Composites Science and Technology

Series Editor

Mohammad Jawaid, Laboratory of Biocomposite Technology, Universiti Putra Malaysia, INTROP, Serdang, Malaysia

Composites Science and Technology (CST) book series publishes cutting edge research monographs (both edited and authored volumes) comprehensively covering topics shown below:

- Composites from agricultural biomass/natural fibres include conventional composites-Plywood/MDF/Fiberboard
- Fabrication of Composites/conventional composites from biomass and natural fibers
- Wood, and Wood based materials
- Chemistry and biology of Composites and Biocomposites
- Modelling of damage of Composites and Biocomposites
- Failure Analysis of Composites and Biocomposites
- Structural Health Monitoring of Composites and Biocomposites
- Durability of Composites and Biocomposites
- Thermal properties of Composites and Biocomposites
- Flammability of Composites and Biocomposites
- Tribology of Composites and Biocomposites
- Bionanocomposites and Nanocomposites
- Applications of Composites, and Biocomposites

To submit a proposal for a research monograph or have further inquries, please contact springer editor, Ramesh Premnath (ramesh.premnath@springer.com).

More information about this series at https://link.springer.com/bookseries/16333

S. M. Sapuan · R. A. Ilyas · M. R. M. Asyraf

Safety and Health in Composite Industry

Springer

S. M. Sapuan
Advanced Engineering Materials
and Composites Research Centre
Department of Mechanical
and Manufacturing Engineering
Universiti Putra Malaysia
Serdang, Selangor, Malaysia

M. R. M. Asyraf
Department of Aerospace Engineering
Universiti Putra Malaysia
Serdang, Selangor, Malaysia

R. A. Ilyas
School of Chemical and Energy
Engineering, Faculty of Engineering
Universiti Teknologi Malaysia
Johor Bahru, Johor, Malaysia

ISSN 2662-1819 ISSN 2662-1827 (electronic)
Composites Science and Technology
ISBN 978-981-16-6135-8 ISBN 978-981-16-6136-5 (eBook)
https://doi.org/10.1007/978-981-16-6136-5

This Springer imprint is published by the registered company Springer Nature Singapore Pte Ltd.
The registered company address is: 152 Beach Road, #21-01/04 Gateway East, Singapore 189721, Singapore

Preface

Safety and Health in Composite Industry provides the latest developments and best possible safety practices utilized in composite manufacturing facilities for students, workers, engineers, and other participants. Composite products range from skateboards to components of the space shuttle. Materials within the composites industry are often called "advanced" if they combine the properties of high strength and high stiffness, low weight, corrosion resistance, and in some cases special electrical properties. Several of the manufacturing processes and potential hazards are common to advanced matrix composites. The materials most frequently used in composite manufacturing, such as matrix (polyester, vinyl ester, phenolic, epoxies, methyl ethyl ketone peroxide, benzoyl peroxide, hardeners, and solvents) and reinforcement materials (carbon, glass and Kevlar fibers, honeycomb, and foams) can be highly toxic to human body. These materials can also be very toxic to the environment when dumped out uncontrollably, creating major future health and environmental concerns. Throughout the manufacturing process, workers inhale vapors of the liquid matrix, hardeners, and solvents/thinners, as well as reinforcement materials (chopped fibers and particles) in airborne. Milling, cutting, and machining of the composites can further increase the toxic inhalations of airborne composite particles, resulting in major rashes, irritation, skin disorders, coughing, severe eye and lung injury, and other serious illnesses. The major portions of these hazardous materials can be controlled using appropriate personal protective equipment for the chemicals and materials used in composite manufacturing and machining. This book covers topics of (1) introduction to safety and health, (2) composites and biocomposites: manufacturing and processing, (3) emission of hazardous air pollution in the composite production, (4) safety in composite laboratory, (5) design for safety in composites, (6) carbon footprint in health care, (7) safety issues in composite materials, (8) fire safety in polymers composites, (9) health hazard from composites, (10) safety and health issues associated with fiber-reinforced polymer composites in various industrial sectors, (11) occupational safety and health administration in composite industry, (12) the role of biocomposites in health issues during COVID-19 pandemic, and (13) safety issues in transportation design. This book includes commentary from leading academic experts in the field who present cutting-edge research on advanced composite materials. Illustrations, figures, and tables will be included in this book in order to make it easier for students, workers, engineers, and other participants to understand the contents of this book. The end user will know that the safety and health should be practiced in composite industry and their right in composite industry. Besides that, the composites industry players can upgrade their current safety system to the recommended practiced system. A lot of problems will be solved by integrating the current system and advanced technology system from extensive research.

Serdang, Malaysia S. M. Sapuan
Johor Bahru, Malaysia R. A. Ilyas
Serdang, Malaysia M. R. M. Asyraf

Contents

Chapter 1
Introduction to Safety and Health

Abstract This chapter discusses on the introductory and background of safety and health in general aspect. In this section, the elaborations on theories and theroretical frameworks of accident and its causes are outlined accordingly. Later, this section narrows toward the foundation of safety and health in term of composites and engineering perspectives, such as safety engineering, chemical engineering, ergonomics and human factor engineering, industrial engineering and fire protection engineering. The preventive concepts and strategies in order to comprehend the accidents occurred in workplaces were also revealed. Lastly, this section identifies the details on significance and importance of safety and health monitoring.

Keywords Composite · Safety and health · Accidents · Control preventions · Safety strategies

1.1 Introduction

Safety and health are the two current vital and debatable subjects, which are the most significant factors building every sector of industrial development. During the years of 1788–1860, the health was significant, by quoting that 'health is not everything, however without it, everything is nothing' (Hassard et al. 2012). Thus, the overall meaning of health and safety can be holistically described as working towards achieving the total wellbeing of the employees at work (Amponsah-Tawiah 2013). World Health Organization (1994) described health as the actions for hygiene, psychology, medicine, physiotherapy, ergonomics, safety, and rehabilitation during in the workplace. On the other hand, safety is the protection of personnel against any physical injury (Hughes and Ed 2015). According to International Occupational Hygiene Association (IOHA), safety and health could also be defined as the scientific approach of recognition, evaluation, anticipation, and control of hazards that occurred in the workplace that later would affect the health and well-being of workers (ILO 2009). Therefore, the term 'safety and health' could be justified by relating to the promotion and maintenance of the highest degree of physical, mental, and social well-being of handling personnel in all workplaces (Westgaard and Winkel 2011).

S. M. Sapuan et al., *Safety and Health in Composite Industry*,
Composites Science and Technology ,
https://doi.org/10.1007/978-981-16-6136-5_1

Safety and health also requires securing human life from being damaged and loss. Globally, human society put the human life and welfare among the most priority aspect (Molamohamadi 2014). This shows that humanitarianism overrides the other issues to put the safety concerns as vital and compulsory items due to moral basis. However, each individual has different levels of regard and standard for the things, which are right or wrong. To reduce these differences, humanity has standardized the codes of conduct among people to ensure safety is grasped accordingly. Via any governing body for formalized standards, constitutionalized laws and policies were established due to this safety reason. Laws and policies recognize that life and the ability to live it fully has worth. The property, too, has worth. In terms of economical view, any values of life, human capabilities, and properties have to be evaluated within the human codes of conduct. Cost is a term referred to as the real outlays to escape from disbursement, or lost properties and abilities. Thus, it can be concluded that there are three main reasons for safety, which are humanitarianism, laws and policies, and cost.

From the above statement, the safety aspects could be branched in various disciplines, including the engineering sector. Engineers and technical officers have been significantly contributing to the safety issues, especially in solving the safety problems. Engineers are required to expect and cooperate with other professionals, such as safety and health officers to rectify hazards, establish efficient solutions to safety problems, and document safe operations and systems (Kim et al. 2016). It is impossible to keep up with all the new laws and regulations at the federal, state, and local levels. In the specific area of safety, several current safety professionals are responsible to assist the activities of other knowledgeable professionals in applying safety principles to particular problems. Documenting and executing safety involve professionals, often from different backgrounds (Sorensen et al. 2018).

In general, there are numerous cases involving accidents and diseases when highly exposed to chemical toxicity, heavy machine tools, and equipment as well as repeating the same work task frequently (Suárez Sánchez et al. 2017). This would cause a major impact on the competitiveness, work quality of an organization, productivity, and livelihoods of individuals and society (Amponsah-Tawiah and Mensah 2016). In addition, any accidents and illness would permit intolerable humanitarian and economic problems, which would disrupt all economic sectors that subsequently inhibiting sustainable economic growth. The prevention effort of implementing safety and health policies in the workplaces would reduce any diseases that have been causing occupational burdens globally.

1.2 Accident and Its Consequences

Accident is a term that clarifies a situation of an unexpected event that causes injury, damage, or any casualties. It usually conjures the ideas of undesirable effects or consequences. The term suggests to most people an immediacy between event and effect. Several problems arise when accidents happen related to safety and health issues, which include the idea chance of occurrence, the duration of events, and the correlation between accident situations and their effects (Brauer 2005). The definition of 'accident' can be elaborated into unintended and unplanned events that result in immediate or long-term undesirable effects due to the unsafe conditions and acts.

Any accident event has one or more identifiable causes or factors. Probability and possibility might significantly contribute to bringing the accident factors to come together. Generally, there are two fundamental backgrounds of accident factors, such as unsafe conditions and unsafe acts (Adane and Abeje 2012). Commonly, the accidents might due to either coming from only one case factor or both. Identification of accident events and their factors is a useful effort to pursue accident prevention. The repeated faulty events are usually associated with the accident events which later contribute to their consequences.

Immediacy is another type of relationship between accident and consequence that can be in error. In some cases, injuries and other types of losses are usually associated with the accident events. They commonly appear right after an event on the victims. However, within safety and health issues, one must also deal with illnesses and diseases. The health issues would lead to a delay or long-term period between the accident and post-event results (Ericson 2005). For instance, in the area of polymeric composite works, the operator is exposed to some cancer diseases which have a latency period of around 20–40 years after exposure from inhaling any toxic chemical substances. Moreover, a repetition of activities such as hand-layup to fabricate composite products that extend over days, weeks or months would also cause fatigues and body injuries. The idea of immediate results implied in the term accident makes it difficult to include illness, diseases, or cumulative injury as accident effects.

1.3 Losses

Losses are subsequent effects after accidents that can take many forms such as injury, disease, illness, damage to property, and death. The damage of property such as equipment, tool, materials, infrastructures, and environment would incur a cost for the repairing and replacement (Borkovskaya and Passmore 2020). Later, any losses could also lead to any loss of sales, time, and production from accident incidents (Koran et al. 2020). All of these losses could contribute to cost which is further defined in the subsequent paragraph.

Two types of costs can be associated with any accidents occurred, such as direct costs and hidden costs (Bayram et al. 2017). These direct costs typically involve those expenses that occurred due to the accident. These include any compensation and medical expenses due to the injured employee since the accidents had caused them to experience casualties.

In terms of indirect cost, it is commonly considered as real expenses associated with accidents, but hard to evaluate based on individual case or situation. Table 1.1 shows the eleven categories of indirect costs, as stated by Heinrich (Basford and

Table 1.1 Indirect costs after accidents (Brauer 2005; Bhushan et al. 2017)

No	Hidden costs
1	Overhead costs that continue during lost work
2	Loss of profit because of lost work time and idle machines
3	Lost time of the injured employees
4	Losses resulting from less than full productivity of injured workers to return to work
5	Losses due to reductions in productivity of coworkers because of concern
6	Losses due to late or unfilled orders, loss of bonuses, or payment of penalties
7	Payments made to the injured employees under benefit programs
8	Damage to tools, equipment, materials, or property
9	Time spent by company first aid, medical, and safety staffs on the case
10	Time lost by a trained advisor to assist the injured worker, investigate the incident, prepare reports, and make adjustments in work and staffing
11	Time lost by other personals to assist injured coworker, to see what is going on, and to discuss events

DePoy Smith 2019). Table 1.1 highlights that there are still other medical fees and compensation needed to be settled by the organization when an accident is took placed. Some researchers suggested the total cost of 4:1 ratio to the direct expenses when accidents occur (Basford and DePoy Smith 2019). However, the ratios of the total cost to the direct cost vary among different organizations and their types of operation.

1.4 Unsafe Acts and Unsafe Conditions

In the past few years, there was a study conducted to determine the ratio of accidents due to unsafe acts with unsafe conditions. Heinrich evaluated around 75,000 accident cases and he discovered that 88% of them happened due to unsafe acts, while 10% and 2% were caused by unsafe conditions and unpreventable factors (Da and Zheng 2016). This can be summarized that the ratio of unsafe acts to unsafe conditions was at 88:10 ratio. Another study carried out by the Pennsylvania Department of Labor and Industry (Brauer 2005) revealed that both unsafe conditions and acts were dominating the contributing factors for more than 98% of the total industrial accidents analyzed of 80,000 cases.

From the above statements, it was found out that both factors (unsafe acts and conditions) were two major causative factors for the occurrence of accidents. Thus, engineers and other technical officers are required to reduce the possibility of unsafe situations (Wong et al. 2019). Designs that reflect an understanding of human error and behavior can limit the range of human behavior that leads to or causes incidents.

1.5 Engineering Knowledge in Safety and Health for Composite Fields

In composite materials and engineering fields, they have their own contributions toward safety and health within their specialization. Occupations in any composite engineering field embrace a significant amount of safety-related tasks (Peciłło 2016). In general, engineers and technologists who develop composite products have the responsibility to ensure that preventive efforts for safety are employed. Their roles are to make sure all the hazards and safety problems are identified during the design process of composite products and eliminate them. They are also required to guarantee the safety of the workplace and activities by preventing any unsafe behavior and environment in designing products. Moreover, the engineers and technologists should ease the impact of unsafe behavior during the designing stage, thus, it could limit any unsafe environment.

1.5.1 Safety Engineering

One of the knowledge that should be embedded in the design and fabricating of composite products is safety engineering. It includes the implementation of engineering, scientific, and fundamentals approaches to reduce and control any possible hazards (Adler et al. 2016). The application of this knowledge is widely covered in order to deal with various circumstances and issues regarding the design and fabrication safety. In addition, the knowledge specializes in recognizing and regulating hazards to the hazard-free environment by integrating the engineering and occupational health and safety disciplines.

1.5.2 Ergonomics and Human Factor Engineering

This scope of knowledge is very similar to safety engineering, however, it focuses on the application of psychological and physiological behaviors of humans in developing any product, equipment, and systems (Hoyos-Ruiz et al. 2017). This ergonomics and human factors engineering would assist the engineers and technologists to improve the products' performance without jeopardizing the safety and satisfaction of consumers. It computes the improvement to bridging the consumers and products, environments, workplaces, and systems in order to achieve safe conditions. This would emphasize the enhancement of performance and safety by reducing task errors and physical stresses. In detail, ergonomics is the scope of knowledge focusing on physiological and biomechanical aspects (da Conceição et al. 2019), meanwhile, human factors engineering is narrowed to the behavioral and cognitive aspects of performance and safety (Bevilacqua and Ciarapica 2018).

1.5.3 Chemical Engineering

Chemical engineering is a corpus of applied science that focuses on designing composite products using chemical substances that directly contribute to safety (Al-Zyoud et al. 2019). It applies the system of safety techniques to process design, which aids to utilize less hazardous chemicals and systematic chemical substances process.

1.5.4 Fire Protection Engineering

This field of engineering is concerned to ensure human and property safety from loss and damage due to fires and explosions (Wegrzynski and Sulik 2016). This field is highly related to chemical engineering, where any explosive or flammable chemical substances could be systematically managed and contained in appropriate apparatus and equipment. Fire protection engineers are specialists in the prevention, protection, detection and alarms, and fire control and extinguishment for structures, equipment, processes, and systems (Östman et al. 2017). They design egress routes to allow for a safe exit from fires.

1.5.5 Industrial Engineering

Since the composite products can be fabricated on large scale with complex instruments and machines, industrial engineering should be embedded by engineers to ensure a higher production rate without compromising safety (Ram et al. 2016). The knowledge focuses on any related methods and work environments to be safe for workers. Thus, the engineers should confirm that they have to understand the knowledge of occupational safety and health, safety engineering, ergonomics, human factors engineering, fire protection engineering, and chemical engineering.

1.6 Preventive Strategies and Approach

Generally, any possible accident is evaluated using investigation by finding their factors and sources of accidents. From this point, a detailed assessment is carried out to evaluate those affected persons and properties from any possible potentials which could lead to accidents. Later, the preventive approach can be taken to avoid the accident from happening. The preventive strategies can be divided into two approaches; reactive and proactive approaches (Mitra et al. 2017). Figure 1.1 shows the overall method in applying data from accidents to avoid them from appearing in the future, which is called the reactive approach. Usually, this method is performed when the

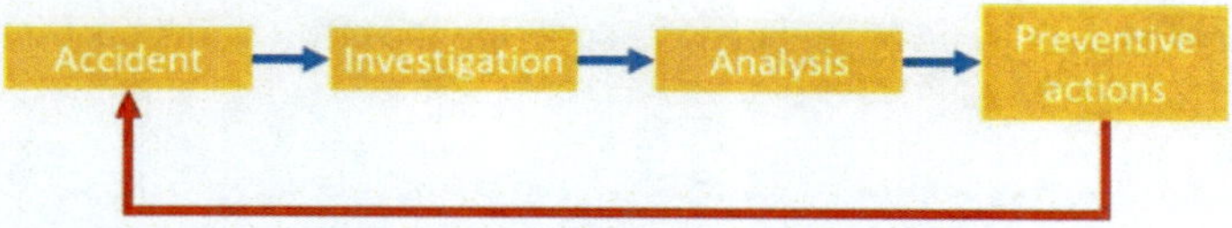

Fig. 1.1 Reactive approach for deriving preventive actions from accidents (Brauer 2005)

Fig. 1.2 Proactive approach for developing preventive actions before accidents (Brauer 2005)

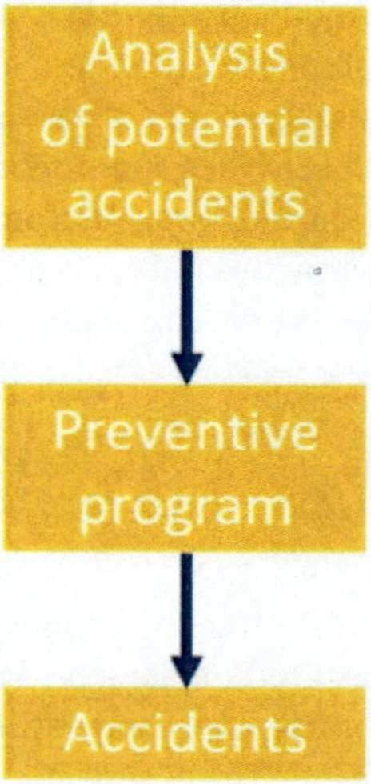

accident has occurred for the later detailed evaluation to identify the preventive actions to be performed (Oyama et al. 2017). Meanwhile, the proactive approach is a preventive method that is carried out by analyzing any possible solutions to create a preventive program to avoid possible accidents (Haslam et al. 2016). Different strategies are possible for this approach based on frequency, severity, and cost, where each has merit, depending on preventive goals (Fig. 1.2).

1.6.1 Frequency Strategy

Frequency strategy is one of those strategies used to evaluate any possible accidents that might occur. Thus, the main components of preventive steps such as investigation, analysis, and preventive actions have to direct towards frequently occurring accidents to reduce the frequency of the occurrence (Suarez et al. 2017).

Identification of these factors would aid direct preventive efforts, which could be the most effective. For instance, around 50% of injuries and casualties happen to workers during the first month of their job. After three months, the amount reduces to half which might be explained by the centering actions, such as formal training on the new workers to cope with their work environments (Brauer 2005). From this action, the accident frequency can be reduced via applying the effort with equal intensity to all workers.

1.6.2 Severity Strategy

Another method can be applied in any serious cases by evaluating the deaths, number of injuries, long-term disabilities, serious illnesses, and number of property losses. A study reported that many serious cases, especially injuries often happen in four types of activities, e.g. nonproductive activities, work involving high health risks, construction, and rarely performed and unusual non-routine works (Kazan and Usmen 2018). From this information and data, any serious illness and casualties could be prevented by formatting a proper strategy.

1.6.3 Cost Strategy

Cost strategy is another approach used to avoid high-cost accidents, where this method implements Parato's law (Obolewicz and Dąbrowski 2018) that follows the basis for measuring the seriousness of incident consequences, not the injury or illness itself. The cost strategy evaluates only for losses that involve other than human ones.

1.6.4 Combined Strategy

A combined strategy is applied from the combination of severity, cost, and frequency approaches. To establish the priority actions, it can be applied by using a number of risk analyses and related techniques (Brauer 2005). The process counts on the seriousness of the event if it does occur, the possibility of the event to happen or the frequency of its occurrence, and the subsequent cost to correct for the post-accident stage.

1.7 The Three Es' of Safety

Aside from those five preventive action strategies, another concept of selection preventive action is the "three Es' of safety", which are engineering, education, and enforcement (Brauer 2005). From the engineering perspective (Ericson 2005), this includes substituting with less hazardous materials, modify processes, and implementing fail-safe devices. Moreover, another action could be taken by designing out hazards, reducing the inventory of hazardous materials, and applying personal protective equipment (PPE) during conducting engineering activities.

In terms of education, proper training has to be embedded for any personals using the equipment and tools that are exposed to hazard (Hadikusumo and Rowlinson 2004). Thus, the training or seminar activities would aid any workers or laboratory personnel to follow safe procedures and practices by correctly and safely performing the job. In addition, a proper training program from trained speakers would guide the user on the appropriate handling of any hazardous products, aware the users of any hazards, and prescribing protective equipment with appropriate protective actions. On top of that, the users such as laboratory personnel could be recognized and evaluated for any hazards, following safety standards and any legal responsibilities.

In these three Es' of safety, enforcement is the last E that elaborates on proper guides to comply with any related policies and laws (Amponsah-Tawiah and Mensah 2016). In specific, every party in the workplace, especially those highly exposed to hazards is required to comply with federal, state, and local laws and regulations, with consensus standards, and with company rules and procedures.

1.8 Significance for Safety and Health Monitoring

There are many advantages when appropriate preventive actions are adopted in safety and health, such as increased productivity and improved product quality. Thus, every worker who directly involves in this field has the legal right to work in a safe and healthy environment. Every organization, including business entities, has to understand that a safe workplace is a key component to developing a positive corporate culture (Lian et al. 2019).

1.8.1 Keeping People Safe

Usually, prevention of accidents by avoiding the potential risk of death of workers is a primary priority. Most safety problems and issues are those being highlighted since they are humanitarian issues (Walters and Wadsworth 2020), thus every corporation should provide training programs for their workers to ensure the safety protocols are followed.

1.8.2 Injury, Cost, Time, and Money

Every injury and death have to deal with compensation fee and medical expenses for the affected victims. A study conducted by Occupational Health and Safety Administration (Brauer 2005) estimated that every employer usually pays around

USD 1 million each week for employees' compensation costs due to accidents. Employers are required to have workers' compensation insurance policies, however, those physical injuries increase the basic costs. Thus, the accidents not only cause casualties for the victims but also increase costs related to the accidents, which could burden an organization when the safety and health are not properly managed. Other than that, the prevention of accidents would provide the workplace to run the operation smoothly without any troubles due to the workers' injury or replacement.

1.8.3 Fewer Injuries Increase Productivity

Any casualties or accidents in any working environment would demoralize the worker spirit in conducting works (Bernard and Patange 2020). This phenomenon would cause workers to divert their attention from completing their daily tasks properly when their state of mind is more focusing on the injury or accident they experienced. An organization that emphasis a safe and healthy workspace and environment would create a stronger relationship with their workers. Subsequently, this could develop team morale among workers, which has a direct correlation to productivity.

1.8.4 Increase Public Perception

A corporation that operates daily in an unsafe environment would diverge it away from its customers (Savelli et al. 2019). For instance, the food processing industry that employs unsafe food preparation practices would avoid consumers from buying their products, since they fear consuming unsafe and unhygienic food products. Thus, any business entity must adopt all good safety and health practices for customer satisfaction. In the end, the positive team morale increases positive customer interactions, which leads to customer retention and loyalty.

1.8.5 Minimizing Legal Liabilities

Accidents which cause casualties could bring the issue into legal liabilities due to technical mistakes and glitch during the operation. Somehow, for any injuries and/or deaths experienced by the workers, the company or the employers could experience court sue from the injured workers and the deceased employees' families (Michaels and Barab 2020). For some small business entities, the resulting lawsuit could result in the bankruptcy of the company.

1.9 Conclusions

All in all, safety and health are big deals for any organization and sector to strive for achieving better economic growth. The safety and health risks are considered by society as moral issues besides being classified as those economic, political, legal, and technical issues. It can be generalized as their acceptability based on the consequences exposed to the victim at what cost. Thus, the application of appropriate actions for prevention aids to reduce many incidents that might cause many injuries, casualties, property loss, and increase in cost. In order to attain an acceptable level of safety and health in the working environment along with the product, technical officers such as engineers and technologists have to identify the existing hazards and implement any current standards of society in terms of regulation, public (consumer) expectation, laws, and judicial interpretation.

References

Adane L, Abeje A (2012) Assessment of familiarity and understanding of chemical hazard warning signs among university students majoring chemistry and biology: a case study at Jimma University, Southwestern Ethiopia. World Appl Sci J 16:290–299

Adler R, Feth P, Schneider D (2016) Safety Engineering for Autonomous Vehicles. Proceedings—46th Annual IEEE/IFIP International Conference on Dependable System and Networks, DSN-W, 200–205 (2016).https://doi.org/10.1109/DSN-W.2016.30

Al-Zyoud W, Qunies AM, Walters AUC, Jalsa NK (2019) Perceptions of chemical safety in laboratories. Safety 5.https://doi.org/10.3390/safety5020021

Amponsah-Tawiah K (2013) Occupational health and safety and sustainable development in Ghana. Int J Bus Adm 4.https://doi.org/10.5430/ijba.v4n2p74

Amponsah-Tawiah K, Mensah J (2016) Occupational health and safety and organizational commitment: evidence from the Ghanaian mining industry. Saf Health Work 7:225–230. https://doi.org/10.1016/j.shaw.2016.01.002

Basford B, DePoy Smith M (2019) Remodeling Heinrich: an application for modern safety management. Prof Saf 64:44–52

Bayram M, Üngan MC, Ardıç K (2017) The relationships between OHS prevention costs, safety performance, employee satisfaction and accident costs. Int J Occup Saf Ergon 23:285–296. https://doi.org/10.1080/10803548.2016.1226607

Bernard E, Patange GS (2020) Productivity improvement through identifying hazardous conditions in steel foundry. IOP Conf Ser Mater Sci Eng 872:012091. https://doi.org/10.1088/1757-899x/872/1/012091

Bevilacqua M, Ciarapica FE (2018) Human factor risk management in the process industry: a case study. Reliab Eng Syst Saf 169:149–159. https://doi.org/10.1016/j.ress.2017.08.013

Bhushan U, Gujarathi R, Banerjee A, Sharma H, Seetharaman A (2017) The impact of hidden costs on production and operations. J Accounting, Bus Manag 24:1–20

Borkovskaya VG, Passmore D (2020) Behavioral engineering model to identify risks of losses in the construction industry. Smart Innov Syst Technol 138:243–250. https://doi.org/10.1007/978-3-030-15577-3_24

Brauer RL (2005) Safety and health for engineers. Wiley, Hoboken, New Jersey, USA

da Conceição CS, Broberg O, Duarte F (2019) A six-step model to transform an ergonomic work analysis into design guidelines for engineering projects. Work a J Prev Assess Rehabil

Da X, Zheng W (2016) A system dynamics model for railway workers' safety behaviors. In: 2016 IEEE International Conference on Intelligent Rail Transportation (ICIRT), pp 409–417. https://doi.org/10.1109/ICIRT.2016.7588762

Ericson CA (2005) Hazard analysis techniques for system safety, 2nd edn. Wiley, Hoboken, New Jersey, USA

Hadikusumo BHW, Rowlinson S (2004) Capturing safety knowledge using design-for-safety-process tool. J Constr Eng Manag 130:281–289. https://doi.org/10.1061/(ASCE)0733-9364(2004)130:2(281)

Haslam C, O'Hara J, Kazi A, Twumasi R, Haslam R (2016) Proactive occupational safety and health management: promoting good health and good business. Saf Sci 81:99–108. https://doi.org/10.1016/j.ssci.2015.06.010

Hassard J, Flintrop T, Clausen K, Muylaert (2012) Motivation for employees to participate in workplace health promotion. Work, Luxembourg

Hoyos-Ruiz J, Martínez-Cadavid JF, Osorio-Gómez G, Mejía-Gutiérrez R (2017) Implementation of ergonomic aspects throughout the engineering design process: human-artefact-context analysis. Int J Interact Des Manuf 11:263–277. https://doi.org/10.1007/s12008-015-0282-3

Hughes P, Ed F (2015) Introduction to health and safety in construction. In: Introduction to health and safety in construction. Elsevier Butterworth-Heinemann, Oxford, UK

ILO (2009) Standards on occupational safety and health: promoting a safe and healthy working environment. Int Labor Conf 98th Sess 1981:162

Kazan E, Usmen MA (2018) Worker safety and injury severity analysis of earthmoving equipment accidents. J Safety Res 65:73–81. https://doi.org/10.1016/j.jsr.2018.02.008

Kim Y, Park J, Park M (2016) Creating a culture of prevention in occupational safety and health practice. Saf Health Work 7:89–96. https://doi.org/10.1016/j.shaw.2016.02.002

Koran S, Sarıhan A, Güllüpınar B, Can Ç, Korkmaz A, Bulut M (2020) Evaluation of occupational accidents-related working day losses in Turkey. Int Med J 27:327–330

Lian J, Cai O, Dong X, Jiang Q, Zhao Y (2019) Health monitoring and safety evaluation of the offshore wind turbine structure: A review and discussion of future development. Sustainability 11:https://doi.org/10.3390/su11020494

Michaels D, Barab J (2020) the occupational safety and health administration at 50: protecting workers in a changing economy. Am J Public Health 110:621–647. https://doi.org/10.2105/AJPH.2020.305597

Mitra S, Geedipally SR, Lord D (2017) Safety analysis of urban signalized intersections in Kolkata , India using a combined proactive and reactive approach. In: Transportation Research Board 96th Annual Meeting. Transportation Research Board, Washington DC, USA

Molamohamadi Z (2014) The relationship between occupational safety, health, and environment, and sustainable development: a review and critique. Int J Innov Manag Technol 5:198–202. https://doi.org/10.7763/ijimt.2014.v5.513

Obolewicz J, Dąbrowski A (2018) An application of the Pareto method in surveys to diagnose managers' and workers' perception of occupational safety and health on selected Polish construction sites. Int J Occup Saf Ergon 24:406–421. https://doi.org/10.1080/10803548.2017.1375781

Östman B, Brandon D, Frantzich H (2017) Fire safety engineering in timber buildings. Fire Saf J 91:11–20. https://doi.org/10.1016/j.firesaf.2017.05.002

Oyama Y, Kashiwagi M, Ogata Y, Hoshishiba Y (2017) Factors associated with the use of the reactive approach to preventing patient safety events. Home Heal Care Manag Pract 29:96–102. https://doi.org/10.1177/1084822316681267

Pecillo M (2016) The resilience engineering concept in enterprises with and without occupational safety and health management systems. Saf Sci 82:190–198. https://doi.org/10.1016/j.ssci.2015.09.017

Ram K, Chaudhary V, Ahmad F, Bajpai PK (2016) Polymer composites for industrial safety helmets: a review. Int J Adv Prod Ind Eng 506:9–10

Savelli E, Murmura F, Liberatore L, Casolani N, Bravi L (2019) Consumer attitude and behaviour towards food quality among the young ones: empirical evidences from a survey. Total Qual Manag Bus Excell 30:169–183. https://doi.org/10.1080/14783363.2017.1300055

Sorensen G, Sparer E, Williams JAR, Gundersen D, Boden LI, Dennerlein JT, Hashimoto D, Katz JN, McLellan DL, Okechukwu CA, Pronk NP, Revette A, Wagner GR (2018) Measuring best practices for workplace safety, health, and well-being: the workplace integrated safety and health assessment. J Occup Environ Med 60:430–439. https://doi.org/10.1097/JOM.0000000000001286

Suarez R, Agbonifo N, Hittle B, Davis K, Freeman A (2017) Frequency and risk of occupational health and safety hazards for home healthcare workers. Home Heal Care Manag Pract 29:207–215. https://doi.org/10.1177/1084822317703936

Suárez Sánchez FA, Carvajal Peláez GI, Catalá Alís J (2017) Occupational safety and health in construction: a review of applications and trends. Ind Health 55:210–218. https://doi.org/10.2486/indhealth.2016-0108

Walters D, Wadsworth E (2020) Arrangements for workers' safety and health in container terminals: corporate core values and concrete practice. Econ Ind Democr. https://doi.org/10.1177/0143831X19893767

Wegrzynski W, Sulik P (2016) The philosophy of fire safety engineering in the shaping of civil engineering development. Bull Polish Acad Sci Tech Sci 64:719–730. https://doi.org/10.1515/bpasts-2016-0081

Westgaard RH, Winkel J (2011) Occupational musculoskeletal and mental health: significance of rationalization and opportunities to create sustainable production systems—a systematic review. Appl Ergon 42:261–296. https://doi.org/10.1016/j.apergo.2010.07.002

Wong IS, Dawson D, Van Dongen HPA (2019) International consensus statements on non-standard working time arrangements and occupational health and safety. Ind Health 57:135–138. https://doi.org/10.2486/indhealth.57_202

World Health Organization (1994) Global strategy on occupational health for all: the way to health at work. World Health Organization. https://www.who.int/occupational_health/globstrategy/en/. Accessed 27 Jul 2020

Chapter 2
Composites and Biocomposites: Manufacturing and Processing

Abstract Currently, the application of composite materials has been ubiquitous in defence, automobile, aerospace, sports, household products and medical equipment. The composite materials were made up from two major phases called as fibre as reinforcement and polymeric matrix. The combination of both phases allows the material to be lightweight and high strength to be implemented in wide applications. Most of composite materials are commercially available in vast sources including synthetic (man-made) and natural-based materials. This chapter elaborates on the background and types of composites which are commercially available. Moreover, the processing and manufacturing of the composites are discussed comprehensively at the end of the manuscript.

Keywords Composite · Biocomposites · Fibres · Polymers · Processing · Manufacturing processes

2.1 Introduction

Along with the historical timeframe till nowadays, humans have been experiencing improvement of life qualities via the research and development activities. The extensive usage of crude oil-based products has generated several issues, including the abundance of plastic wastages smearing in ecosystem affecting the food chain and running down crude oil's resources (Jumaidin et al. 2019). The rapid growth of plastic waste from human activities has worsened the pollution, affecting the fishery sector, air quality, limiting the freshwater resources, flora, and fauna. These environmental issues have raised the concern among the environmentalists and material engineers to pledge the effort in developing biodegradable plastics and biocomposites (Mansor et al. 2015; Bajuri et al. 2016; Ilyas et al. 2018b; Nurazzi et al. 2019; Asyraf et al. 2020b). One of the efforts to tackle the issues is by developing renewable resources originated products that can compete and subsequently substitute the existing petroleum-based products in the current markets. However, the production of biobased materials as alternatives to petroleum-based products is cost-ineffective. A more feasible effort could be performed is by combining the biobased materials with

S. M. Sapuan et al., *Safety and Health in Composite Industry*,
Composites Science and Technology ,
https://doi.org/10.1007/978-981-16-6136-5_2

the petroleum resources to produce economical products with numerous applications (John and Thomas 2008). Natural fibres reinforced in either synthetic or biodegradable polymers are forecasted as alternatives to current carbon and glass fibres. The effort and development of many possibilities in biocomposite applications in various sectors have been continued days by days to ensure the biocomposite revolution is a reality (Mohanty et al. 2002b).

Generally defined, biocomposites are composite materials formed by the post-curing process of biopolymer or synthetic polymer resin and reinforced with natural/biofibre. Apart from that, green composites are likely to be derived from both biopolymers, such as PLA and PHBA with plant-derived fibre (natural/biofibre).

One of the building blocks making up many biocomposite products is lignocellulosic fibre, such as flax, sugar palm, kenaf, pineapple leaf, oil palm, hemp, jute, and corn that have drawn substantial attention as alternatives to the current man-made fibres, such as glass and carbon fibres. These fibres have been implemented in various applications including aerospace, packaging, civil construction, automobile, and even household products. From the statistics and forecasted data recorded by Department of Agriculture and Department of Energy, United States of America (USA), around 50% of basic chemical building blocks coming from plant resources by the year of 2050 (Mohanty et al. 2002a, b). This could happen due to the biofibres themselves that have many promising merits over man-made fibres. Table 2.1 illustrates the comparison between the merits of biofibres and synthetic fibres (Sanyang et al. 2016; Shahroze et al. 2019). Recently, the demands of biofibres have increased tremendously in many sectors due to their affordability, availability, biodegradability, recyclability, renewability, and manufacturability (Ilyas et al. 2020c, d; Asyraf et al. 2021). To be specific, biofibres exhibit several benefits, e.g. easy disposal, fewer health hazards and energy consumption during manufacturing, low density, and better insulation properties with comparable tensile performance.

The most popular renewable resources to produce biodegradable plastic are cellulose and starch (Zhong et al. 2020). Based on the economical perspective, starch is considered as the least expensive biopolymer in the current market. The biopolymer

Table 2.1 Comparison between the merits of biocomposite and synthetic composite (Sanyang et al. 2016; Shahroze et al. 2019)

Attributes	Biofibre/natural fibre	Synthetic fibre
Biodegradability	Yes	No
Carbon dioxide emission	No	Yes
Energy usage	Low	High
Recyclability	Yes	No
Renewability	Yes	No
Physical property	Low density	Double that of biocomposite
Prices	Low	Higher than biocomposite
Respiratory problem affect	No	Yes

is a high potential for non-food products, e.g. plastic bag, table covers, and plastic wrapper. For cellulose, it is obtained from plant components and is useful in the manufacturing of eco-friendly cellulose plastics. Apart from that, vegetable oils (e.g. peanut, sunflower, walnut, sesame, and soybean oils) have also grown the attention internationally to be applied in biodegradable plastics. Soybean oil has been widely used with natural fibre in many applications, such as housing, rigid packaging, and transportation (Fei et al. 2019). In another study, fish oil polymers have shown promising properties, since they possess a high degree of unsaturation and unique good damping and shape memory properties (Li et al. 2000).

Previously, the development of the biocomposite product by using cellulose fibre reinforced in phenolic has begun since 1908 (John and Thomas 2008). Later, it has extended to reach commodity status to the implementation of urea and melamine by reinforcing glass fibre. Even though the discovery of biocomposite potential was long back in time, plastic products are now showing an ambiguous reputation due to their respective mechanical performances and qualities. Many industries are started to adopting this material as one of the vital components, especially in kitchenware, automobile, construction, household and safety products. Thus, in this chapter, it provides a state-of-art review on composites and biocomposites, focusing on processing and manufacturing.

2.2 Polymers

The polymeric matrix is an essential element used in composites to transfer load from the surface of the laminate to fibre for protection from the outside environment and provide mechanical abrasion and chemical resistance. Polymers are the most abundant materials as matrix components in composite materials. It has various applications, especially in structural applications in many sectors such as sports, aerospace, automotive, defence, naval, and construction (Ilyas and Sapuan 2020).

2.2.1 Synthetic Polymers

In general, the polymers are categorized into thermosets and thermoplastics, which they form a linking chain of monomers that undergo polymerization process to form the polymer. In the early twentieth century, synthetic polymer was created and used for many applications, especially in structural usage. The properties of polymers vary depending on their availability, monomer composition, molecular mass, branching structures, molecular mass, flexibility, crystallinity, chemical compounds, chain flexibility, and chemical processes. They usually are tailored according to their processes including copolymerization, blending, and macromolecular architecture alternation (Yu et al. 2006; Jawaid and Abdul Khalil 2011).

For the thermoset polymers, the monomers were cross-linked via chemical bonds, which they are cannot be remelted. They are very useful for the ideal high-temperature applications, such as in aerospace, energy, and electronics. The thermoset polymers exhibit good chemical, temperature resistance as well as possessing high mechanical strength, stiffness, and structural integrity (Johari et al. 2019). These are the major characteristics that are required for the industrial design of products, including the thick and thin-walled structures. These thermosets products are usually having a good appearance with high flexibility and low-cost production (Stochioiu et al. 2018). However, the main drawbacks of the thermosets are those wastages produced from the manufacturing process that cannot be remoulded and recycled (Yang et al. 2012). They also demonstrated less impact resistance and strength. The finished surface of the thermosets is difficult to produce. In most studies, there are numerous forms of fibres and additives particles can be mixed with thermoset resins to form different composite materials. The thermosets polymers include silicon, urethane, polyester, vinylester, polyimides, polycarbonate, and epoxy.

On the other hand, thermoplastic polymers displayed viscoelastic properties, since it can be heated to make them flow (Marques 2011). The curing process for a thermoplastic polymer is fully reversible, which leads to the remouldable and recyclable characteristics of this type of polymer. The remoulding process can be executed without affecting the physical behaviours of thermoplastics. These thermoplastics depict high mechanical strength as well as shrink resistance. Overall, the thermoplastics demonstrated good chemical and high impact resistance and eco-friendlier manufacturing process whereby the leftovers can be recyclable and remoulded (Bashir and Manusamy 2015). For the application point of view, they can be subjected to various structural applications, specifically in lower forces and stresses mode. The downsides of the thermoplastic polymers are low temperature resistance and the costly raw material than thermoset (Chevali et al. 2009). Several examples of thermoplastic polymers are polyvinyl chloride (PVC), polystyrene, low-density polyethylene (LDPE), high-density polyethylene (HDPE), polyether ether ketone (PEEK), Polypropylene (PP), and polyethylene.

The processing stage of thermosets (thermoforming) can be performed via compression moulding, hand-layup, spray up, pultrusion, resin transfer moulding, and filament winding. For those thermoplastics, they can be formed by blow moulding, rotational moulding, extrusion, screw extrusion, and injection moulding.

2.2.2 Natural Polymers

Natural polymers are usually extracted from biodegradable resources, e.g. plant cellulose and animal proteins. The natural polymers are considered more sustainable in comparison to man-made polymers. These polymers are divided into three types: (1) microbial fermentation, (2) natural polymers obtained from natural resources protein, starch and cellulose, and (3) synthetic polymers obtained from natural monomers such as polylactic acid (PLA). Since the biodegradable polymers are attained fully

Table 2.2 Advantages and disadvantages for both natural and synthetic polymers (Azeredo et al. 2009; Shalini and Singh 2009; Jawaid and Abdul Khalil 2011)

Polymers	Advantages	Disadvantages
Naturally-derived polymers	– Well-known as cell-binding sites which give structural support for cell attachment and proliferation	– Expensive processing
	– Do not implicate with the involvement of toxic chemicals during processing	– Limited in the ability to customize with specific properties
	– Non-carcinogenic material	– High rate of degradation
	– Biodegradable	– Less mechanical strength
Synthetic polymers	– Controllable degradability by manipulating their crystallinity, copolymer ration, and molecular weight	– Limited resources since they are petroleum-based products, which could incur the raw material cost
	– High strength to weight ratio	– Possible local toxicity resulting from acidic degradation products
	– Cheaper processing and manufacturing costs	– Non-biodegradable material which could harm the ecosystem

from natural resources, they are considered as green materials or biopolymers. Polyglycolic acid has the highest strength and modulus values, which made it a desirable matrix material for medical applications (Athanasiou et al. 1996). Advantages and disadvantages for both natural and synthetic polymers are as illustrated in Table 2.2.

2.3 Fibres

Fibres are among other constituents used as reinforcements in polymeric composites (Shahroze et al. 2019; Ilyas et al. 2020a). In general, they are branched into synthetic and natural fibres, as displayed in Fig. 2.1.

2.3.1 Synthetic Fibres

Synthetic fibres are not obtainable naturally, they are mainly derived from petroleum-based polymers either in laboratories or industrial scales. They are manufactured from the combination of various chemical compounds that form very long strand and retain high thermal, physical, and mechanical performances. The performances of the synthetic fibres vary based on their compounds and processes took place earlier. There are numerous examples of the polymer-derived fibres, e.g. polyesters, polyurethanes,

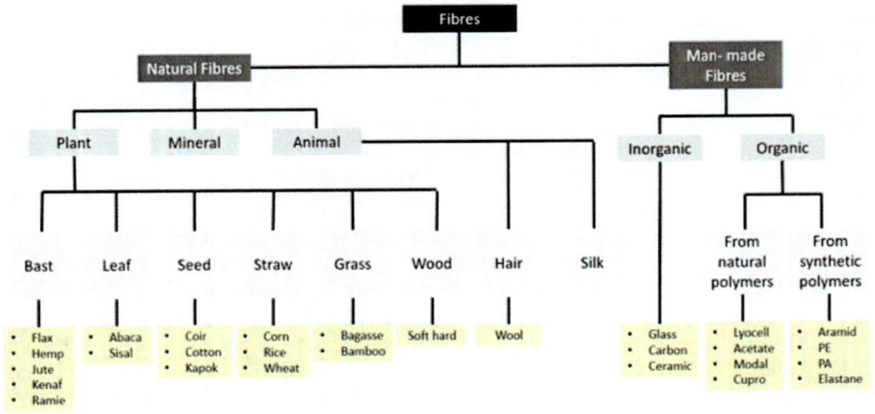

Fig. 2.1 Natural and synthetic fibres classification

nylon, and acrylics. Apart from that, carbon, glass, and aramid (Kevlar) fibres are those three main synthetic fibres used in composite products such as defence, naval, sports, automobile components, energy, and household products. Table 2.3 summarises the mechanical and physical properties of synthetic fibres. Figure 2.2 displays the illustration of glass, aramid, and carbon fibres.

Table 2.3 Mechanical and physical properties of the three most common synthetic fibres (Mohanty et al. 2000; Alves et al. 2010)

Fibre	Tensile strength (MPa)	Elastic modulus (GPa)	Elongation at break (%)	Density (g/cm^3)
Aramid	2700–4500	130	3.3–3.7	1.45
Carbon	3500–5000	260	1.4–1.8	1.8
E-glass	1800–2700	73	2.5	2.6

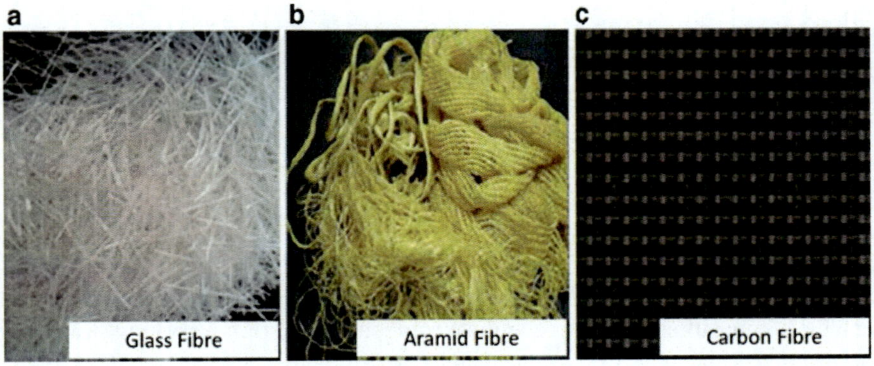

Fig. 2.2 Illustration of **a** glass; **b** aramid; and **c** carbon fibres

2.3.2 *Natural Fibres*

On synthetic fibres counterpart, natural fibres offer benefits to manage the crop and farm wastes by recycling these materials into many useful products, which later create local micro-economies in the developing countries. Moreover, the capabilities of natural fibres such as high stiffness and strength would encourage scientists and manufacturers to utilise the fibres in polymeric composite applications. On the top of that, the natural fibres entail significantly less energy consumption for processing these fibres as compared to their counterparts (synthetic fibres) (Ilyas et al. 2020b). They are considered as renewable materials since the sources are mostly from agricultural wastes, which will be never depleted and very resourceful (Ilyas et al. 2018a). This has led to the researches on natural fibre-based composites or bio-composites.

Typically, natural fibres can be categorized into three types, including mineral, plant, and animal fibres. For those obtained from animal sources, these fibres are usually known as wool, silk, hair, and feather fibres, while leaf, bast, fruit, and grass are plant-derived fibres. Those fibres attained from mineral and animal sources are commonly not used in the manufacturing of composites, instead, they are applied in clothes, handicrafts, and papers productions. Basalt fibre is one of the mineral-based natural fibres used in various composite applications since it has a high ability in terms of chemical, mechanical, and thermal behaviours. At this point of view, most material researchers are conducting various studies to use basalt fibre, since it is an effective alternative to synthetic fibres. Specifically for plant fibres, they are subdivided into wood-based and non-wood based categories. Recently, natural fibres have been widely implemented in many applications, e.g. automotive components (Ishak et al. 2018), safety equipment (Asyraf et al. 2020b) and aerospace (Khan et al. 2018).

2.4 Extraction of Fibres

According to the previous subsection, the natural fibres are commonly extracted from different methods, either mechanically or manually depending on their applications afterwards. These fibres are processed via retting or decortication approach before being further processed into composite laminates. In some cases, both retting and decortication processes are implemented simultaneously to fabric the yarn. Apart from that, mechanical defulling is executed in many fibre retting to remove hull/husk fibres. The subsequent sections discuss the retting or decortication retting processes, mechanical dehulling, and decortication further. Different natural fibres extracted from different plants are represented in Fig. 2.3.

Fig. 2.3 Post-extracted natural fibres

2.4.1 Retting Process

In natural fibres extraction, the retting process is an effective way to extract leaf and bast fibres that produces good quality fibres. The process can be accomplished via four ways, e.g. mechanical, biological, chemical, and physical retting. Specifically, biological retting consists of natural and artificial retting.

Natural retting can be conducted in two methods, such as cold-water and dew/field retting process. For cold-water retting, the process disintegrates the pectin part in plant bast bundles when anaerobic bacteria and water are present. In the process, the outermost layer of the plant is broken down via hydrolysis that infiltrates the central stalk portion. Later, the inner cells are swollen to increase the absorption of moisture and bacteria (Thomsen et al. 2005). The process is highly depending on various factors, e.g. water type, temperature, and bacterial resistance which takes around 1–2 weeks to be completed. Good quality of fibres can be obtained by employing this method. However, the disadvantages of the cold retting process are it is a time-consuming process as well as producing unnecessary pollutions into the environment through the disposal of wastewater (Magnusson and Svennerstedt 2007). The contaminated waste usually resulted from organic fermentation by the anaerobic bacteria.

Dew/field retting is another retting procedure for fibre extraction with limited water sources. In the process, it uses the dew to moisten the bast or leaves to break the fibre (Morrison et al. 2000). Usually, it takes about 3–6 weeks to complete, hence, is considered a very slow process (Bleuze et al. 2020). The retting is mainly dependent on the weather, environmental humidity, and bacterial growth. This retting process induces darker fibres that are reflected as low-quality fibre production compared to water retting approach. It has also typically considered an unpredictable process and also leads to pollution due to the fermentation to the environment.

Apart from biological and natural retting, physical retting is well-known as clean and fine fibre extraction with high and consistent quality. The physical retting or wet retting could alter the characteristics of the fibres depending on their applications by adjusting the processing parameters. Generally, three ways of physical retting are conducted, which include enzyme retting, stem explosion methods (STEX), and ultrasonic retting (Potluri 2019). The enzyme retting is the most expensive retting process in comparison with other retting methods, since the whole process takes only 2–24 h. The extracted fibres are usually long, and most of the fibres have high strength (Tian et al. 2016). Meanwhile, STEX retting produces great properties and fineness of fibres that are par with cotton fibres. The process is highly suitable for the textile industry (Song et al. 2017). Lastly, for the ultrasound retting, the process is employed for the stalks of the plant, which is usually useful for non-textile grade fibres (Tian et al. 2017).

Green retting process is a cost-effective retting process which implements a mechanical approach to separate fibre from xylem and straw of plant (Sisti et al. 2018). The green/mechanical retting entails the drying of leaves or stalks before mechanical extraction using machines to skin off and yield the fibres. The process does not require bacteria and water to successfully perform the fibre extraction. Usually, the fibres produced are coarser than those obtained by biological retting. As aforementioned statements, it shows that the course fibres do limit the usefulness of fibres via this process to be used as reinforcements in the composite applications.

Chemical retting carries out the extraction of fibres using heated water along with the addition of various chemical compounds to it (Jose et al. 2016). The process is time-effective, which dissolve the pectin in the stalks to separate the fibres. The chemical compounds use in this process are chlorinated lime, sulphuric acid, potassium hydroxide, sodium carbonate, and sodium hydroxide to chemically ret the fibre. The process could produce consistent and good properties fibres. However, the major drawback of this process is its expensive cost.

2.4.2 Mechanical Decortication

Mechanical decortication is applicable to extract most of the leaf fibres after harvesting the crops. To decorticate fibres easily, water retting is introduced that takes only about 1–5 days to complete (Hums et al. 2018). However, the process does not cover all types of leaf fibres. In the process, the leaves are crushed under pressure by the rotating drum consisting of blunt blades. From the crushing action, the fibres are broken apart from the other plant's hard structure. After decortication of the leaf, the fibres are washed and then dried.

Then, the fibres are either sun-dried or oven-dried under controlled temperature (Boyko et al. 2020). Specifically for this step, it is essential to determine the mechanical performance of the extracted fibre. For the subsequent step, the fibres are separated based on their specific dimensions, especially length and diameters. For specific fibres such as flax, additional step such as hackling is performed to separate the small and coarse fibres by combing the fibre using hackle (comb-like structure). This step is followed by mechanical carding to untangle, clean, and mix the fibre for future processing. Later, the drawing and spinning processes are carried out to extract plant fibres such as sisal, PALF, bamboo, hemp, and banana.

2.4.3 Mechanical Dehulling

Mechanical dehulling is a separation process of the extracted fibres which involves rolling the fibres to separate the hull from the seed or grain. Hull is a by-product from the seed or grain, which later is used as reinforcement in composite materials. The process is very useful, especially to extract green coconut coir fibre, wheat husk, and rice husk. Figure 2.4 shows the mechanical dehulling machine using rotary machine pulley and hopper (Olaniran et al. 2020).

2.5 Fibre Treatments

Interaction between fibre-matrix interfaces is essential in identifying the mechanical performance and thermal stability of the composite. In a general view, natural fibre such as plant cellulosic fibre has poor interfacial adhesion interaction between the matrix and fibre surface. Good interfacial adhesion of fibre-matrix interface is crucial to transfer the load from matrix to fibre in a composite laminate (Alsubari et al. 2021). The inhabitant components in plant fibre such as pectin, lignin, hemicellulose, and wax would reduce the strength of the cellulosic fibre (Omran et al. 2021). There are various treatments such as physical, chemical, and physicochemical to improve fibre

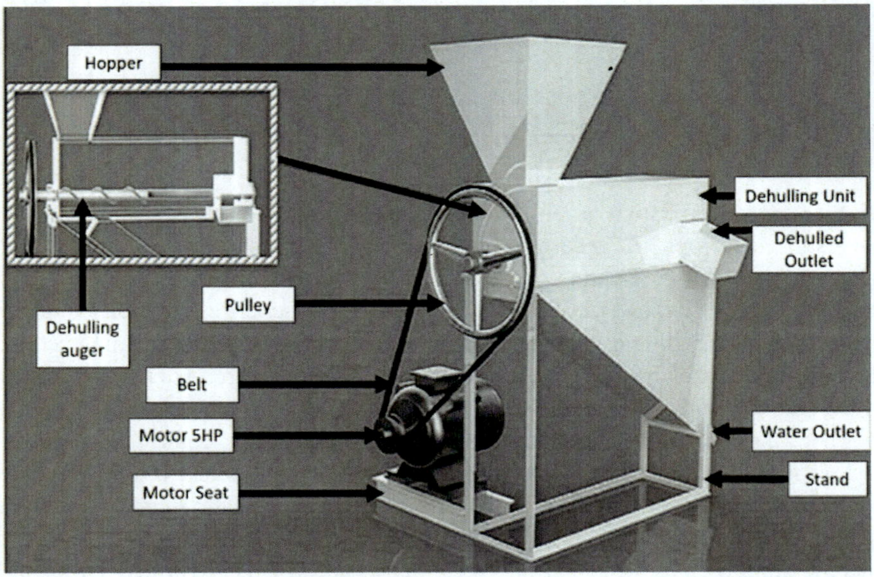

Fig. 2.4 Mechanical dehulling machine using rotary machine pulley and hopper (Olaniran et al. 2020)

surface by removing these components to provide compatibility between fibre-matrix (Hamidon et al. 2019).

Physical treatments are one of the methods used to alter the structural and interface of lignocellulosic fibres. These physical modification methods include fibre stretching, gamma-ray projection, electric discharge, calendaring, swaging or rolling, thermal treatments, and solving attraction (Hamidon et al. 2019). Chemical treatments are other approaches to treat fibres by using numerous chemical compounds to modify the fibre surface by removing several unwanted plant components like lignin and pectin. For instance, the alkaline, grafting, acetylation, silane treatment, benzoylation and bleaching are used as coupling agents for chemical treatments. Meanwhile, Physicochemical treatment or in other words combination treatments of physical (thermal or steam explosion) with chemical treatments produces better fibre bundles. This treatment aids in enhancing chemical reactions.

2.6 Manufacturing Process of Composites and Biocomposites

To fabricate any composite products, an appropriate manufacturing process is required to translate the concept design to a finished product. The product can be forged in many ways including resin transfer moulding (RTM), compression moulding, hand lay-up, vacuum bagging, pultrusion, filament winding, and injection moulding (Vijay et al. 2016). Among these methods, the most widely used manufacturing process techniques for thermoset are compression moulding and hand lay-up, while thermoplastic matrix composites' manufacturing processes are injection moulding and screw extrusion. A brief discussion about some of the processing methods used for the production of hybrid composites is mentioned in the following paragraphs.

2.6.1 Hand Lay-Up

A hand lay-up process is a famous open moulding method used in manufacturing polymeric composites that require high skill operator to command this foremost process. Generally, both thermoplastics and thermosets composites can be used this process to fabricate the specific composite products. An optimum amount of the raw material determination for the required density of composite is done by estimating the mixing ratio of fibre to the matrix (Singh and Kumar 2017). Initially, the mould surface is prepared with any releasing agents to detach composite laminate from the surface of the mould. This process has largely fabricated various automotive components, e.g. door panels, interior engine covers (Rosli et al. 2013), parking lever brake (Mansor et al. 2014), and automobile spoiler (Mansor et al. 2015) since

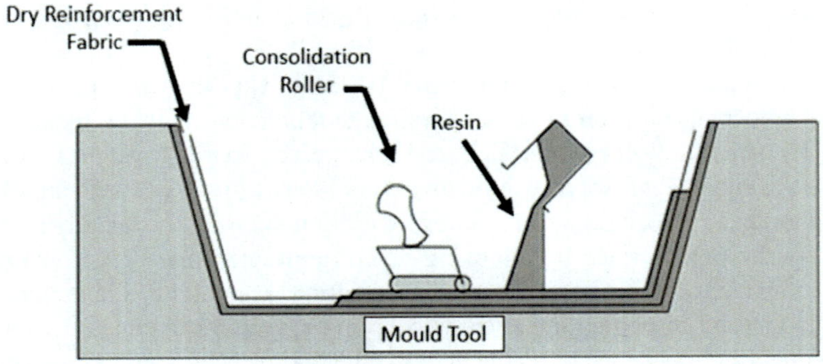

Fig. 2.5 Hand lay-up process (Sharma et al. 2016)

the method is efficient and accurate to extrude complex shape. Figure 2.5 shows the hand lay-up process of composite with specific mould.

2.6.2 Pultrusion

Another manufacturing technique of composite products is pultrusion which combines the process of both pulling and extruding. The process started when the fibres are impregnated with any thermoset or thermoplastic resins to form impregnated fibre composite. At this point, the fibres are passed through the resin bath before getting by through heated die to obtain final product shape (Asyraf et al. 2020a). There are various shapes of the final products, such as rectangular, I and H shapes, circular, and shape to form long and symmetrical products. For instance, the pultruded composite products include cross arms, pipes, and structural beams (Nadhirah et al. 2017). Figure 2.6 displays the pultrusion of impregnated fibre with polymeric resin in a single processing line.

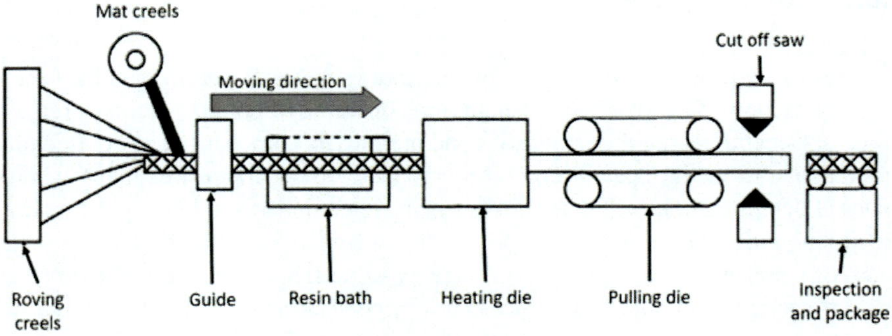

Fig. 2.6 Pultrusion of the symmetrical composite product (Fairuz et al. 2014)

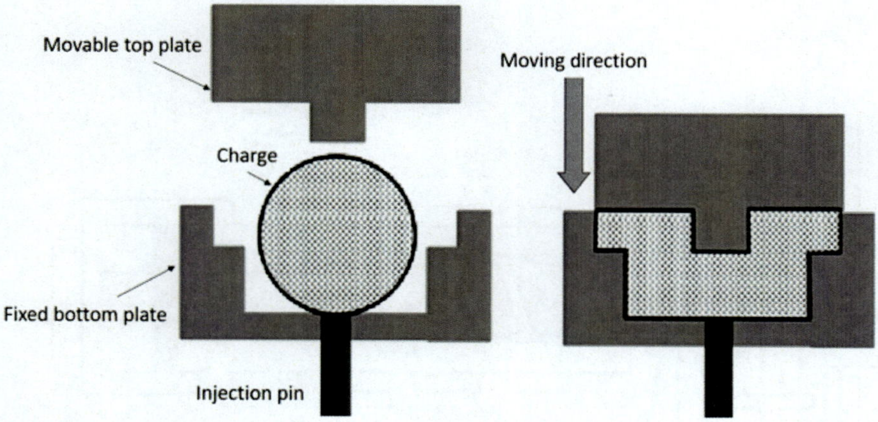

Fig. 2.7 Process of compression moulding of composite (Ketabchi et al. 2015)

2.6.3 Compression Moulding

Both thermosets and thermoplastics polymeric composites used compression moulding to produce composite laminates at a faster rate (Curvelo et al. 2001). This manufacturing method is divided into two processes, which are cold and hot presses. Both presses implement pressure during the moulding process of polymeric composites, but hot press method executes both heat and pressure to fabricate the composite laminate. Application of heat in hot press moulding is to initiate the curing of resin during the process. Figure 2.7 depicts the compression moulding using both upper and bottom plate moulds. The compression moulding is a well-known method implement in the automotive industry due to the process that is suitable for large production volumes (Kasiviswanathan et al. 2015). Additional equipment and machines such as internal mixers and twin screw extruders usually are introduced to attain better distribution of fibres in the polymeric resin.

2.6.4 Injection Moulding

The most well-known manufacturing plastics components, natural, and synthetic-based fibre composite products are injection moulding (Pickering et al. 2007). At first, the fibres should be either in form of small or powder form, while the polymers have to be in the form of small pellets. As shown in Fig. 2.8, the screw extruder is applied in the manufacturing process to mix the resin granules with short/powdered fibres particles. Later, the mixed compounds are flowed via hopper to heat chamber to melt the granules. Then, the melted granules are injected to the shape-desired mould at very high pressure. This method is essentially applicable to produce intricate parts and products for thermoplastic polymers (Liu et al. 2009).

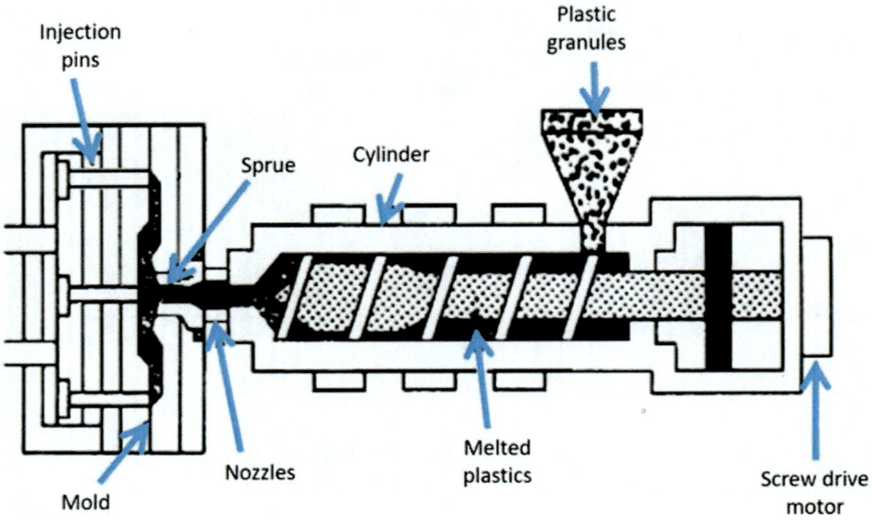

Fig. 2.8 Injection moulding of composite products

2.6.5 *Vacuum Bagging/Moulding*

This moulding requires preparation of prepreg, which can be divided into two techniques, which is autoclave and vacuum bagging methods. Vacuum bagging and vacuum moulding differ from each other in terms of the method of curing of the composite's polymeric resin. The manufacturing process takes place when the plies of the composite laminate are layered and consolidated. The compaction pressure is applied by vacuum technique, which is known as a vacuum bagging process (Bajuri et al. 2017). In the first place, the prepreg materials are employed on a horizontal mould, and it is enclosed with a vacuum bag that is later sealed to mould, forcing the air to withdraw out from the bag, creating a necessary vacuum to cure the resin as shown in Fig. 2.9 (Mallick 2010; Fiore et al. 2012). Various types of resins are used in this technique, such as epoxies, phenolic resins, and polyimides.

2.6.6 *Resin Transfer Moulding*

Resin transfer moulding (RTM) is a renowned manufacturing technique for high capacity production at low-cost (Saba et al. 2015). As shown in Fig. 2.10, the process requires fine particles of resins to be injected on the fibres at the stationary platform. Next, a chamber is used to dry the mould. Post curing is also required for this process. Usually this method is executed to produce various composite products for the automotive and aerospace applications.

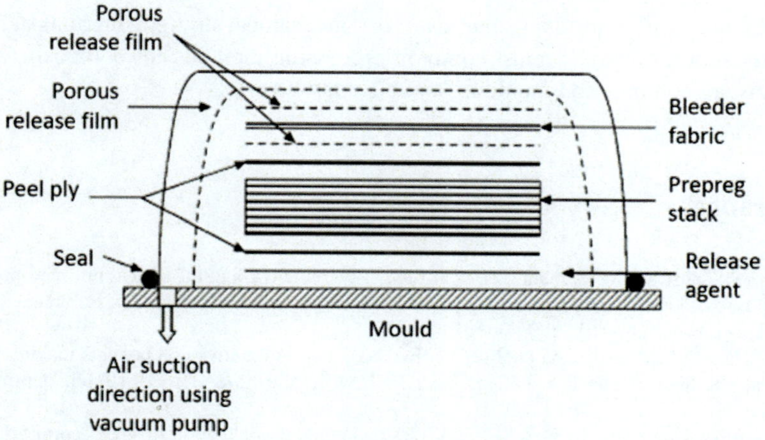

Fig. 2.9 Vacuum bagging technique to fabricate composite product (Mallick 2010)

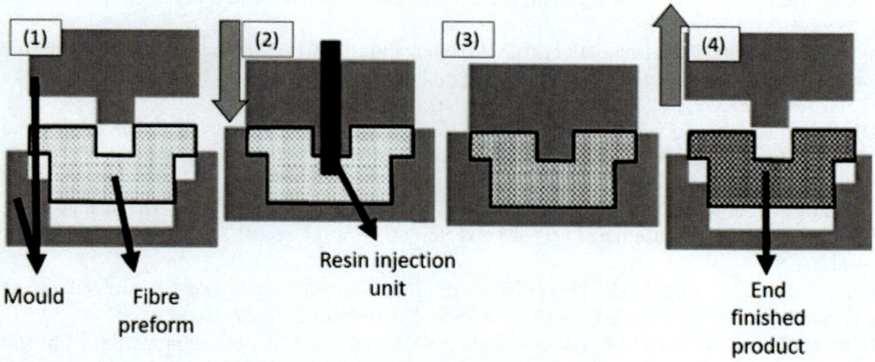

Fig. 2.10 Fabrication process of resin transfer moulding (RTM) (Ketabchi et al. 2015)

2.7 Conclusions

Many studies have been performed to put forward the composite products in many sectors, especially automotive, aerospace, construction, marine, sports, and entertainments. Moreover, recent progress and development were carried out by all society pillars, including government bodies, educational institutions, and corporate industries to shift the current synthetic composites towards biocomposites for greener and better environmental sustainability. They have worked together to improve the current biocomposites by enhancing the performance of natural fibres via retting, treatment, and hybridising with suitable thermoset and thermoplastic resins. The manufacturability of both synthetic composites and biocomposites were those highlighted by most researchers for large volumes productions and to obtain complex shape products. In this case, also, a specific manufacturing process of composite products is selected based on the product's purpose, cost, mechanical strength, thermal stability, and production time. The most common processing methods for composite-based products are compression moulding and hand lay-up.

References

Alsubari S, Zuhri MYM, Sapuan SM et al (2021) Potential of natural fiber reinforced polymer composites in sandwich structures: a review on its mechanical properties. Polymers (basel) 13:423. https://doi.org/10.3390/polym13030423

Alves C, Ferrão PMC, Silva AJ et al (2010) Ecodesign of automotive components making use of natural jute fiber composites. J Clean Prod 18:313–327. https://doi.org/10.1016/j.jclepro.2009.10.022

Asyraf MRM, Ishak MR, Sapuan SM et al (2020) Potential application of green composites for cross arm component in transmission tower: a brief review. Int J Polym Sci. https://doi.org/10.1155/2020/8878300

Asyraf MRM, Rafidah M, Azrina A, Razman MR (2021) Dynamic mechanical behaviour of kenaf cellulosic fibre biocomposites: a comprehensive review on chemical treatments. Cellulose. https://doi.org/10.1007/s10570-021-03710-3

Asyraf MRM, Rafidah M, Ishak MR et al (2020) Integration of TRIZ, Morphological Chart and ANP method for development of FRP composite portable fire extinguisher. Polym Compos 41:2917–2932. https://doi.org/10.1002/pc.25587

Athanasiou KA, Niederauer GG, Agrawal CM (1996) Sterilization, toxicity, biocompatibility and clinical applications of polylactic acid/polyglycolic acid copolymers. Biomaterials 17:93–102. https://doi.org/10.1016/0142-9612(96)85754-1

Azeredo HMC, Mattoso LHC, Wood D et al (2009) Nanocomposite edible films from mango puree reinforced with cellulose nanofibers. J Food Sci 74. https://doi.org/10.1111/j.1750-3841.2009.01186.x

Bajuri F, Mazlan N, Ishak MR (2017) Effect of silica nanoparticles in kenaf reinforced epoxy: Flexural and compressive properties. Pertanika J Sci Technol 25:1029–1038

Bajuri F, Mazlan N, Ishak MR, Imatomi J (2016) Flexural and compressive properties of hybrid kenaf/silica nanoparticles in epoxy composite. Procedia Chem 19:955–960. https://doi.org/10.1016/j.proche.2016.03.141

Bashir ASM, Manusamy Y (2015) Recent Developments in Biocomposites Reinforced with Natural Biofillers from Food Waste. Polym—Plast Technol Eng 54:87–99. https://doi.org/10.1080/03602559.2014.935419

Bleuze L, Chabbert B, Lashermes G, Recous S (2020) Hemp harvest time impacts on the dynamics of microbial colonization and hemp stems degradation during dew retting. Ind Crops Prod 145. https://doi.org/10.1016/j.indcrop.2020.112122

Boyko G, Tikhosova H, Ternova T (2020) Optimization of the decortication process of industrial hemp stems by mathematical planning method. INMATEH—Agric Eng 60:53–60. https://doi.org/10.35633/INMATEH-60-06

Chevali VS, Dean DR, Janowski GM (2009) Flexural creep behavior of discontinuous thermoplastic composites: Non-linear viscoelastic modeling and time-temperature-stress superposition. Compos Part A Appl Sci Manuf 40:870–877. https://doi.org/10.1016/j.compositesa.2009.04.012

Curvelo AAS, De Carvalho AJF, Agnelli JAM (2001) Thermoplastic starch-cellulosic fibers composites: preliminary results. Carbohydr Polym 45:183–188. https://doi.org/10.1016/S0144-8617(00)00314-3

Fairuz AM, Sapuan SM, Zainudin ES, Jaafar CNA (2014) Polymer composite manufacturing using a pultrusion process: A review. Am J Appl Sci 11:1798–1810. https://doi.org/10.3844/ajassp.2014.1798.1810

Fei M, Liu T, Fu T et al (2019) Styrene-free soybean oil thermoset composites reinforced by hybrid fibers from recycled and natural resources. ACS Sustain Chem Eng 7:17808–17816. https://doi.org/10.1021/acssuschemeng.9b04308

Fiore V, Valenza A, Di Bella G (2012) Mechanical behavior of carbon/flax hybrid composites for structural applications. J Compos Mater 46:2089–2096. https://doi.org/10.1177/0021998311429884

Hamidon MH, Sultan MTH, Ariffin AH, Shah AUM (2019) Effects of fibre treatment on mechanical properties of kenaf fibre reinforced composites: a review. J Mater Res Technol 8:3327–3337. https://doi.org/10.1016/j.jmrt.2019.04.012

Hums ME, Moreau RA, Yadav MP et al (2018) Comparison of bench-scale decortication devices to fractionate bran from sorghum. Cereal Chem 95:720–733. https://doi.org/10.1002/cche.10087

Ilyas RA, Sapuan SM (2020) Biopolymers and biocomposites: chemistry and technology. Curr Anal Chem 16:500–503. https://doi.org/10.2174/157341101605200603095311

Ilyas RA, Sapuan SM, Asyraf MRM et al (2020a) Introduction to biofiller reinforced degradable polymer composites. In: Sapuan SM, Jumaidin R, Hanafi I (eds) Biofiller reinforced biodegradable polymer composites. CRC Press, Boca Raton, USA, pp 1–23

Ilyas RA, Sapuan SM, Asyraf MRM et al (2020b) Mechanical and dynamic mechanical properties of macro-nanosized natural fibre reinforced polymer composite. In: Kumar SMK (ed) Mechanical and dynamic mechanical analysis of biocomposite. Wiley, West Sussex, UK

Ilyas RA, Sapuan SM, Atiqah A et al (2020c) Sugar palm (*Arenga pinnata* [Wurmb.] Merr) starch films containing sugar palm nanofibrillated cellulose as reinforcement: water barrier properties. Polym Compos 41:459–467. https://doi.org/10.1002/pc.25379

Ilyas RA, Sapuan SM, Ishak MR et al (2018a) Characterization of sugar palm nanocellulose and its potential for reinforcement with a starch-based composite. In: Sugar palm biofibers, biopolymers, and biocomposites. CRC Press, First edition. CRC Press/Taylor & Francis Group, Boca Raton, FL, pp 189–220

Ilyas RA, Sapuan SM, Norrrahim MNF et al (2020d) Nanocellulose/starch biopolymer nanocomposites: Processing, manufacturing, and applications. In: Al-Oqla FM, Sapuan SM (eds) Advanced Processing, Properties, and Applications of Starch and Other Bio-Based Polymers, 1st edn. Elsevier Inc., Amsterdam, Netherland, pp 65–88

Ilyas RA, Sapuan SM, Sanyang ML et al (2018b) Nanocrystalline cellulose as reinforcement for polymeric matrix nanocomposites and its potential applications: a review. Curr Anal Chem 14:203–225. https://doi.org/10.2174/1573411013666171003155624

Ishak NM, Sivakumar D, Mansor MR (2018) The application of TRIZ on natural fibre metal laminate to reduce the weight of the car front hood. J Brazilian Soc Mech Sci Eng 40:1–12. https://doi.org/10.1007/s40430-018-1039-2

Jawaid M, Abdul Khalil HPS (2011) Cellulosic/synthetic fibre reinforced polymer hybrid composites: a review. Carbohydr Polym 86:1–18. https://doi.org/10.1016/j.carbpol.2011.04.043

Johari AN, Ishak MR, Leman Z et al (2019) Fabrication and cut-in speed enhancement of savonius vertical axis wind turbine (SVAWT) with hinged blade using fiberglass composites. In: Seminar Enau Kebangsaan. Bahau, Negeri Sembilan, Malaysia, pp 978–983

John MJ, Thomas S (2008) Biofibres and biocomposites. Carbohydr Polym 71:343–364. https://doi.org/10.1016/j.carbpol.2007.05.040

Jose S, Mishra L, Basu G, Samanta AK (2016) Study on reuse of coconut fiber chemical retting bath. Part 1: Retting efficiency. J Nat Fibers 13:603–609. https://doi.org/10.1080/15440478.2015.1093441

Jumaidin R, Khiruddin MAA, Asyul Sutan Saidi Z et al (2019) Effect of cogon grass fibre on the thermal, mechanical and biodegradation properties of thermoplastic cassava starch biocomposite. Int J Biol Macromol. https://doi.org/10.1016/j.ijbiomac.2019.11.011

Kasiviswanathan S, Santhanam K, Kumaravel A (2015) Evaluation of mechanical properties of natural hybrid fibers, reinforced polyester composite materials. Carbon—Sci Technol 7:43–49

Ketabchi MR, Hoque ME, Khalid Siddiqui M (2015) Critical concerns on manufacturing processes of natural fibre reinforced polymer composites. In: Sapuan SM, Jawaid M, Yusoff N, Hoque

ME (eds) Manufacturing of natural fibre reinforced polymer composites, 1st edn. Springer International Publishing, Cham, Switzerland, pp 125–138

Khan T, Hameed Sultan MT, Bin AAH (2018) The challenges of natural fiber in manufacturing, material selection, and technology application: a review. J Reinf Plast Compos 37:770–779. https://doi.org/10.1177/0731684418756762

Li F, Marks DW, Larock RC, Otaigbe JU (2000) Fish oil thermosetting polymers: synthesis, structure, properties and their relationships. Polymer (guildf) 41:7925–7939. https://doi.org/10.1016/S0032-3861(00)00030-6

Liu T, Yao K, Gao F (2009) Identification and autotuning of temperature-control system with application to injection molding. IEEE Trans Control Syst Technol 17:1282–1294. https://doi.org/10.1109/TCST.2008.2006746

Magnusson K, Svennerstedt B (2007) Influence of temperature on the water retting process of hemp (*Cannabis sativa* L.) cultivated under Swedish climate conditions. J Ind Hemp 12:3–17. https://doi.org/10.1300/J237v12n02_02

Mallick PK (2010) Thermoset–matrix composites for lightweight automotive structures. In: Mallick PK (ed) Materials, design and manufacturing for lightweight vehicles, 1st edn. Woodhead Publishing Series, Boca Raton, USA, pp 208–231

Mansor MR, Sapuan SM, Hambali A (2015) Conceptual Design of Kenaf Polymer Composites Automotive Spoiler Using TRIZ and Morphology Chart Methods. 761:63–67. https://doi.org/10.4028/www.scientific.net/AMM.761.63

Mansor MR, Sapuan SM, Zainudin ES, Nuraini AA (2014) Conceptual design of kenaf fiber polymer composite automotive parking brake lever using integrated TRIZ-Morphological Chart-Analytic Hierarchy Process method. Mater Des 54:473–482. https://doi.org/10.1016/j.matdes.2013.08.064

Marques AT (2011) Fibrous materials reinforced composites production techniques. Fibrous Compos Mater Civ Eng Appl, 191–215. https://doi.org/10.1533/9780857095583.3.191

Mohanty AK, Drzal LT, Misra M (2002a) Engineered natural fiber reinforced polypropylene composites: influence of surface modifications and novel powder impregnation processing. J Adhes Sci Technol 16:999–1015. https://doi.org/10.1163/156856102760146129

Mohanty AK, Misra M, Drzal LT (2002b) Sustainable Bio-Composites from renewable resources: opportunities and challenges in the green materials world. J Polym Environ 10:19–26. https://doi.org/10.1023/A:1021013921916

Mohanty AK, Misra M, Hinrichsen G (2000) Biofibres, biodegradable polymers and biocomposites: an overview. Macromol Mater Eng 276–277:1–24. https://doi.org/10.1002/(SICI)1439-2054(20000301)276:1%3c1::AID-MAME1%3e3.0.CO;2-W

Morrison WH, Archibald DD, Sharma HSS, Akin DE (2000) Chemical and physical characterization of water- and dew-retted flax fibers. Ind Crops Prod 12:39–46. https://doi.org/10.1016/S0926-6690(99)00044-8

Nadhirah A, Mohamad D, Zainoodin M et al (2017) Properties of fiberglass crossarm in transmission tower—a review. Prop Fiberglass Crossarm Transm Tower Rev 12:15228–15233

Nurazzi NM, Khalina A, Sapuan SM et al (2019) Thermal properties of treated sugar palm yarn/glass fiber reinforced unsaturated polyester hybrid composites. J Mater Res Technol. https://doi.org/10.1016/j.jmrt.2019.11.086

Olaniran AF, Okonkwo CE, Erinle OC et al (2020) Optimum boiling duration and its effect on nutritional quality and acceptability of mechanically dehulled unfermented locust bean seeds. Prev Nutr Food Sci 25:219–224. https://doi.org/10.3746/pnf.2020.25.2.219

Omran AAB, Mohammed AABA, Sapuan SM et al (2021) Micro- and nanocellulose in polymer composite materials: a review. Polymers (basel) 13:231. https://doi.org/10.3390/polym13020231

Pickering KL, Beckermann GW, Alam SN, Foreman NJ (2007) Optimising industrial hemp fibre for composites. Compos Part A Appl Sci Manuf 38:461–468. https://doi.org/10.1016/j.compositesa.2006.02.020

Potluri R (2019) Natural fiber-based hybrid bio-composites: processing, characterization, and applications. In: Muthu SS (ed) Green composites: processing, characterisation and applications for textiles, 1st edn. Springer Nature Singapore Pte Ltd., Singapore, pp 1–46

Rosli MU, Ariffin MKA, Sapuan SM, Sulaiman S (2013) Integrated AHP-TRIZ innovation method for automotive door panel design. Int J Eng Technol 5:3158–3167

Saba N, Paridah MT, Jawaid M et al (2015) Manufacturing and processing of kenaf Fibre-reinforced epoxy composites via different methods. In: Sapuan SM, Jawaid M, Yusof N, Hoque ME (eds) Manufacturing of natural fibre reinforced polymer composites. Springer, New York, USA, pp 101–124

Sanyang ML, Sapuan SM, Jawaid M et al (2016) Recent developments in sugar palm (*Arenga pinnata*) based biocomposites and their potential industrial applications: a review. Renew Sustain Energy Rev 54:533–549. https://doi.org/10.1016/j.rser.2015.10.037

Shahroze RM, Chandrasekar M, Senthilkumar K et al (2019) A review on the various fibre treatment techniques used for the fibre surface modification of the sugar palm fibres. In: Seminar Enau Kebangsaan. Bahau, Negeri Sembilan, Malaysia, pp 48–52

Shalini R, Singh A (2009) Biobased packaging materials for the food industry. J Food Sci Technol 5:16–20

Sharma SD, Sowntharya L, Kar KK (2016) Polymer-based composite structures: processing and applications. In: Kar KK (ed) Composite materials: processing, applications, characterizations, 1st edn. Springer-Verlag Berlin Heidelberg, Berlin, Heidelberg, pp 1–36

Singh S, Kumar V (2017) Review of current technology in glass fibre composite. Int J Mech Prod Eng Res Dev 7:109–118. https://doi.org/10.24247/ijmperdaug201712

Sisti L, Totaro G, Vannini M, Celli A (2018) Retting process as a pretreatment of natural fibers for the development of polymer composites. In: Kalia S (ed) Lignocellulosic composite materials, 1st edn. Springer International Publishing AG, Cham, Switzerland, pp 97–135

Song Y, Han G, Jiang W (2017) Comparison of the performance of kenaf fiber using different reagents presoak combined with steam explosion treatment. J Text Inst 108:1762–1767. https://doi.org/10.1080/00405000.2017.1285200

Stochioiu C, Chettah A, Piezel B et al (2018) Study of the time varying properties of flax fiber reinforced composites. AIP Conf Proc 1932:https://doi.org/10.1063/1.5024191

Thomsen AB, Rasmussen S, Bohn V et al (2005) Hemp raw materials: the effect of cultivar, growth conditions and pretreatment on the chemical composition of the fibres. Risø-R Rep 1507:6–30

Tian Y, Liu X, Zheng X et al (2017) Effects of ultrasonic wave on flax retting and fiber properties. Wool Text J 45:40–42. https://doi.org/10.19333/j.mfkj.2016070080503

Tian Y, Liu X, Zheng X, Wang L (2016) Antimicrobial properties of flax fibers in the enzyme retting process. Fibres Text East Eur 24:15–17. https://doi.org/10.5604/12303666.1172082

Vijay N, Rajkumara V, Bhattacharjee P (2016) Assessment of composite waste disposal in aerospace industries. Procedia Environ Sci 35:563–570. https://doi.org/10.1016/j.proenv.2016.07.041

Yang Y, Boom R, Irion B et al (2012) Chemical Engineering and Processing : Process Intensification Recycling of composite materials. Chem Eng Process Process Intensif 51:53–68. https://doi.org/10.1016/j.cep.2011.09.007

Yu L, Dean K, Li L (2006) Polymer blends and composites from renewable resources. Prog Polym Sci 31:576–602. https://doi.org/10.1016/j.progpolymsci.2006.03.002

Zhong Y, Godwin P, Jin Y, Xiao H (2020) Biodegradable polymers and green-based antimicrobial packaging materials: a mini-review. Adv Ind Eng Polym Res 3:27–35. https://doi.org/10.1016/j.aiepr.2019.11.002

Chapter 3
Emission of Hazardous Air Pollution in the Composite Production

Abstract Composite production contributes to air pollution by releasing hazardous pollutants into the atmosphere. This study reviews the emission of styrene from the fibre reinforced plastics composites and thermoset composites manufacturing, formaldehyde emission in wood-based composites production, the national standard for hazardous air pollution in composites production, environmental issues related to health and safety, and the control measures of the composites pollution. Based on the review on styrene emission in fibre reinforced plastics composites and thermoset composites, the styrene emission by composites production through the process of open mould process and closed mould process. Other than that, the production of the wood-based composite is used to produce furniture components, support building as well as interior and exterior uses. Those wood-based composites release volatile organic compounds, VOCs that consist of formaldehyde that is carcinogenic to humans and deteriorating indoor air quality, contributing to environmental issues related to health and safety. For example, the emissions of VOCs and formaldehyde cause cancer and other respiratory diseases, due to the main route of those emissions entering the humans' bodies is through inhalation. Thus, some of the studies have shown that the emission of chemicals also might cause skin and eye irritations and allergic reactions. In addition, the national standard for hazardous air pollutants in composites production need to be complimented to reduce the air pollutant and control the hazards. Some of the regulations and acts are applying to these issues, including the Clean Air Act (CAA) Amendments 1990, Industrial Code of Practice (ICOP) on Indoor Air Quality, Occupational and Safety (Use and Standard of Exposure Chemical Hazardous to Health) Regulation 2000 (USECHH Regulation), and International Regulations and Guidelines. The controls are made to reduce air pollution or to improve the effect of emissions from composite production to comply with the regulations.

Keywords Composite · Air pollution · National standard · Control · Health · Formaldehyde · Styrene

S. M. Sapuan et al., *Safety and Health in Composite Industry*,
Composites Science and Technology ,
https://doi.org/10.1007/978-981-16-6136-5_3

3.1 Introduction

Composites production is extensively releasing hazardous air pollutants in the working environment and atmosphere. A composite material resulted from a combination of two or more materials with different chemical and physical properties. Di Tomasso et al. (2014) stated that composite manufacturing production consists of three moulding processes; open, closed, and cast polymer. Examples of composites include wood, thermoset, fibre reinforced plastics, and ceramic-matrix.

In this book chapter, researchers had studied the emission of styrene and formaldehyde from fibre reinforced plastics, thermoset composites, and wood-based composite productions, respectively, the national standard for hazardous air pollution in composite manufacturing, environmental issues related to health and safety, and control of composite pollution.

Baley et al. (2006) found that the fibre reinforced (FRP) fabrication in the United States is the main source of styrene emission. According to the Clean Air Act Amendments, styrene is considered as ozone precursors and delegated as a toxic air pollutant. These composite productions have the potential to release hazardous air pollutants, e.g. styrene, methyl methacrylate (MMA), methylene chloride (dichloromethane), and formaldehyde. Moreover, Crawford and Lungu (2011) also found that the styrene is also released from the thermoset composite manufacturing using the environmental test chamber method. Besides, formaldehyde (HCHO) is the simplest, naturally occurring chemical in the woods but a highly reactive organic compound is still detected in a small amount from combustion process.

Composite production emits hazardous chemicals to human health through the air that comprises VOCs, styrene, and formaldehyde. Most of the air pollutants have higher to harm humans' health when indoor than outdoor since they enter humans' bodies through respiration. They cause lung cancer, ischemic heart disease, asthma, and other respiratory diseases. Other than that, some of these chemicals such as formaldehyde also cause irritation and sensitization on skin and eye.

A National Emission Standard was introduced for these hazardous air pollutants as an effort to minimize the environmental issues related to health and safety by the implementation of these hazardous air pollutants (HAPs). For example, the Maximum Achievable Control Technology (MACT) was introduced by the United States Environmental Protection Agency (USEPA) to control the emission of HAPs. Meanwhile, to control and reduce the release of formaldehyde emissions from imported and domestic wood production, the Formaldehyde Emission Standard was introduced.

Besides, being the source of composite pollution, the industry also plays an important role to control composite pollution. There are a few measures that can be taken by the company or manufacturer to reduce the composite pollution and comply with the regulations. These include the usage of low-styrene materials, VOC abatement system, and lowering the mol ratio of urea–formaldehyde resin.

3.2 Styrene Emissions

3.2.1 *What Is Styrene?*

In general, styrene which goes by the name of phenylethene, etenylbenzene, and vinylbenzene is an organic compound with the chemical formula of $C_6H_5CH{=}CH_2$. In spite of the fact that high concentrations of styrene have a less pleasant odour, but it also evaporates easily in the air (EPA 2018a, b). According to Crawford and Lungu (2011), the unsaturated polyester resin, a volatile organic compound that is extensively used in fibre reinforced plastic composite manufacturing is a component of styrene monomer. Moreover, the monomer of styrene is also a component of vinyl ester resin that is used in the production of thermoset composite material. Di Tomasso et al. (2014) claimed that styrene can affect humans health as it is classified as hazardous air pollution. The processes in composite production that emit styrene emission are open mould and closed mould. This issue raised the interest of more researchers to study styrene emission. Therefore, this review paper compiles and discusses the previous studies' findings on the emission of styrene in composite production.

3.2.2 *Importance and Uses of Styrene*

Two million tonnes of composite per year are produced by the fibre reinforced plastic composite production in both Europe and North America regions. Di Tomasso et al. (2014) reported that styrene's primary source of emission is open mould composite processes that possibly spread some VOCs into the working environment and atmosphere. Then, Hammond et al. (2011) clarified that when the resin process is completely restored, styrene evaporates into the environment. In another study, Nunez et al. (1999) also explained that styrene can reduce the viscosity and increase the performance of resin as it acts as a reactive diluent in FRP production. Figure 3.1 shows the consumption of styrene around the world in 2017. The involved countries are China, Japan, Taiwan, South Korea, the United States, and Western Europe which have higher styrene consumption in the market than others IHS Marlit (2018).

Styrene is mainly used in the production of glass fibre reinforced plastic and thermoset composites that used unsaturated polyester resin and vinyl ester resin. The glass fibre reinforced plastic products possess a wide range of applications, including storage tanks, transportation rail, electrical equipment, construction industry, automobile and aircraft, boat construction, machine components, bathroom fittings, and recreation items. Furthermore, the products from thermoset composite are used extensively in many applications such as stairs, aircraft, furniture, ceiling, wall, and components of toys and games. According to Clean Air Act Amendments 1990, styrene is considered as ozone precursors and delegated as a toxic air pollutant. It is also being

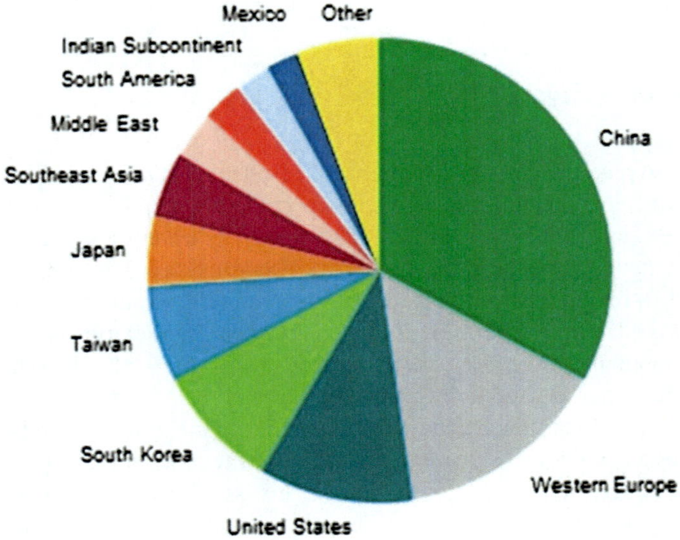

Fig. 3.1 The world consumption of styrene in year 2017

categorized under Group 2A, which is probably carcinogenic to humans according to animal experiments on its carcinogenicity.

3.3 Formaldehyde Emissions

3.3.1 Wood Based Composites

In the new global economy, the forest products industry, especially wood based panels have become a center for environmental issues. After World War II in 1954 to 1975, solid wood in homes and construction are hugely being replaced from plywood (Weschler 2009). Any type of wood being bonded with adhesives or other bonding methods is called a composite. Today, engineered wood products, or commonly known as wood-based composite (WBC) or composite wood products (CWP) stand out as unique and versatile that are ideal for their application in construction, industry, and home purposes, since they are available in a variety of classifications according to size, thickness, and grade.

There are a few categories of wood-based composites (WBC); the panel products such as plywood, oriented strand board (OSB), particleboard, and medium-density fibreboard (MDF); the structural timber products, e.g. glued-laminated timber (glulam), laminated veneer lumber (LVL), and laminated strand lumber; and the wood-non wood composites, e.g. wood fibre-thermoplastics and inorganic bonded composites (Cai and Ross 2010). The commonly used adhesives in the industry are the urea-based adhesives, such as urea–formaldehyde resin, the phenolic-based adhesive such as phenol resorcinol adhesives, the isocyanate-based adhesive, and the adhesives from inexhaustible resources, such as from soybean and lignin (Shi and Walker 2006). However, the industries have traditionally used urea–formaldehyde resins in the manufacturing of wood-based panels (Athanassiadou et al. 2009).

3.3.2 *Building Materials and Indoors Products*

Wood-based composites (WBC) are being used nowadays in numerous product lines, which include both interior and exterior uses, structural and non-structural applications, parts of furniture, and as support structures in buildings (Cai and Ross 2010). The chemical properties of wood are complex, but humans had modified it so that it can be used in building unlimited structures. These extraordinary materials are commonly used in the construction of buildings and boats, and also were massively used as home furniture and in the decor industry.

These innovations of wood have some benefits, such as lighter weight, greater thermal insulations, good fireproofing, have acoustic properties, and high energy-dissipation capacity, especially as an earthquake-proof (Clark et al. 2012). Since the incidence of earthquakes in Abruzzo (L'Aquila) during 2009, the usage of innovative wood products has increased dramatically in Italy. Figure 3.2 shows buildings that were constructed with wood-based composites (WBC) have their own environmental values. Clark et al. (2012) claimed that in multi-storey buildings built with WBC, their greenhouse gas emissions were 10% lower than other similar buildings without WBC.

According to Weschler (2009), the most obvious sources of airborne formaldehyde indoors are composite wood products (CWPs), such as kitchen cabinets and furniture made from urea–formaldehyde (UF) and melamine–formaldehyde (MF) resin. It is reported that when plywood first came into the market, its adhesive resin used urea–formaldehyde and its formaldehyde emission was high during that time, which was approximately more than 1000 $\mu g\ m^{-2}\ h$ (Weschler 2009).

3.4 Volatile Organic Compound (VOC) Emissions—Formaldehyde

Volatile organic compounds (VOCs) are one of the organic compounds that have high volatility and low water solubility. The VOC emissions could be detected as early as the wood is still in the forest and continues to present throughout the processes of making composite wood products and lasts in the products themselves (Roffael 2006).

Fig. 3.2 The largest solid timber building located in the UK, named Bridport House (Clark et al. 2012)

Formaldehyde (HCHO) is the most simple but highly reactive organic compound. It is a naturally occurring chemical in wood and can be detected with a small amount of free formaldehyde. It is one of the aldehydes that emitted the most from the composite wood products, due to the fact that they are bonded with formaldehyde-based resins such as urea–formaldehyde (UF) resins and phenol–formaldehyde (PF) resins (Kim et al. 2006). Formaldehyde might also be released from composite wood products that use thermosetting adhesives, such as amino plastic resins. The IARC had reclassified formaldehyde as Group 1, which is carcinogenic to humans (IARC 2006).

In industries that manufacturing wood products, their exposure to formaldehyde is likely to occur during glue mix preparation, lying off the mat, hot pressing work, and sanding (Fig. 3.3) (Shi and Walker 2006). They are primarily emitted from works that used hot pressing in wood products, including wood products that were bonded with urea–formaldehyde (UF) resins, for examples, particleboard and medium-density fibreboard (MDF); and were bonded with phenol–formaldehyde (PF) resins, such as softwood plywood and oriented strand board (OSB).

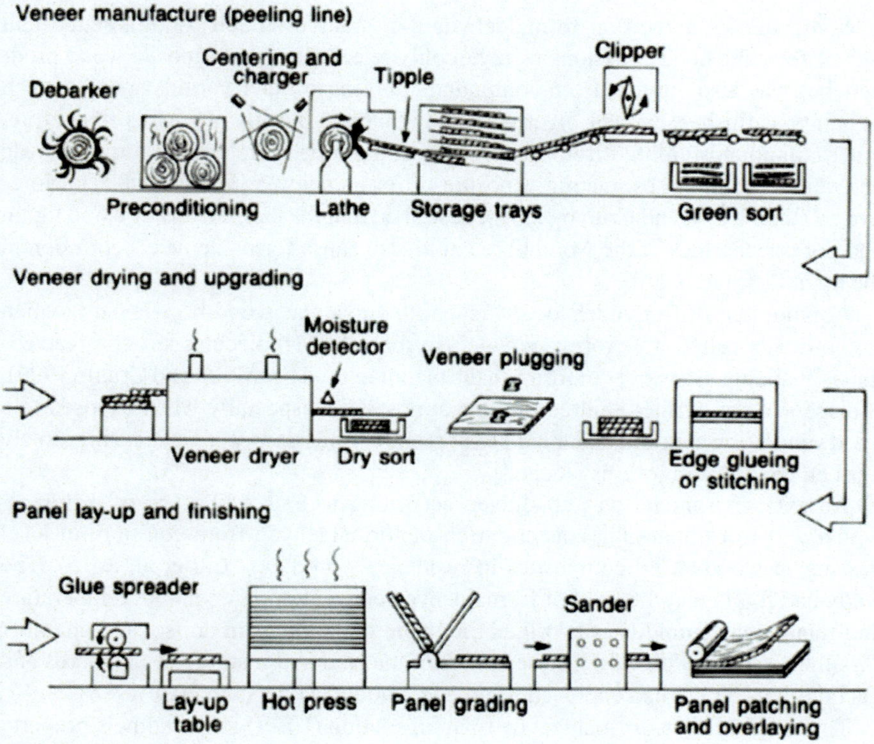

Fig. 3.3 Illustrations of the process of plywood production (Shi and Walker 2006)

3.5 Indoor Air Quality

One of the first pollutants measured indoors is formaldehyde (Athanassiadou et al. 2009). Although the concentration of formaldehyde in the atmosphere is only ranging between 0.1 and 0.2 ppb, which is generally low, however, the formaldehyde emission from composite wood products could give a counter effect on the indoor air quality (Roffael 2006).

According to Athanassiadou et al. (2009), the level of formaldehyde emissions from composite wood products is influenced by external and internal factors. When measuring indoor formaldehyde emissions, all the external and internal factors have to be taken into consideration. External factors are the surrounding temperature, air exchange rate, and humidity, meanwhile, the types of wood and resin used, parameters and conditions of wood panel production as well as the age of the wood panel are the internal factors of formaldehyde emissions. Moreover, a study by Kelly et al. (1999), also agreed on several factors that significantly affect the level of formaldehyde emissions, which are temperature, humidity, and ventilation rate.

Some studies argued on formaldehyde emissions; Mendell (2007), stated that indoor formaldehyde emissions were not only released from composite wood products but was also emitted from compounds present in other various sources, such as paints, varnishes, carpets, cigarettes, and other combustion processes. Therefore, other compounds emitted from these sources might be correlated to formaldehyde and countered in the estimating exposure of formaldehyde (Mendell 2007). Moreover, a study by Yu and Crump (1999), described that the thermolysis of wood lignin and polysaccharides of the wood fibre could also cause formaldehyde's emission in the manufactured board.

A major part of formaldehyde is first emitted in a new wood-based board, which is commonly called 'free formaldehyde' (formaldehyde molecules left non-reacted) that is available in a variety of forms in the manufactured board (Yu and Crump 1999). It dispersed and further emitted into the atmosphere, especially when exposed to a good ventilation area and under high temperature. Emissions were decreasing slowly until an equilibrium level is reached.

An extensive model was conducted according to Fick's first law of diffusion, which was to estimate the concentration of formaldehyde from equilibrium level gassing in wood-based composites in residence's products (Liu et al. 2015). The study had found that the level of formaldehyde concentration on indoor temperature and relative air humidity was originating from one emitter in an isolated chamber (Matthews et al. 1986, 1987). Researchers of the National Institute of Standards and Technology (NIST) had endorsed Matthews' model (Silberstein et al. 1988).

The usage of urea–formaldehyde foam insulation (UFFI) in the home's property was once in demand, mainly in difficult-to-access wall spaces. However, in the mid-1980s, indoor formaldehyde concentrations were decreasing significantly due to restrictions on permissible formaldehyde emissions from composite wood products and the prohibition on the use of UFFI in homes and schools (Weschler 2009).

3.5.1 Subsequent Formaldehyde Release Regulations

Germany was the earliest European country to come out with regulations regarding the formaldehyde emission from composite wood productions in actual industrial practice (Athanassiadou et al. 2009; Roffael 2006).

German Chemical Prohibition Ordinance 1994 banned the usage of composite wood products that emit formaldehyde greater than 0.1 ppm concentration in the air when put into the test in an environmental chamber. German Institute of Building Technology incorporated this limit of 0.1 ppm in a guideline that only permitted the 'E1' and 'E1b' emission classes of wood based products (coated or uncoated) to operate in building's construction and structure (Yu and Crump 1999).

Japan has one of the strictest formaldehyde emission standards in the world. In Japan, the Building Standards Law (BSL) Sick House regulations were intended to solve health issues caused by formaldehyde emissions in residential buildings. Wood based panels are categorized with their formaldehyde emissions, by using the

Table 3.1 F* rating system according to its formaldehyde emission rate (Eastin and Mawhinney 2011)

Formaldehyde emission rating	Formaldehyde emission rate, X (mg/m^{2}h)
F*	More than 0.12
F**	Between 0.02 and 0.12
F***	Between 0.005 and 0.02
F****	Equal and less than 0.005

desiccator reference values, which are F****, F***, F**, and F* (Roffael 2006). The more stars the wooden product rated, the less the amount of formaldehyde it emits (Table 3.1). The emission from F**** boards can be used in interior spaces without limitations. When building materials have a F* rating, it cannot be applied in building residential buildings at all (Eastin and Mawhinney 2011).

On July 7, 2010, President Obama had signed the Formaldehyde Standards for Composite Wood Products Act (S.1660) (Liu et al. 2015). It urged the United States Environmental Protection Agency (US EPA) to publicize regulations to execute specific standards to restrict formaldehyde emissions from composite wood products. It also highlighted the US EPA's attentiveness in controlling indoor formaldehyde exposure sources and the component that impacted its indoor fate of emissions.

In Malaysia, the issue of formaldehyde emissions has received considerable critical attention. It is clearly stated in Section 15 of Occupational Safety and Health Act 514 (Act 1994), the primary duties of both employers and self-employed persons to their workers. Hence, the Industry Code of Practice on Indoor Air Quality (2010), has been enforced to make sure both workers and users are protected from unhealthy indoor air quality that could give counter effect to their health and well-being, and thereby decrease their productivity. For chemical contaminants such as formaldehyde, the limits are according to the time-weighted average eight-hour (TWA8) concentration of airborne. The Permissible Exposure Limit (PEL) for formaldehyde is 0.1 ppm.

Thus, formaldehyde is a carcinogenic chemical that has critical attention worldwide. A good understanding of the usage of formaldehyde together with formaldehyde emission exposure limits and standards are crucial for safeguarding the workers' and users' safety and quality of health.

3.6 Environmental Issues Related to Health

3.6.1 Emission of Composite Production

The emissions from the composite production into the air cause air pollution that is either indoor or outdoor and both have the tendency to affect humans' health and safety. VOCs are the main elements produced by composite that consist of many chemicals that contribute to air pollution, e.g. formaldehyde and styrene. For example, the wood composites produce formaldehyde that causes respiratory problems in children and allergic sensitization (Si 2018).

3.6.2 *Styrene Emissions Effect*

IARC Monographs Vol 121 Group (2018) stated that some countries like Europe, Denmark, the USA, Washington State, and the UK had the highest styrene exposure level produced by the reinforced plastic industry. The findings on lymph hematopoietic malignancies were assessed in several studies, where the mortality of subtypes of leukemia and lymphomas did not change as a result of increasing of leukemia diseases number, specifically on myeloid leukemia. One large cohort of reinforced plastic workers was founded to have sinonasal adenocarcinoma, which is rare cancer but the cases were limited and as the confounding and chance could not be counted. The experiment on mice was done in O20 mice and they found the prevalence of lung adenoma or carcinoma in both males and females rat and lung carcinoma in females by transplacental exposure. Styrene exposure through inhalation increased the prevalence of bronchioloalveolar carcinoma in males, and in another study, high bronchioloalveolar adenoma or carcinoma was found in both females and males for CD-1 mice (Cruzan et al. 2017). The same effect of styrene exposure in B6C3F1 mice was observed where females had hepatocellular adenoma, while males had high of brochioloalveolar adenoma and carcinoma. Based on one of the two inhalation studies, female rats had higher malignant mammary tumors than the male rats.

Styrene was categorized in Group 2A as "probably carcinogenic to humans" because of the adequate evidence of animals' experiments for carcinogenic and restricted evidence in humans. By inhaling styrene, it breaks down into styrene-7,8-oxide as it reacts directly with DNA. Styrene-7,8-oxide might cause DNA damage, sister chromatid exchanges, micronucleus formation, chromosomal aberrations, and gene mutations in human cells in vitro. When rodents were exposed to styrene-7,8-oxide, the results were positive for DNA damage in multiple tissues but ambiguous for cytogenetic effects. Besides, styrene reduced cell proliferation in cultured human lymphocytes, and both styrene and styrene-7,8-oxide increased proliferation in various rodent tissues.

Matanoski and Tao (2003) stated that exposure to styrene also had might have some acute effects on the human cardiovascular system and probability threat on ischemic heart disease. Based on this risk, the case-cohort study was performed during the years of 1943–1984 on the United States in 2 styrene-butadiene rubber-manufacturing plants' employees. The study did not discrete acute from chronic ischemic heart disease and no data on deaths were recorded in active workers or workers who had left the plants for many years before they died. There was no explanation for the phenomenon of reverse risk with the period of employment. The result of the study had shown that both short-term and long-term workers were influenced by styrene exposure. Tables 3.2 and 3.3 show styrene exposure among

Table 3.2 Styrene exposure among workers with the risk of ischemic heart disease (Matanoski and Tao 2003)

Model	Disease	Styrene exposure variable (ppm)[b]	Employed for > 0 years		Employed for ≥ 2 years		Employed for ≥ 5 years	
			RH[c]	95% CI[a, c]	RH	95% CI	RH	95% CI
1	Acute IHD[b, c]	For recent 2 years	3.26	1.09,9.72	5.86	1.59, 21.64	6.60	1.78, 24.54
		For 2–5 years ago	0.94	0.85, 1.04	0.92	0.82, 1.03	0.90	0.79, 1.01
2	Acute IHD	For recent 2 years	2.34	0.72, 7.57	4.99	1.07, 23.34	4.27	0.84, 21.79
		For 2–5 years ago	0.98	0.87, 1.11	0.95	0.82, 1.09	0.94	0.81, 1.10
3	Chronic IHD	For recent 2 years	1.22	0.08, 18.13	1.45	0.09, 23.93	1.52	0.09, 24.65
		For 2–5 years ago	0.96	0.72, 1.22	0.93	0.72, 1.20	0.93	0.72, 1.20
4	Chronic IHD	For recent 2 years	1.92	0.08, 46.74	2.13	0.07,62.97	2.18	0.07, 68.27
		For 2–5 years ago	0.89	0.66, 0.70	0.88	0.64, 1.19	0.88	0.65, 1.20

[a] Using age as timescale, stratified by plant, controlled for race and birth year. Adjustment is made for butadiene in models 2 and 4

[b] Time-dependent, time-weighted average intensity prior to the time of event

[c] *RH* relative hazard; *CI* confidence interval; *IHD* ischemic heart disease

Table 3.3 Different styrene exposure to workers (Matanoski and Tao 2003)

Time-weighted styrene intensity (ppm) for the most recent 2 years	Employed for > 0 years		Employed for ≥ 2 years		Employed for ≥ 5 years	
	RH[a]	95% CI[a]	RH	95%CI	RH	95% CI
< 0.10	1.00		1.00		1.00	
0.10–< 0.20	0.94	0.29, 3.04	1.24	0.36, 4.33	1.25	0.36, 1.35
0.20–< 0.30	2.14	0.80, 5.72	2.95	1.02, 8.57	3.00	1.03, 8.73
≥ 0.30	3.10	1.25, 7.67	4.30	1.56, 11.84	4.24	1.54, 11.68

[a] *RH* relative hazard, using age as timescale, stratified by plant, controlled for race and birth year, *CI* confidence interval

workers employed for 2 or more years than workers who were employed for less than 2 years at the same styrene level exposure. The result showed that styrene impacted the health of both short and long-term workers.

The importance of this study was due to the data suggested that chemicals found in the air and industrial settings can be associated with an increased risk of heart disease. At low industrial exposures, the risk appeared as acute events. These chemicals might react once adsorbed onto particulates related to ischemic heart disease mortality. Besides, some other chemicals might also have similar reactions in both outdoor and indoor environments that might consequently contribute to increasing deaths from ischemic heart disease (Matanoski and Tao 2003).

3.6.3 Formaldehyde Emissions Effect

Formaldehyde indoor air pollution is much more dangerous than outdoor as it typically resulted from low air exchange rates and stronger formaldehyde sources. The release of formaldehyde can happen at any stage of the production, use, storage, transportation, or disposal of products containing the remaining formaldehyde. The formaldehyde emissions come from wood based composite products, and the information on wood based composite products are presented in Table 3.4. In the 1960s and

Table 3.4 Possible formaldehyde sources in the indoor environments from wood based composite products (Salthammer et al. 2010)

Sources	Details
Wood and wood based products, solid wood	Oak, Douglas fir, beech, spruce, pine
Particle board	Effect of hot-pressing
Particle board	Recycled wood-waste sprayed with PMDI/PF
Particle board, MDF	Comparison of standard methods
Particle board	Effect of aging
Particle board	Effect of humidity and temperature
Particle board, oriented-strand board	Comparison of analytical techniques
Wood-based composites	Laminate, engineered flooring, MDF, particle board
Wood-based panels	Effect of loading and ventilation
Wood panels	Interlaboratory comparison
Particle board, plywood	With carpet and insulation
Pressed wood product, wood-based flooring material	Effect of ozone, infrared, sunlight, UV-A, UV-B
Insulation materials, mineral wool	Interlaboratory comparison
Flooring material, carpet	Interaction of ozone

1970s, the UF bonded particleboard under living conditions released a high concentration of formaldehyde. Many residents complained and were concerned about the bad odour and adverse health effects that were being used extensively (Salthammer et al. 2010).

The International Agency for Research on Cancer (IARC) had categorized formaldehyde as Group 1 human carcinogenic because it affects nasopharyngeal cancer in humans and inhalation causes squamous cell carcinoma in rats. It also causes leukemia and sinonasal cancer in humans but with limited evidence. Formaldehyde is a major contributor to indoor air pollution through inhalation (Si 2018). In 2010, an indoor air guideline was developed by the World Health Organization (WHO). In the guideline, the limit value of formaldehyde exposure of 100 $\mu g/m^3$ should not be exceeded for any 30-min period of the day. The sensory irritation of the eyes, upper airways, and portal-of-entry effects were considered as the critical effects (Mendell et al. 2008). Besides that, formaldehyde also might contribute to sick-building syndrome (SBS) symptoms. In Japan, the concentration of formaldehyde is knowingly higher in homes with occupiers reporting sick-building syndrome (SBS) symptoms than that in homes among non-SBS reporting participants (67 $\mu g/m^3$ vs. 56 $\mu g/m^3$) (Takigawa et al 2010). By inhaling 1% of formaldehyde, it will be transported into the blood and reacted with blood proteins as well as haemoglobin with valine as one of the binding sites. The half lifespan of formaldehyde in the blood is about 1–1.5 min. As the metabolism of formaldehyde in the blood is fast, there was no formaldehyde accumulation reported over long-term exposure. This has also resulted from the reversibility of the formaldehyde-blood reaction (Nielsen et al 2013).

3.6.4 VOCs Emissions Effect

The main route of VOCs is through inhalation. Jensen et al. (2001) stated that most of the toxicological works of literature were based on oral studies in animals. Some of the VOCs are well-known skin sensitizers. Animals were used to conduct most of the experimental studies and the doses used were greater than those significant in indoor air. Through direct contact or adherence to particles in the air, this sensitizing chemicals exposure to the skin might cause allergic contact dermatitis. The clinical data in humans and animals on direct skin contact were existing. The result from VOCs in the wood had caused health effects among workers in the wood industry, but it was unsure whether people can develop allergies as well as skin and airway sensitizations as the VOCs exposure effects from the production of wood or wood based materials in the indoor environment. Children with airway diseases such as asthma can develop an exacerbation of symptoms. Table 3.5 shows the sum of the lowest concentration for VOCs from wood and wood based products. This evaluation method is suitable for more general application. Several countries had enacted regulations to control the release of VOCs from the material, as some of the VOCs are directly or indirectly affect the humans' health and the environment (Väisänen et al. 2018).

Table 3.5 The sum of the lowest concentration for VOCs from wood and wood based products (Väisänen et al. 2018)

Material	S-values (comfort and health evaluation by load 0.4 m^2/m^3 and 2.2 m^2/m^3)			Main effect	Essential substances relative to effects
	3 or 4 days	9–11 days	27 or 28 days		
Solid ash	0.2–1.3	0.2–1.0	0.03–0.2	Irritation	Acetone
Solid oak	0.2–1.2	0.2–1.5	0.1–0.4	Irritation	Acetone
Solid beech	0.1–0.6	0.2–0.8	0.04–0.2	Irritation	Acetone
Solid spruce	0.4–2.4	0.3–1.8	0.1–0.7	Irritation	Alpha-pinene, limonene
Particle board (MUPF glue)	1.4–8.0	1.4–8.2	0.5–2.8	Irritation, allergy	Formaldehyde, acetone
Particle board (UF glue)	1.2–6.8	1.0–5.4	0.8–4.6	Irritation, allergy	Formaldehyde, acetone
Particle board (PU glue)	1.5–8.3	0.9–5.0	0.2–1.2	Irritation, allergy	Acetone, acrolein
Plywood (phoenic-glue)	0.3–1.9	0.2–1.0	0.1–0.8	Irritation, allergy	Acetone
MDF (UF glue)	0.9–5.1	0.9–5.0	0.8–2.6	Irritation	Formaldehyde
Beech veneered particle board (PVA glue)	2.4–12.6	1.8–10.4	1.7–9.5	Irritation	Benzyl alcohol
Beech veneered particle board (UF glue)	2.4–12.7	1.5–8.4	1.0–5.6	Irritation	Formaldehyde
Solid pine-heartwood from northern Finland	> 32–> 180	> 28- > 152	16.1–88.4	Irritation, allergy	Alpha-pinene, limonene
Solid pine-sapwood from northern Finland	> 5–> 28	> 3- > 6	3.6–19.5	Irritation, allergy	Alpha-pinene, 3-carene
Solid pine-heartwood from southern Sweden	18.4–102	13–71.4	17.5–96.2	Irritation, allergy	Alpha-pinene, 3-carene
Solid pine-sapwood from southern Sweden	> 13.9–> 77	> 18- > 98	9.8–54	Irritation, allergy	Alpha-pinene, 3-carene
Oriented strain board (OSB) (phenolic glue)	4.4–24.1	2.9–15.9	0.7–3.7	Irritation, allergy	Furfural, acrolein

(continued)

Table 3.5 (continued)

Material	S-values (comfort and health evaluation by load 0.4 m^2/m^3 and 2.2 m^2/m^3)			Main effect	Essential substances relative to effects
	3 or 4 days	9–11 days	27 or 28 days		
Solid beech (alkyde and linseed oil)	8.0–44.2	1.8–10.1	0.3–1.4	Irritation, allergy	Vinyl ethyl ketone, C2-alkybenzenes
Solid beech (resin and linseed oil)	0.7–3.7	0.4–2.2	0.2–1.3	Irritation, allergy	Acetone, C2-alkylbenzenes
Beech veneered particle board (nitrocellulose lacquer)	20.7–113.8	9.5–52.2	1.4–7.4	Irritation	2-Octenal, 2-Decenal
Beech veneered particle board (UV curing lacquer)	0.1–0.7	0.1–0.6	0.1–0.7	Irritation	Acetone
Beech veneered particle board (acid-curing lacquer)	8.5–46.7	5.0–27.5	3.7–21.3	Irritation	Butanol, formaldehyde
Beech veneered particle board (water-borne lacquer)	6.0–33.4	3.3–18.3	1.6–8.7	Irritation	2-(2-Buthoxy-ethoxy-ethanol), 1-butoxy-2-propyl acetate
Beech veneered particle board (polyurethane lacquer)	2.9–15.7	1.8–9.8	0.9–4.8	Irritation	1-Methoxy-2-propyl acetate, C2-alkylbenzenes

In conclusion, VOCs are the main elements contributing to air pollution, either in indoor or outdoor environments. VOCs include chemicals such as formaldehyde and styrene, giving acute or chronic health effects. Most of these chemicals are associated with respiratory diseases in humans through inhalation. Styrene was studied for its risk that might cause lung cancer, ischemic heart disease, asthma, and DNA damage as it metabolises as styrene-7,8-oxide. Styrene is categorised under Group 2A, which is possibly carcinogenic to humans. For formaldehyde, it might result in skin and eyes irritations as well as respiratory diseases. Besides, it was proven to contribute to sick building syndrome (SBS) symptoms, as found in most VOCs.

3.7 National Standard for Hazardous Air Pollutants in Composite Production

3.7.1 Clean Air Act (CAA) Amendments 1990

Clean Air Act (CAA) Amendments 1990 were introduced to rein four major hazards, which are urban air pollution, acid rain, ozone depletion, and hazardous emissions that are considered as threats to the health and the environment. The amendments also inaugurated a national operating permits program to make the regulations more practicable and intensified enforcement to ensure better compliance with the regulation.

According to Environmental Protection Agency (2004), Title I of CAA Amendments includes the provisions in achieving the National Ambient Air Quality Standards (NAAQS) for pollutants criteria and provisions for curbing the emissions of hazardous air pollutants (HAPs). Section 112 in Title I stated that the provisions that addressed the curb of HAP emissions are also known as toxic air. In Section 112 of the CAA, the provisions for the proclamation of National Emission Standards for Hazardous Air Pollution (NESHAP), or maximum achievable control technology (MACT) standards, as well as any other interconnected programs were designated to elevate and assist the NESHAP programs (EPA 2004).

Two types of static sources produced routine emissions of hazardous air pollutants (HAPs) or air toxins, which are major and area sources (EPA 2004). The major source is defined as a source that releases 10 tonnes per year (10 tpy) of any of the listed hazardous air pollutants, or 25 tpy or more ($\geq$25 tpy) of any combination of toxic air. Meanwhile, an area source is defined as a source that emits less than 10 tpy of a single air toxic, or less than 25 tpy ($\leq$25) of any mixture of the toxic air (EPA 2004). Basically, the area sources consist of smaller size facilities that emit lesser quantities of hazardous air pollutants into the atmosphere.

The need to comply with Section 112 of the CAA depends on the facility itself; either the facility is an existing major source, area source, or new source.

If the facility is an existing major source, it shall comply with Section 112 of the CAA by 21 April 2006 or it must reach the HAPs emission limit below the major source threshold before 21 April 2006 (EPA 2004). If the facility initially emitted less than 100 tpy of organic HAP from the mixture of all continuous lamination operations and centrifugal casting prior to or on 21 April 2003, and then the emission rises to 100 tpy or more from the operations mentioned in Table 3.6, the owner of the facility shall comply with the 95% of reduction requirement within three years after the semiannual compliance report and shows the evidence that the facility reached or overreached tpy threshold (EPA 2004).

If the facility is an existing area source and then becomes a major source after 21 April 2006, the owner of the facility shall comply with Section 112 of the CAA within three years after the facility becoming a major source.

Table 3.6 Emission limits for existing new sources emitting less than 100 tpy (EPA 2004)

Type of material and/or application	Limits by type of operation (lb/ton)						
	Mechanical	Filament	Manual	Centrifugal casting	Gelcoat	Continuous lamination/casting	Pultrusion
Corrosion-resistant and/or high strength	112	171	123	25	605	–	–
Non-corrosion-resistant and/or non-high strength	87	188	87	20	–	–	–
Tooling	254		157	–	437		–
Low-flame/LOW SMOKE RESIn	497	270	238	–	854	–	–
Shrinkage control	354	215	180	–		–	–
White/off-white pigmented gel coat	–	–	–	–	267	–	–
All other pigmented gel coat	–	–	–	–	377	–	–
Clear gel coat	–	–	–	–	522	–	–
All resins	–	–	–	–	–	15.7 lb/ton OR 58.5% reduction	60% reduction

Table 3.7 Work practice standards (existing and new sources) (EPA 2004)

Operation	Requirement
Compression/injection molding	Uncover only one charge per mould cycle per machine. One charge means sufficient material to fill all moulds for multiple-mould machines or to fill the hopper for hopper-fed machines
Cleaners	Use non-HAP containing cleaners, except styrene can be used to clean closed systems and organic HAP cleaners can be used to clean cured resin from the equipment
Containers	Keep closed when storing HAP-containing materials
Mixing/BMC manufacturing	Use covers with no visible gaps except for up to one shift around mixing shafts and instrumentation. Close mixer vents that are not vented to a 95 percent (or more) efficient control device when mixing (except when adding materials or as needed for safety reasons)
SMC manufacturing	Enclose resin delivery system to the doctor box and use nylon-containing film to wrap SMC
Pultrusion (larger parts)	Reduce air velocity; limit ambient air across the wet-out area; no point suction of ambient air (unless directed to a control device); cover vessels containing HAP-containing materials

The owners of the facility shall comply with Section 112 of the CAA upon the start-up if their facility is a new and a major source from the start-up of the facility. If it subsequently became one of the major sources, the owner of the facility shall comply with Section 112 as soon as it became a major source. The owner of the facility shall comply with the 95% reduction requirements if the facility is a new source that originally only subjected to Tables 3.6 and 3.7, and later elevates its initial organic HAP emissions to a level where the requirements apply. The owners need to comply with such requirements within three years after the semiannual compliance report to show that the facility reached or overreached the 100 tonnes per year threshold. Lastly, the new source facility shall show evidence of their compliance within 180 days after its start-up if the additional controls were initially used.

The emission limits depend on the type of material and/or the manufacturing technique of a particular composite product as shown in Tables 3.6 and 3.7.

3.7.2 Industrial Code of Practice (ICOP) on Indoor Air Quality (IAQ)

Indoor air quality or known as IAQ, refers to the quality of air within and around the buildings and structures, specifically as it relates to the health and comfort of the building occupants (EPA 2018a, b). One of the regulations concerning the reduction in the risk of indoor health concern was the Industrial Code of Practice (ICOP) on IAQ 2010 that was approved on 30 August 2010.

According to the Department of Occupational Safety and Health (2016), good indoor air quality (IAQ) is required to ensure a healthy indoor work environment as poor IAQ can cause various short term and long term health problems, e.g. allergic reactions, respiratory problems, bronchitis, sinusitis pneumonia, and eye irritation. IAQ problems that occur in buildings could be caused by insufficient control of indoor air pollutants or by poor ventilation.

There are plenty of causes of pollutants that can be found in the environment, e.g. environmental tobacco smoke (ETS) that are released from the burning of tobacco products. Other than that, furnishings can also be one of the sources that emit a variety of pollutants, for example, formaldehyde. Next, volatile organic compounds (VOCs) can also be released from the use and application of solvent and lastly, laser printers and photocopiers are also known as the sources of ozone released into the atmosphere.

The acceptable limit of formaldehyde stated in ICOP on IAQ 2010 is 0.1 parts per million (ppm) for an eight-hour time weighted-average (TWA_8) for airborne concentrations. Table 3.8 shows the list of indoor air contaminants and their acceptable limits from the ICOP.

In addition, according to Use and Standards of Exposure of Chemicals Hazardous to Health (USECHH) Regulations 2000, it is stated that the ceiling limits the airborne concentration of formaldehyde of 0.3 ppm or 0.37 mg/m^3.

Table 3.8 List of indoor air contaminants and the acceptable limits (DOSH 2016)

Indoor air contaminant	Acceptable limits		
	ppm	mg/m^3	cfu/m^3
Chemical contaminants			
Carbon monoxide	10	–	–
Formaldehyde	0.1	–	–
Ozone	0.05	–	–
Respirable particulates	–	0.15	–
Total volatile organic compounds (TVOC)	3	–	–
Biological contaminants			
Total bacteria counts	–	–	500*
Total fungal counts	–	–	1000*
Ventilation performance indicator			
Carbon monoxide	C1000	–	–

Table 3.9 The occupational exposure limit for styrene (DOSH 2000)

Chemical	[CAS]	Eight-hour time-weighted average airborne concentration	
		ppm	mg/m^3
Stearates		–	10
Stibine	[7803-52-3]	0.1	0.51
Stoddard solvent	[8052-41-3]	100	525
Strontium chromate, as Cr	[7789-06-2]	–	0.0005
Strychine	[57-24-9]	–	0.15
Styrene, mononer- (skin)	[100-42-5]	20	85.2

3.7.3 *Occupational and Safety (Use and Standard of Hazardous Chemical Exposure to Health) Regulation 2000 (USECHH Regulation)*

USECHH Regulation 2000 was implemented to provide a constitutional structure to curb the exposure of hazardous chemicals to the health of workers in the workplace (DOSH 2016). USECHH was made to set the hazardous chemical exposure standards for workers in the workplace. DOSH (2016) also stated that if a chemical is classified as a hazardous chemical under CLASS Regulations but was not listed in Schedule I or II of USECHH Regulations, the chemical will still be incorporated under USECHH Regulations if it is used by the workers in the workplace. Styrene falls under Schedule I of USECHH Regulation.

Based on Schedule 1 of USECHH Regulation, the eight-hour time-weighted average (TWA_8) airborne concentration for styrene is 20 ppm or 85.2 mg/m^3, as shown in Table 3.9.

3.7.4 *International Regulations and Guidelines*

The differences in exposure limits in countries or organizations are presented in Tables 3.10 and 3.11:

Based on the standards, it can be concluded that the limits of the standard, e.g. occupational exposure limits (OEL) or permissible exposure limits (PEL), short-term exposure limit (STEL), and ceiling limits differ between countries based on their organizations. The emission limits of a particular hazardous chemical are also different based on the type of material and/or applications, and the work practice standards (EPA 2004).

Table 3.10 Occupational exposure limits for formaldehyde (EPA 2005)

Country or region	Concentration (mg/m^3) [ppm]	Interpretation	Carcinogen classification
Australia	1.2 [1]	TWA	2; Sen
	2.5 [2]	STEL	
Belgium	0.37 [0.3]	Ceiling	
Brazil	2 [1.6]	Ceiling	
China	0.5	Ceiling	
Canada			
Alberta	2.5 [2]	Ceiling	
Ontario	0.37 [0.3]	Ceiling	
Quebec	2.5 [2]	Ceiling	A2[2]
Denmark	0.4 [0.3]	STEL	L, K
Finland	0.37 [0.3]	TWA	
	1.2 [1]	Ceiling	
France	0.6 [0.5]	TWA	
	1.2 [1]	STEL	
Germany	0.37 [0.3]	TWA (MAK)	A; Sh; I
		STEL	
	0.7 [0.6]	Ceiling	
	1.2 [1]		
Hong Kong	0.37 [0.3]	Ceiling	A2[b]
Ireland	2.5 [2]	TWA	
	2.5 [2]	STEL	
Japan	0.6 [0.5]	TWA	
Malaysia	0.37 [0.3]	Ceiling	
Mexico	2.5 [2]	Ceiling	A2[b]
Netherlands	1.2 [1]	TWA	
	2.5 [2]	STEL	
New Zealand	1.2 [1]	Ceiling	A2[b]
Norway	0.6 [0.5]	TWA	Ca[c]; Sen
	1.2 [1]	Ceiling	
Poland	0.5	TWA	
	1	STEL	
South Africa	2.5 [2]	TWA	
	2.5 [2]	STEL	
Spain	0.37 [0.3]	STEL	
Sweden	0.6 [0.5]	TWA	Ca[d]; Sen

(continued)

Table 3.10 (continued)

Country or region	Concentration (mg/m^3) [ppm]	Interpretation	Carcinogen classification
Switzerland	0.37 [0.3]	TWA	Sen
United Kingdom (MEL)	2.5 [2]	TWA	
	2.5 [2]	STEL	
USA			
ACGIH (TLV)	0.37 [0.3]	Ceiling	A2[b]; Sen
NIOSH (REL)	0.02 [0.016]	TWA	Ca[d]
	0.12 [0.1]	Ceiling	
OSHA (PEL)	0.9 [0.75]	TWA	Ca[d]
	2.5 [2]	STEL	

3.8 Control of Composite Pollution

3.8.1 Law Regarding the Control of Composite Pollution

There are many forces established in order to control the pollution from composite, as it can affect the environment, human health, and safety. The efforts came whether from the governments, organizations, or the manufacturers themselves. United States Environmental Protection Agency (USEPA) plays a significant role in controlling composite production. According to Grande (2007), in 2006, the Maximum Achievable Control Technology (MACT) Standard was introduced by the United States Environmental Protection Agency, a minimum standard that industry needs to meet and after a year of it come into operation, the MACT standard was complied by most of the composite producing firms. Not only focusing on technology, for a facility to reduce the Hazardous Air Pollution (HAP) emissions, control technology also includes procedure, techniques, systems, and processes (Chamberlian 2018). In MACT, major sources are required to emit toxic pollutants at least 10 tons per year (tpy) when the toxic pollutants are mixed, consistently comply with the maximum achievable control technologies and processes in order to lessen the release of HAP. This is the reason for MACT's effectiveness as it lowers exposure to toxic pollutants which results in increasing rates of cancer diseases (Barreto 2018).

Other than MACT, USEPA also took the initiative to control formaldehyde emissions. In 2016, Formaldehyde Emission Standard was introduced to control the emissions dissipated from some wood products, whether they are being imported into the United States or they are made in the country (United States Environmental Protection Agency 2018a, b). The standard controls over laminated wood products, including more detailed labeling on covered products and the need for the compliance details to be passed down through the distribution chain from importers or manufacturers (National Association of Music Merchants 2018). The final rule guaranteed that medium-density fibreboard, hardwood plywood, and particleboard products being

Table 3.11 Occupational exposure limits for styrene across Europe (European Composite Industry Association 2017)

Country	8-h TWA (ppm)	15 min STEL (ppm)
Austria	20	80 (4 × 15 min)
Belgium	25	50
Bulgaria	20	50
Czechia	24	94[a]
Denmark	–	25[a]
Estonia	2.4	7
Finland	20	100
France	23.3	46.6
Germany	20	40 (4 × 15 min)
Greece	100	250
Hungary	12	12
Ireland	20	40
Italy	20	40
Latvia	2.4	7
Lithuania	20	50
Luxembourg	20	40 (30 min)
Netherland	25	50
Norway	25	37.5
Poland	12	47
Portugal	20	40
REACH DNEL	20	68
Romania	12	35
Slovakia	20	40[a]
Slovenia	20	80
Spain	20	40
Sweden	10	20
Switzerland	20	40 (4 × 10 min)
United Kingdom	100[b]	250

[a] Ceiling limit
[b] Obligation to reduce as much as possible

manufactured, supplied, sold, imported to, or sold in the country follow the emission standard by imposing provisions for laminated products, labeling, record keeping, product-testing requirements, and import certification (United States Environmental Protection Agency 2018a, b). To guarantee the final national law was compatible with the similar wood composite product of California's requirements, EPA joined the forces together with California Air Resources Board (CARB).

3.8.2 Usage of Low-Styrene Materials

Other than the effort of the organizations, manufacturers or industry players are also working on some ways to control the composite pollution to comply with the regulations. One of the methods is the usage of low-styrene materials in their production. Within the last 10 years, to obey the highly tight regulations, especially in the fulfillment of boat industry, there exists a wide usage of glass fibre that strengthens the low styrene polyester resins (Perrot et al. 2007).

The suppliers adapted their formulations to meet the regulations to the health and safety of the amounts of tolerable volatiles of I boatyards and proposed different kinds of resin. First, polyester in the low styrene decreased the styrene content, but to maintain the low viscosity, a reduction of the prepolymer molecular weight were made. Second, low styrene emissions resins were used. In order to form a film while limiting the styrene emission, additives were added that moved to the surface, forming a layer on the surface of the resin. For example, the addition of small quantities of paraffin to the resin lowered the amount of styrene emitted from the polyester during the curing process. The types and levels of paraffin used, including the resin composition and curing characteristics affect the extent of the reduction of styrene emission, as shown in Table 3.12 (Savage 1985a). Lastly, the produced composites possess low styrene and emit less styrene (Perrot et al. 2007). The effect of curing conditions, e.g. temperature and time on the properties of the mechanical of resins with unstrengthened polyester was showed in previous research (Baley et al. 2006). The significant difference between polyester design and resins with standard polyester to reduce styrene emissions was shown in that study (Perrot et al. 2007).

Not only the initial content of the resin styrene monomer was reduced to 20%, but maintenance in certain situations was also allowed. It also allowed the fabricated reinforced plastic ("RP") laminate physical properties to be improved. Significantly lowering the emissions of styrene monomer without other properties being notably sacrificed is the major advantage of this new family of LSE resins. From the comparison, it can be observed that the obtained emission of the new LSE resin was lowered by 70% to only 30 gm/m^2 compared to 99 gm/m^2 measured from the RP test laminates, where the resin was used in the preparation (Savage 1985b).

Table 3.12 The results of styrene emissions and interlaminate bonding (Savage 1985a)

Styrene	Without low styrene emission resin	With low styrene emission resin
Paraffin (weight %)	0.0	0.15
Adhesion promoter (weight %) (from Example 1)	0.0	0.30
Styrene emission (grams/square mere one hour after gelation)	103	18
Interlaminate adhesion (lbs.)	41	38

3.8.3 VOC Abatement System

The utilization of VOC-abatement technology is a more capital-intensive way of reducing the release of VOCs and is usually used to incinerate paint fumes in automatic plants. This technology requires an early investment of at least a million dollars including continuous operating costs. According to ACMA's Lacovara, this approach only was used by approximately, twelve composites fabricators. This abatement method is considered a burden to the manufacturers and is too expensive in most situations (Grande 2007).

Over the last 20 years, there was a growing interest in the abatement of VOCs from scientists who were fascinated with the NTP technology (Vandenbroucke et al. 2011). Accelerating electrons are consuming most of the energy transferred to the system. While the room temperature of the background gas is being maintained, a normal temperature of 10,000 to 250,000 K (1–20 eV) is achieved. The whole treated gas flow is not necessarily being heated due to the non-equilibrium state. Secondary photons, radicals, electrons, and ions were produced from the collision of the background molecules (N_2, O_2, H_2O) with the accelerated primary electrons. The oxidation of VOC molecules comes from these latter species, even though ionic reactions are also possible (Vandenbroucke et al. 2011). For example, there was a large amount of attention was given to the chlorinated olefin, which is trichloroethane. This was due to the fact that NTP can easily remove the TCE without requiring considerable energy added to the system. This happens because the plasma discharge produced reactive radicals which are easy to be added to the carbon–carbon double bond which initiated the oxidation process (Vandenbroucke et al. 2011). A variety of NTP reactors have been investigated by researchers for environmental sustainability purposes. These reactors are classified depending on their multiple characteristics. Figure 3.4 shows the illustrations of different NTP reactor configurations (Vandenbroucke et al. 2011).

The emissions of VOC can be treated in various ways, including the destruction or recovery based methods that can be applied depending on the quantity and quality of the emissions. Adsorption and condensation are among the recovery based methods. These methods are suitable if an uncomplicated emission mixture is released and some economic value is possible to be produced from the recovered chemicals, either by selling the recovered chemicals (Ojala et al. 2011).

3.8.4 Lowering the Urea–Formaldehyde mol Ration in the Synthesis of Urea–Formaldehyde Resin

Another alternative to control composite pollution is by reducing the emission of formaldehyde. This can be achieved when the mol ratio of the urea–formaldehyde is reduced in the synthesis of urea–formaldehyde resin. It is known that by lowering the mole ratio of formaldehyde/urea (F/U), these requirements for producing resins, especially Formaldehyde Standard can be obeyed (Nuryawan et al. 2017). There

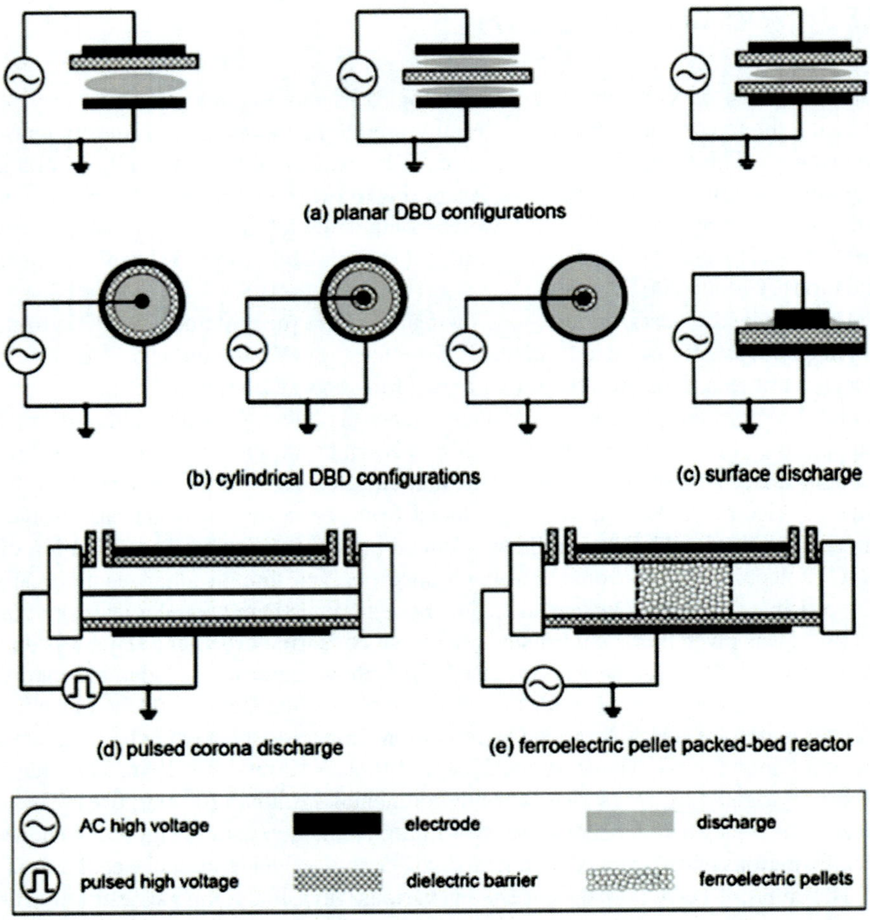

Fig. 3.4 Illustrations of different NTP reactor configurations (Vandenbroucke et al. 2011)

were a lot of studies that reported the mole ratio of F/U that significantly influenced the properties of UF resins and the incorporation of excess urea was done in the condensation step. During the incorporation of excess urea in condensation step (pressing stage), most hemiformals and free formaldehyde species can be removed (Nuryawan et al. 2017). To prove this statement, a total of 18 panels with a size of 16 (T) × 500 (W) × 500 (L) mm were prepared by different resin types. Six of them were using 12 different commercial UF resins while the other 12 were using commercial MUPF. Table 3.13 shows the conditions employed in the study (Que et al. 2007). The physical and mechanical properties of the UF resin's bonded wood panels might be associated with the low mole ratio of UF resins (Nuryawan et al. 2017). The result of the synthesis of UF resin is shown in Table 3.13. Furthermore, according to Park and Causin (2013), for interior applications, the use of bonded wood based panels and UF resin adhesives can also be limited with lower resistance of UF resins. In

Table 3.13 Resin and board variations (Que et al. 2007)

Resin designation	Resin					Content[a] (%)	Additive (%)		Emission class[d]	Press temperatur[e] (°C)
	Type	Molar ratio (F:U)	Viscosity 25 °C (cp)	Solids content (%)	pH		Wax[b]	catalyst[c]		
A	UF	1.27	300	65	8.3	Face: 11 Core: 7.9	0.4	Face: 0.26 Core: 2.0	E2	150
B	UF	1.19	270	65	8.3	Face: 11 Core: 7.9	0.4	Face: 0.26 Core: 2.0	E2	150
C	UF	Face:1.16 Core:1.14	260 140	66 66	7.4 7.7	Face: 11 Core: 7.9	0.4	Face: 0.70 Core: 2.5	E2	150
D	UF	1.05	320	65	8.3	Face: 11 Core: 7.9	0.4	Face: 0.70 Core: 3.0	E1	150
B	UF + FA Scavenger[f]	1.01	–	–	–	Face: 11 Core: 7.9	0.4	Face: 0.70 Core: 3.0	–	150
C	UF + FA Scavenger[f]	0.97	–	–	–	Face: 11 Core: 7.9	0.4	Face: 0.70 Core: 3.0	–	150
E	MUPF	1.05	300	63	9.2	Face: 13 Core: 13	1.0	Face: 0.70 Core: 3.0	E1	180

[a] Surface particles: 80% roundwood (spruce/pine = 50/50), 20% sawdust; Core particles: 70% roundwood (spruce/pine/birch = 45/45/50), 30% sawdust. 7–10% moisture content after spraying. The resin used in this study was supplied by Degussa AG

[b] Percent of dry furnish

[c] Percent of resin solution added to resin before spraying. The catalyst is ammonium chloride (NH_4Cl)

[d] Emission class referred to the German emission class for particleboard. E1 class: ≤9 mg/100 g, E2 class: >9 mg/100 g ≤ 30 mg/ 100 g. The emission class for resin is provided by resin company

[e] Press time 10 s/mm

[f] Formaldehyde-bonding substances added. The Scavenger is sodium sulphite, UF, urea–formaldehyde, FA, formaldehyde absorb, MUPF, melamine–urea–phenol formaldehyde

Fig. 3.5 UF resin synthesis and its results in the laboratory (Nuryawan et al. 2017)

addition, it is well known that the major causes of sick building syndrome come from the emissions of formaldehyde from the panels used for interior applications as shown in Fig. 3.5 (Nuryawan et al. 2017).

Thus, the major important topic of the research of UF resin is the formaldehyde emission issues. In placing more emphasis, Que et al. (2007) claimed that in their study, the formaldehyde released from the board was considerably influenced by the UF resin's mole ratio during pressing and after production. However, they stated that scientific published studies that focus on the role of mole ratio that includes the whole scope of the particleboard's pertinent measurable properties and the comparative measurement of formaldehyde with different methods are difficult to find (Que et al. 2007).

In conclusion, there are many ways to control composite pollution, and it requires the efforts of the manufacturers themselves to take action into their hands or vice versa. However, agencies such as the government already play their roles by implementing the regulations for the industry to comply. The manufacturers must realize the importance of controlling composite pollution for their workers' and consumers' health as well as environmental sustainability.

3.9 Conclusions

In all studies reviewed here, a tremendous proliferation of useful composite wood products, with advanced ones are being formulated and invented. However, some of these products might release some hazardous air pollutants into the atmosphere. Most of the VOCs emitted to the atmosphere give counter effects on the environment and humans health in either direct or indirect ways. Hence, the efforts from the manufacturers, governments, authority bodies, and consumers are needed to make the world a better place to live in despite growing composites production along with the advancement in manufacturing technologies.

- Clean Air Act Amendments 1990 had considered styrene as ozone precursors and delegated as a toxic air pollutant. It is also being categorized under Group 2A, which is possibly carcinogenic to humans according to the confirmation on animal experiments on its carcinogenicity.
- Formaldehyde is recognised as 'carcinogenic to human' (Group 1) according to the latest reclassification by IARC and has critical attention worldwide. The level of formaldehyde emissions is based on a variety of internal and external factors, including the types of resins used, production factors, and the surrounding atmosphere. A good understanding of the usage of formaldehyde together with formaldehyde emission exposure limits and standards is crucial for safeguarding the workers' and users' safety and quality of health.
- The styrene emission from wood composite products will affect the environment and human health. Styrene is probably carcinogenic to humans that might cause lung cancer, ischemic heart disease, asthma, and DNA damage. Irritation on skin and eye as well as respiratory diseases might result from formaldehyde exposure.
- Many standards had been implemented by some countries to control and reduce the hazardous air pollutants from composites production. For example, EPA had introduced Clean Air Act Amendments 1990 to control major hazards that are considered as threats to health and the environment.
- Hazardous air pollutants (HAPs) emissions can be controlled through a range of tests developed to identify the products that might release HAPs to the atmosphere. Great efforts and various controls had been adopted to minimize the emission of HAPs to the workers and consumers around the world, which include Maximum Achievable Control Technology (MACT) by US EPA, VOC abatement system, and lowering the urea–formaldehyde mol ratio in the synthesis of urea–formaldehyde resin as an effort from the manufacturers' side.

Acknowledgements This chapter was written with Nur Intan Shafiqah Roslan, Nurul Syafiqah Illyana Shafie, Nur Shabrina Azreen Mohd Shabri, Fauziatul Syuhada Mansor, and Aimi Athifah Suhaimy.

References

Athanassiadou E, Tsiantzi S, Markessini C (2009) Producing panels with formaldehyde emission at wood levels, pp 2–6. https://www.researchgate.net/publication/242310214

Baley C, Perrot Y, Davies P, Bourmaud A, Grohens Y (2006) Mechanical properties of composites based on low styrene emission polyester resins for marine applications. Appl Compos Mater 13. https://doi.org/10.1007/s10443-005-9000-9

Barereto, J. D. (2018). EPA Quietly Moves to Allow More Toxic Air Pollution that Causes Cancer. Retrieved from https://blog.ucsusa.org/juan-declet-barreto/epa-quietly-moves-to-allow-more-toxic-air-pollution-that-causes-cancer

Cai Z, Ross RJ (2010) Chapter 12. Mechanical properties of wood-based composite materials. In: Wood handbook—wood as an engineering material, pp 1–12. https://doi.org/10.1038/nchembio.2217

Chamberlian A (2018) MACT regulations summary: breaking down NESHAP & MACT regulations. Understanding MACT and NESHAPs could be essential to your business. Retrieved from https://info.era-environmental.com/blog/bid/39728/MACT-Regulations-Summary-Breaking-Down-NESHAP-MACT-Regulations

Clark D, Aurenhammer P, Spear M (2012) Innovative wood-based products. UNECE/FAO Forest Products Annual Market Review 2011–2012:141–151

Crawford S, Lungu CT (2011) Influence of temperature on styrene emission from a vinyl ester resin thermoset composite material. Sci Total Environ 409(18):3403–3408. https://doi.org/10.1016/j.scitotenv.2011.05.042

Cruzan G, Bus JS, Banton MI et al (2017) Editor's highlight: complete attenuation of mouse lung cell proliferation and tumorigenicity in CYP2F2 knockout and CYP2F1 humanized mice exposed to inhaled styrene for up to 2 years supports a lack of human relevance. Toxicol Sci 159:413–421

Department of Occupational Safety and Health (2000) Federal Subsidiary Legislation. Retrieved from http://www.dosh.gov.my/index.php/en/legislation/regulations-1/osha-1994-act-154/522-pua-131-2000-1/file

Department of Occupational Safety and Health (2016) Indoor air quality. Retrieved from http://www.dosh.gov.my/index.php/en/chemical-management/indoor-air-quality

Di Tomasso C, József Gombos Z, Summerscales J (2014) Styrene emissions during gel-coating of composites. J Clean Prod 83:317–328. https://doi.org/10.1016/j.jclepro.2014.07.051

Eastin IL, Mawhinney DE (2011) Working Paper 120 Japanese F-4Star Formaldehyde Rating Process for Value-Added Wood Products. Center for International Trade in Forest Products (C i n t r a f o r), 7–9

Environmental Protection Agency (2004) An overview of the final rule. Retrieved from https://www.epa.gov/technical-air-pollution-resources

Environmental Protection Agency (2005) Formaldehyde. Retrieved from https://monographs.iarc.fr/wp-content/uploads/2018/06/mono88-6A.pdf

Environmental Protection Agency (2018), Formaldehyde emission standards fir composite wood products. Retrieved from https://www.epa.gov/formaldehyde/formaldehyde-emission-standards-composite-wood-products

Environmental Protection Agency (2018) Introduction to indoor air quality. Retrieved from https://www.epa.gov/indoor-air-quality-iaq/introduction-indoor-air-quality

European Composite Industry Association (2017) Occupational exposure to styrene. Retrieved from http://www.upresins.org/wp-content/uploads/2017/09/170731_UPR_SHG2_EN.pdf

Grande JA (2007) Composites and VOCs: plain and fancy ways to meet emissions rules. Retrieved from https://www.ptonline.com/articles/composites-and-vocs-plain-and-fancy-ways-to-meet-emissions-rules

Hammond D, Garcia A, Feng HA (2011) Occupational exposures to styrene vapor in a manufacturing plant for fibre-reinforced composite wind turbine blades. Ann Occup Hyg 55(6):591e600

IARC (2006) Monographs Vol 88: formaldehyde, 2-butoxyethanol and 1-tert-butoxypropan-2-ol. IARC Monographs, Lyon, France

IARC Monographs Vol 121 Group (2018) Carcinogenicity of quinoline, styrene, and styrene-7,8-oxide. Lancet Oncol 19(June):728–729. https://doi.org/10.1016/S1470-2045(18)30316-4

Kelly TJ, Smith DL, Satola J (1999) Emission rates of formaldehyde from materials and consumer products found in Califomia homes. Environ Sci Technol 33(1):81–88. https://doi.org/10.1021/es980592+

Kim S, Kim JA, Kim HJ, Do Kim S (2006) Determination of formaldehyde and TVOC emission factor from wood-based composites by small chamber method. Polym Testing 25(5):605–614. https://doi.org/10.1016/j.polymertesting.2006.04.008

Liu X, Mason MA, Guo Z, Krebs KA, Roache NF (2015) Source emission and model evaluation of formaldehyde from composite and solid wood furniture in a full-scale chamber. Atmos Environ 122:561–568. https://doi.org/10.1016/j.atmosenv.2015.09.062

National Association of Music Merchants Inc (2018) Formaldehyde emissions from composite wood and wood laminate. Retrieved from https://www.namm.org/issues-and-advocacy/regulatory-compliance/formaldehyde-emissions

Matanoski GM, Tao XG (2003) Styrene exposure and ischemic heart disease: a case-cohort study. Am J Epidemiol 158(10):988–995. https://doi.org/10.1093/aje/kwg247

Matthews TG, Fung KW, Tromberg BJ, Hawthorne AR (1986) Impact of indoor environmental parameters on formaldehyde concentrations in unoccupied research houses. JAPAC 36:1244–1249

Matthews TG, Wilson DL, Thompson AJ, Mason MA, Bailey SN, Nelms LH (1987) Interlaboratory comparison of formaldehyde emissions from particle-board underlayment in small-scale environmental chambers. JAPCA 37:1320–1326

Mendell MJ, Mirer AG (2008) Dampness, mould, and health—a review of epidemiologic evidence for the upcoming WHO guidelines for indoor air quality. Epidemiology 19(6):S136–S137

Mendell MJ (2007) Indoor residential chemical emissions as risk factors for respiratory and allergic effects in children: a review. Indoor Air 17(4):259–277. https://doi.org/10.1111/j.1600-0668.2007.00478.x

Nielsen GD, Larsen ST, Wolkoff P (2013) Recent trend in risk assessment of formaldehyde exposures from indoor air. Arch Toxicol 87(1):73–98

Nunez CM, Ramsey GH, Bahner MA, Clayton CA (1999) An empirical model to predict styrene emissions from fiber-reinforced plastics fabrication processes. J Air Waste Manag Assoc 49(10):1168–1178. https://doi.org/10.1080/10473289.1999.10463912

Nuryawan A, Risnasari I, Sucipto T, Heri Iswanto A, Rosmala Dewi R (2017) Urea-formaldehyde resins: production, application, and testing. IOP Conf Ser Mater Sci Eng **223**(1). https://doi.org/10.1088/1757-899X/223/1/012053

Ojala S, Pitkäaho S, Laitinen T, Koivikko NN, Brahmi R, Gaálová J, Matejova L, Kucherov A, Päivärinta S, Hirschmann C, Nevanperä T (2011) Catalysis in VOC abatement. Top Catal 54(16–18):1224–1256. https://doi.org/10.1007/s11244-011-9747-1

Perrot Y, Baley C, Grohens Y, Davies P (2007) Damage resistance of composites based on glass fibre reinforced low styrene emission resins for marine applications. Appl Compos Mater, 67–87. https://doi.org/10.1007/s10443-006-9033-8

Que Z, Furuno T, Katoh S, Nishino Y (2007) Evaluation of three test methods in determination of formaldehyde emission from particleboard bonded with different mole ratio in the urea-formaldehyde resin. Build Environ 42(3):1242–1249. https://doi.org/10.1016/j.buildenv.2005.11.026

Roffael E (2006) Volatile organic compounds and formaldehyde in nature, wood and wood based panels. Holz Als Roh—Und Werkstoff 64(2):144–149. https://doi.org/10.1007/s00107-005-0061-0

Salthammer T, Mentese S, Marutzky R (2010) Formaldehyde in the indoor environment. Chem Rev 110(4):2536–2572. https://doi.org/10.1021/cr800399g

Savage AEO (1985a) United States Patent (19), 19(54), 1–6.

Savage AEO (1985b) United States Patent (19), 19(54), 5–8

Shi S, Walker JCF (2006) Wood-based composites: plywood and veneer-based products. Primary Wood Process Principles Pract 9781402043:391–426. https://doi.org/10.1007/1-4020-4393-7_11

Si H (2018) Indoor air pollution, lung cancer and solutions. Cancer Cell Res 19:464–470

Silberstein S, Grof RA, Ishiguro K, Mulligan JL (1988) Validation of models for predicting formaldehyde concentrations in residences due to press-wood products. JAPCA 38:1403–1411

Takigawa T, Wang BL, Saijo Y et al (2010) Relationship between indoor chemical concentrations and subjective symptoms associated with sick building syndrome in newly built houses in Japan. Int Arch Occup Environ Health 83(2):225–235

Väisänen T, Laitinen K, Tomppo L, Joutsensaari J, Raatikainen O, Lappalainen R, Yli-Pirilä P (2018) A rapid technique for monitoring volatile organic compound emissions from wood–plastic composites. Indoor and Built Environment 27(2):194–204. https://doi.org/10.1177/1420326X16669976

Vandenbroucke AM, Morent R, De Geyter N, Leys C (2011).Non-thermal plasmas for non-catalytic and catalytic VOC abatement. J. Hazard. Mater. **195**(x), 30–54. https://doi.org/10.1016/j.jhazmat.2011.08.060

Weschler CJ (2009) Changes in indoor pollutants since the 1950s. Atmos Environ 43(1):153–169. https://doi.org/10.1016/j.atmosenv.2008.09.044

Yu CWF, Crump DR (1999) Review: testing for formaldehyde emission from wood-based products—a review. Indoor and Built Environment 8(5):280–286. https://doi.org/10.1177/1420326X990080050

Chapter 4
Safety in Composite Laboratory

Abstract In this chapter, it reveals on the review of safety in composite laboratory. Throughout this topic, it elaborates on the steps and safety measures needed to be taken in the laboratory in three periods such as before, during and after handling composites. Several safety and protective equipment also have been mentioned to allow reader to understand the functions and post-effect if not wearing or using it appropriately. On the top of that, specific composite processes such as pyrolysis would potentially cause several hazards, which could harm the operator during its handling process. Thus, this part deliberates comprehensively on the composite processes step-by-step and its hazards which could harm the users. At the end, the emergency safety equipment such as fire extinguisher and emergency showers explains thoroughly in the manuscript, which functions during post-accident events. This manuscript also presents the guidelines of important facts and basic concepts that can be used as a guide to preventing accidents and injuries in Biocomposite Technology Laboratory, Universiti Putra Malaysia (UPM). The objective of these guidelines is to be used to prevent accidents and injuries in biocomposite laboratories in particular while increasing the safety and health of students/customers.

Keywords Safety and health · Accidents · Composite · Composite laboratory · Safety and emergency equipment

4.1 Introduction

Safety is a state of situation or event that is protected from any danger, risk, injury, or fatal. It can be also referred to as the management of any recognised hazards and harms to attain a satisfactory condition of risk. The topic of 'safety' is frequently seen as one of a branch of disciplines related to quality, reliability, availability, maintainability, and safety itself (Niu et al. 2019). According to a recent database recorded by the U.S. Bureau of Labour Statistics, around 2.9 million nonfatal injuries and 5190 fatal injuries were recorded only in the year 2016 (USDOL 2018). In conjunction with these facts, the estimated annual cost from serious injuries was projected around \$151.1 billion. In response to these numbers and trends, international bodies,

S. M. Sapuan et al., *Safety and Health in Composite Industry*,
Composites Science and Technology ,
https://doi.org/10.1007/978-981-16-6136-5_4

government organisations, commercial industries, and standard organisations have to intensively focus on the efforts to eliminate or reduce the injuries and death rates by applying any policies and schemes to increase workers safety in their workplace.

Since then, many discussions are focusing on this subject matters, especially on their fundamentals, principles, and approaches to achieve better health and safety quality for world communities (Bucksch and Schlicht 2006; de Hartog et al. 2010). Therefore, many interventions were proposed internationally and locally as a guideline in order to handle any potential risks in any sector (Ladewski and Al-Bayati 2019). For instance, ISO 45001:2018 is an international standard that has been adopted by engineering and manufacturing sectors to manage their workers' safety and health qualities (Rostykus and Baker 2018). This international standard postulates the requirements on occupational health and safety management system to provide safe and healthy workplaces by preventing work-related injury and ill-health.

In recent years, safety issues in the composite industry, especially composite laboratories have raised many concerns from researchers, technical staff, and even students. These concerns have been gained since numerous potential hazards and harms were identified in the composite laboratories (Brown et al. 2002). However, currently, no occupational exposure limits governing workplace exposure to engineered composite materials were reported. These exposures might lead to any hazards, including physical injuries and health problems such as suffocation, exposure to cancer risks, and other related diseases (Maynard and Kuempel 2005; Shvedova et al. 2005). For these reasons, composite materials substances such as glass fibre, carbon fibre, and synthetic resins revealed new challenges to manage potential safety and health risks to the mentioned laboratory users.

Having a strong set of laboratory safety rules and guidelines which can be strictly adhered to by students/customers is an important component to avoid accidents and injuries in the laboratory. Overall workplace injury cases were increased from 12,810 cases in 2018 to 13,779 cases in 2019. The number of workplace fatal injury cases were decreased from 41 cases in 2018 to 39 cases in 2019, resulting in a lower fatal injury rate of 1.1 per 100,000 workers. The number of workplace major injuries were increased by 5.5% from 596 in 2018 to 629 in 2019. Likewise, the number of workplace minor injuries were increased by 7.7% from 12,173 in 2018 to 13,111 in 2019. The confirmed cases of the occupational disease were decreased from 563 cases in 2018 to 517 cases in 2019, while the number of dangerous occurrences was decreased from 23 incidents in 2018 to 21 incidents in 2019 (Table 4.1).

Safety procedure usually involves chemical hygiene plan, waste disposal procedures, and significant physical and health hazards associated with the specific type of research and instruction in specific procedures that researchers should use to prevent and limit exposure to the health hazards in that workplace (Shrivastava 2017).

This chapter focuses on the principles, guidelines, and approaches of safety in the composite laboratories to recognise practical methods to improve safety cultures in research laboratories within the theme of hazard and risk management. To achieve this aim, all parties and laboratory sers related to the composite laboratories such as staff, educators, and students have to prepare necessary actions, emergency responses, and equipment during handling activities related to composite-based products. The safety

Table 4.1 Number of workplace injuries, dangerous occurrences, and occupational diseases, 2018 and 2019 (Ministry of Manpower 2020)

Year	2018	2019
Workplace injuries	12,810	13,779
(i) Fatal	41	39
(ii) Major	569	629
(iii) Minor	12,173	13,111
Dangerous occurrence	23	21
Occupational diseases	563	517

guidelines, cultures, and practices are also highlighted such as to conduct effective training modules, develop safety guidelines and signs for safety awareness, prepare proper personal protective equipment, provide safety emergency response guides, and regular maintenance services for the tool and equipment. Those practices are explained in the following subtopic of this section.

4.2 Occupational Safety and Health Act (OSHA) 1994

The Occupational Safety and Health Act (OSHA) 1994 was enacted to provide for the safety, health, and welfare of workers and to protect against the risk of accidents involving workers' environmental activities. In OSHA 1994, section IV is for the employer and Part VI is for the employee (Occupational Safety and Health Act 1994).

The objectives of the acts are:

(a) To secure the safety, health, and welfare of persons at work against risks to safety or health arising out of the activities of persons at work.
(b) To protect persons at the workplace other than persons at work against risks to safety or health arising out of the activities of persons at work.
(c) To promote an occupational environment for persons at work that is adapted to their physiological and psychological needs.
(d) To provide the means whereby the associated occupational safety and health might be progressively replaced by a system of regulations and approved industry codes of practice operating in a combination of the provisions of this act that is designed to maintain or improve the standards of safety and health.

4.3 Safety

Students/customers need to make sure the strategic plan is in place before carrying out the tasks and work in the laboratory. This is because students/customers are exposed to risks while in the laboratory. Among the risks involved are exposure to chemicals, use of high-temperature laboratory tools, hydraulic stops that can cause cluttered hands and feet without proper handling, and machinery in the laboratory. The Biocomposite Technology Laboratory, Universiti Putra Malaysia (UPM) is installed with exhaust fans to get rid of chemicals or odours as students/customers conduct experiments in the laboratory.

In the event of an emergency or injuries among the students while performing work in the laboratory, the supervisor will perform follow-up actions such as obtaining an injury report, assisting the student in follow-up treatment, as well as contacting the most relevant contacts including the University Health Centre and the Office of Occupational Management and Occupational Health (OSH) UPM.

4.4 Laboratory Safety Knowledge

Laboratory safety knowledge is constantly emphasised by the officer in charge of the laboratory as well as the Biocomposite Technology Laboratory, Universiti Putra Malaysia (UPM) showing the safety aspects inside and outside the laboratory to guide students/customers. This is to create awareness to students/customers to be careful when carrying out assignments and works in Biocomposite Laboratory Laboratory, Universiti Putra Malaysia (UPM).

Knowledge of the use of laboratory tools is also provided in each of the tools as guidelines for instrumentation and safety. The knowledge to use the personal protective equipment (PPE) in accordance with the requirements of the equipment or machinery is available at the Biocomposite Technology Laboratory, Universiti Putra Malaysia (UPM).

4.5 General Laboratory Safety Rules and Guidelines

Following are the rules and guidelines related to most laboratories. Compliance with safety rules and guidelines while in the laboratory is very important to adhere to as students/customers are at high risk when using tools or machines in the Biocomposite Technology Laboratory, Universiti Putra Malaysia (UPM).

The rules and guidelines are as detailed out below (Hanton 2017):

1. Be sure to read all fire alarm and safety signs and follow the instructions in the event of an accident or emergency.
2. Ensure you are fully aware of your facility's/building's evacuation procedures.
3. Make sure you know where your lab's safety equipments are located, which includes first aid kit(s), fire extinguishers, eyewash stations, and safety showers, and how to properly use them.
4. Know emergency phone numbers to call for help in case of emergency.

5. If there is a fire drill, be sure to turn off all electrical equipments and close all containers.
6. Do not, drink, or eat while working in the laboratory.
7. Never use laboratory equipment that you are not approved or trained by your supervisor to operate.
8. If an instrument or piece of equipment fails during use or is not operating properly, report the issue to a technician right away. Never try to repair an equipment problem on your own.
9. Do not work alone in the laboratory.
10. Never leave an ongoing experiment unattended.
11. Make sure to always follow the proper procedures for disposing laboratory waste.
12. Report all injuries, accidents, and broken equipments or glasswares right away, even if the incident seems small or unimportant.
13. If you are injured, yell out immediately and as loud as you can to reach out for help.
14. In the event of a chemical splashing into your eye(s) or on your skin, immediately flush the affected area(s) with running water for at least 20 min.
15. If you notice any unsafe conditions in the laboratory, let your supervisor know as soon as possible.

4.6 Awareness, Attitude, and Action Toward Safety

Perceptions, attitudes, and actions toward safety aspects are those critical aptitudes needed to be implemented in order to maintain a safe environment in the composite laboratory. That safety awareness should be embedded in each laboratory users who uses the laboratories for their purpose. In composite laboratories, many hazards and risks could cause casualties and injuries to a person without proper following of the established safety guidelines and regulations. These hazards including materials that can be irritant, explosive, flammable, radioactive, or a health threat, which requires special attention and considerations during conducting the activities (Lunar et al. 2014). In general, the factors of casualties and accidents in any chemical laboratories, including composite laboratories are reported by Adane and Abeje (2012):

1. Absence of personal protective equipment (PPE);
2. Lack of experience during conducting experiments;
3. Mishandling of chemicals; and
4. Lack of knowledge about the proper actions to be taken in emergency cases.

According to Faller et al. (2010), the accident could happen on university campuses, such as laboratories during the presence of students. This could be explained by the students' lack of safety awareness and attitude of taking things for granted. Hence, it is compulsory to apply all safety procedures for more professional safety education, coherent risk, and safety climate management.

A research work conducted by Al-Zyoud et al. (2019) on strengths and weaknesses of student awareness of chemical safety found out that in most cases of accidents in laboratories, there were weaknesses during the staff's handling and dealing a specific emergency. For instance, some of the staff were lack of knowledge in the proper use of fire extinguishers. Moreover, the outcomes from the project also highlighted and recommended the improvement of the culture of safety ethics and risk management among the staff members and students. On the other hand, according to their descriptive statistics, students demonstrated fair to good familiarity and understanding of chemical hazard warning signs, when the sign was placed at the proper position. Another approach suggested from their work was a specific organisation, where the Environmental Health and Safety Office should be available on the campus, which is responsible for applying and following up on compliance with safety rules and procedures. Lastly, it was proposed that a compulsory course program regarding hazardous waste and risk management has to be developed before any laboratory works. Hence, these efforts would alert and make every laboratory user more conscious and aware of their safety in the composite laboratory.

4.7 Safety Practices

Currently, there is a rising understanding of chemical hazards and risk management which contribute to the rise of a new culture of concern for laboratory safety. It is important to note that composite educators including trainers, technicians, and lecturers have to maintain and establish a safe and secure environment surroundings (Ladewski and Al-Bayati 2019). This effort can be subjected to fostering the users' and students' "good attitudes toward rational risk assessment and safe habits" throughout the academic sessions and training programs. This task is vital to embed good practices and cultures among the laboratory users to keep safety as the number one priority. Moreover, an effective safety education can also minimise instructor liability in the event of an accident. A proper safety culture would provide self-awareness among laboratory members to conduct appropriate practices to conduct the composite preparations and experimental works safely (Alaimo et al. 2010).

4.8 General Rules and Policies for General Users of the Advanced Composites Laboratory

Laboratory operators that enter the laboratory must follow the rules very strictly. Users might be prohibited from using the laboratory if they handle chemicals and/or nano-particles carelessly. Some of the general rules and policies for general users of the advanced composites laboratory are listed by Tate (Tate and Cook 2020) from Ingram School of Engineering, Texas State University, San Marcos:

1. Every laboratory user should read the 'Hazardous Communication' presentation on TRACS before start work in the laboratory.
2. Every laboratory user also should sign-in and sign-out as well as book the dates when using any instruments within the laboratory space area.
3. The instructions in the material specification data sheets (MSDS) related to handling and storing for the specific chemical must be read and followed.
4. Any experiment must be performed under the supervision of the laboratory coordinator or staff and students are stringently not allowed to conduct the laboratory works alone.
5. All laboratory users should acknowledge all chemicals around the laboratory, chemical disposable bins, emergency contacts, MSDS documents, emergency tools (first-aid kit, fire extinguisher, and safety shower), personal protective equipment, as well as emergency exits at the laboratory.
6. For the clothing ethics, all laboratory personals should wear full-length pants, closed tows shoes, and a laboratory coat/workshop jacket.
7. No food or drink is allowed in the laboratory.
8. The appropriate PPE must be worn when handling liquid chemicals, extreme temperature, extreme pressures, and spark releasing equipment such as gloves, aprons, safety glasses, face shields, nose masks, and respirators.
9. A half-mask respirator must be worn when managing nanofibres and volatile chemical compounds (VOCs) resins.
10. Always ensure a clean environment workspace in the laboratory.
11. All equipment and tools that have been used should be placed and stored back at their designated stored location.
12. Any broken equipment and tools should be reported to laboratory supervisors immediately.
13. All VOCs resins should be prepared under the fume hood and should be labelled to designate their content.
14. Washbasin in the lab is only for washing hands. Do not clean glassware or plastics ware in the washbasin.
15. The equipment should be not left unattended.
16. All machines, tools, and equipment should be switched off once the work has been completed.
17. Notes or indications when an equipment's operation is to be kept/continued and stopped should be placed during operation for long hours.
18. Any cured, uncured, rages, and VARTM bagging materials should not be thrown in the designated satellite waste containers, not the common garbage.
19. All nanoparticle contaminant wastes such as wipes, rags, beakers, napkins, respirator, and cartridges must be put in a Zip-lock bag and kept in Nanoparticle Containment Room.

4.9 Preliminary Training Courses on Safety

Generally, this subtopic is narrowed down on the aspect of conducting a training workshop on composite laboratory safety. Any person who works in a laboratory must receive training to become knowledgeable about potential hazards in the laboratory. The main objective of this laboratory safety training course is to explain, clarify, and provide knowledge to the laboratory members and users in terms of safety requirements. This allows them to ensure that the laboratory personnel knows about handling laboratory activities to prevent accidents, injuries, and illnesses (Simplify Compliance 2020).

The courses are commonly prepared and conducted by professional trainers with a background in safety. During the program, the trainees would be provided with a written module, training kit, and physical demonstration on the guidelines of safety and emergency response during an accident. On the other hand, the training workshops usually outline and cover several topics, including national laws and policies regarding safety, incidents related to safety accidents, emergency preparedness, procedures and postings, laboratory and equipment inspections, pollution prevention and waste disposal, personal protective equipment, hazardous substances, and processes (2020a).

4.10 Raw Material Handling and Preparation

In a composite laboratory, raw materials such as glass fibre, carbon nanotube, and volatile organic compounds (VOCs) are the materials required to prepare a composite laminate. However, as up to our current knowledge, these materials would cause health problems to the operators when it is not handled properly. A proper method of storage, movement, and handling of raw material for composite preparation could improve and reduce the risks and injuries to a person. Thus, manual material handling is of major concern in health and safety aspects, thus, practical ways of reducing health risk have to be identified accordingly.

In the manual of material handling, the activities cover the preparation of raw materials, progress of work, finishing goods, scraps, and packaging materials. Commonly, there are various sizes, shapes, weights, and properties of raw materials used in composite product preparation. The term 'material handling' is usually referred to as a systematic and scientific method of moving, packing, and storing materials in an appropriate and suitable location. The main objectives of material handling are as follows (MSG 2020):

1. To determine the appropriate distance of raw material to the working area.
2. To govern the reduction of material damage by improving the quality of the raw material.
3. To create safe and hazard-free work conditions.
4. To improve productivity and efficiency.

5. To manage schedule efficiently.
6. To reduce the overall manufacturing time by designing efficient material movement.

It is essential to determine the importance of material handling to improve and reduce safety and health hazards for the operator from lower back injuries, breathing difficulties, and other physical injuries and pains. In the current competitive and globalised environment, it is important to control cost and reduce time in material handling along with promoting a safe environment. An efficient material handling process promotes the proper design of facility layout, development of a method that improves and simplifies the work process, increases productivity and efficient material handling, and reduces the total cost of production (Heragu 2018). Hence, a suitable safety action assists to cut down the number of accidents during handling raw materials and equipment.

4.11 Personal Protective Equipment

Personal protective equipment or PPE is one of the important ethics needed to be cultivated in the composite laboratory work area when working with chemical and physical hazards. The PPE is a good approach when handling composite materials in a layup operation. Personal protective tools should be carefully chosen to safeguard the human body from chemicals and the process used (Newill et al. 1989; Zhang et al. 2020). The PPE includes safety glasses, eye protection, face shields, glove, helmets, and respirators as shown in Fig. 4.1.

4.11.1 Eye Protection

Eye protection tools are that equipment that is useful to protect human vision from volatile chemicals and fine fibre entering the eyes (Jaafar et al. 2018b; Ilyas et al. 2020b). The common equipments are safety glasses and chemical goggles as displayed in Fig. 4.2. In practice, the goggles have to be worn before entering any wet bench laboratory, including cell culture labs. This applies to laboratory visitors, technical maintenance staff, custodial workers, as well as staff and students. Commonly, the safety glasses and goggles function to shield chemical splash and are good enough for impact resistance.

Fig. 4.1 Personal protective equipment to protect the operator from hazards (Loibner et al. 2019)

Fig. 4.2 Wearable safety glass goggles for eye protection

4.11.2 Laboratory Coats and Jackets

Other safety clothes used to cover and shield the user from chemical danger are the laboratory coat and jacket. They are used to protect personnel from chemicals, biological, or unsealed radiological sources that might cause diseases. Normally, the laboratory coat does cover the wearer to the knees and it usually applies a long sleeve to protect arms from any harm. To be specific, the laboratory coat fabric is made up from poly-cotton blends, which is a flame-resistant material. Figure 4.3 depicts the laboratory coats and jackets that could be used in the composite laboratory.

Fig. 4.3 Laboratory coat and jacket

4.11.3 Face Protection

A face shield is one of the face protection tools used when cutting a composite laminate that might produce sharp dust particles by-product. These sharp fine particles can cause injury to facial skin and eye redness. Typically, the face shield used in the composite laboratory is made up of synthetic resin, which is polypropylene with a 3-D printed adjustable bracket (Roberge 2016). Face shields must always be worn over safety glasses or goggles, not instead of safety glasses or goggles. Figure 4.4 exhibits a face shield used with safety goggles.

Fig. 4.4 Application of face shield with safety goggles

4.11.4 Hand Protection

A hand protection kit is necessary to guarantee the hand is resisted from corrosive chemicals being used, which would cause loss of dexterity, risk of ergonomic injury, or even loss of the user's hands. There is no single glove material that provides 100% protection from all chemicals, a good all-purpose glove is the nitrile exam gloves made up of latex (Dave et al. 1999). Many institutions such as hospitals have banned the use of powdered latex glove because it can cause latex allergies and there is circumstantial evidence that the powder used might increase environmental bacterial contamination (Dave et al. 1999).

There are several general rules regarding chemical resistant glove use. The general rules can be found as followed (EHS Today 2000):

1. Nitrile exam gloves are the general-purpose glove of choice in the composite and chemical laboratories.
2. Powder free-latex glove should be used when handling composites sample fabrication, laminate cutting, and joining composite products.
3. Appropriate gloves should be chosen that are suitable for a specific chemical liquid.
4. Any physical damages on the gloves such as tears or pinholes and for previous chemical damage have to be checked before use, especially when dealing with a high concentration of acid.
5. The external surface of the gloves have to be washed often during managing composite fabrication.
6. Keep hands away from unprotected flames or any heat sources of high temperature to avoid the gloves to melt, since the latex is combustible.
7. When removing gloves, the contaminated exterior must be avoided from contacting the skin as shown in Fig. 4.5.
8. The contaminated gloves must be properly disposed inside an appropriate container as shown in Fig. 4.5.
9. Do not ever attempt to reuse the contaminated gloves after their removal.
10. Do not wear possibly contaminated gloves outside the laboratory or touch any daily use devices and personal items such as telephones, computer keyboard, or wallet.
11. The user's hands should be immediately washed using soap immediately after removing the gloves.

In terms of handling hot or cold conditions works, a specialised glove should be used. These specialised gloves are commonly made up of leather, Kevlar, and insulating foam (Awais et al. 2015). The thermal protective gloves should be selected according to their length, the required level for protection, and the level of dexterity. For this case, gloves made of asbestos cloth are not allowed.

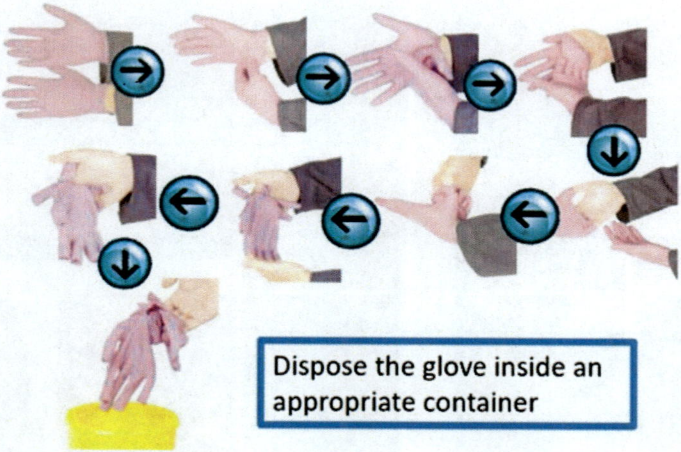

Fig. 4.5 Method to remove gloves after use

4.11.5 Respiratory Protection

Respirators are the last resort when it comes to protecting people in the workplace. A respirator is a device designed to protect the wearer from inhaling harmful substances. It can protect the wearer from harmful gases, mists, vapors, fumes, and fine particulates.

Respirators fall into the following two general classifications according to the modes of operation such as atmosphere-supplying respirators and air-purifying respirators. For atmosphere-supplying respirators, it can be further classified into supplied air (SA) or air-line respirator and self-contained breathing apparatus (SCBA). For air-purifying respirators comprise an N-95 filtering facepiece respirator, half-face air-purifying respirator (HF APR) with an elastomeric facepiece, and full-face air-purifying respirator (FF APR) with an elastomeric facepiece (2020d). Figure 4.6 shows the classification of the respiratory protections used in the laboratory to reduce the airborne hazard risks.

4.12 Waste Disposal Management

A composite waste is usually the leftover from composite fabrication, which involves toxic substances that might affect the living organisms as well as the environment (Shahroze et al. 2019). Composite wastes are typically disposed via incineration process. Recent studies displayed that there are large quantities of hazardous materials could be absorbed or particulates the wastes are improperly disposed. This was due to several chemical compounds that are mutagens and carcinogens toward living organisms especially humans that enters the body via inhalation, possessing

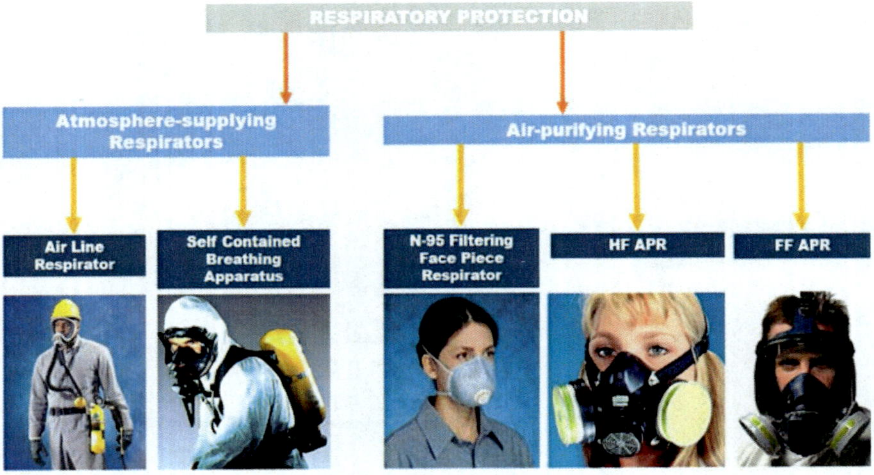

Fig. 4.6 Classification of the respiratory protections used in the laboratory (2020b; d)

acute or chronic effects. For instance, a work carried out by Gupta (2009) found that consistent and continuous exposure to a single high dose fibrous particulates and any synergistic interactions with the organic chemical causes bad long-term health effects. Thus, the waste products from composites such as fibre and VOCs have to be collected in proper containers before post-processes take place.

The composite manufacturing process generates composite material wastes during manufacturing as well as at the end of life (Jaafar et al. 2018a; Johari et al. 2019; Asyraf et al. 2020a; Ilyas et al. 2020a). Due to the environmental legislation and to protect the earth from hazardous substances, an attempt was made to find out a suitable method to dispose the composite material (Vijay et al. 2016). The current and future waste management and environmental legislations require all engineering materials to be properly recovered and recycled from end-of-life (EOL) products (Yang et al. 2012). To have a clearer picture of the recycling process of waste composite, Fig. 4.7 depicts the overall concept for the composite processes.

There are several post-processes for water disposal initiatives including mechanical recycling, thermal recycling, and landfilling. The specifications of composite post-processes can be found in Table 4.2.

4.13 Equipment Inspection and Maintenance

Regular inspection and maintenance of instrument and equipment in the laboratory is mandatory to ensure the safety of the operator. Numerous casualty incidents happened due to the lack of maintenance and improper use of these instruments. One of the typical instrument-related hazards in composite laboratories come from devices powered by electricity. These devices might harm users such as driller, cutter,

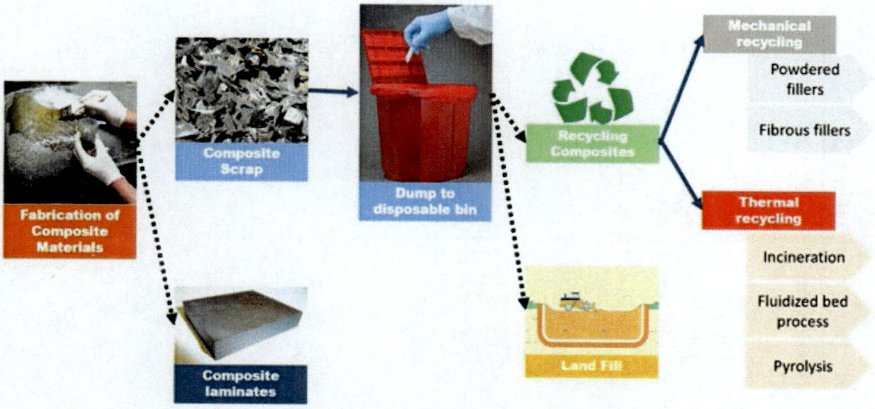

Fig. 4.7 Overall concept for composite product processes and outputs (Yang et al. 2012)

grinder, paint spray, and hot-press machines (Yarahmadi et al. 2016). Seemingly ordinary hazards such as physical accidents with rotating equipment and machines or tools for cutting and drilling, noise extremes, slips, trips, falls, lifting, and poor ergonomics account for the greatest frequency of laboratory accidents and injuries (Barclay et al. 2018). Thus, this subtopic is narrowed to the prudent practices for handling frequently used equipment in the composite laboratory.

4.13.1 Electrical Powered Laboratory Equipment

In a composite laboratory, there are many appliances powered by alternating current (AC) plug to operate heating, cooling, mixing, and pumping. These electrically powered devices include fluid and vacuum pumps, lasers, power supplies, stirrers, and hot plates. High-voltage and high-power requirements are increasingly prevalent; thus sensible practices for handling these devices are increasingly necessary.

In the application of electrical devices, major risk that might be induced is electrical shock. Aroundu 80–100 mA of electrical current can lead to fatalities and 10 mA would cause some serious injuries (Bellini et al. 2016). Furthermore, the improper use of electrical equipment could ignite explosive vapours, which cause fire accidents. Those hazards could be minimised by regular appropriate maintenance operation and the correct use of these devices. Before beginning any work, all personnel should be informed and trained on the use of all electrical power sources and the location of emergency shutoff switches.

Several precautions steps should be practiced especially during installation, modification, repair, and the use of electrical appliances. In the beginning, the operator should ensure all electrical equipment are installed and maintained in accordance with the provisions. In this case, well-trained and qualified personnel should work on repairing and calibrating this equipment before use by any operator. To ensure the

Table 4.2 Specifications of composite post-processes

Post-process	Specific process	Materials	Methodology	Advantages	References
Mechanical recycling	Powdered fillers	Glass fibre	– The mechanical recycling process involved the reduction of composite scrap size at 50–100 mm using low-speed cutting/crushing – Specific process of scrap reduction via hammer mill or other high-speed millings for fine grinding – The produced particles from previous process were encountered with cyclone and sieves chamber to filter the fibre particles with specific size: fibre-rich particle (coarser) and matrix-rich particle (finer)	– The end product from the recycling could be reused as reinforcement material in new thermosetting composites	Van Kets et al. (2019)
	Fibrous fillers				

(continued)

Table 4.2 (continued)

Post-process	Specific process	Materials	Methodology	Advantages	References
Thermal recycling	Incineration	Carbon fibre	– It did not involve the materials recovery process – Cannot be incinerated without proper precautions due to the potential release of small electrically conductive fibres into the environment, which contribute to electrical interference issues	– The by-product in the form of inorganic residues could be used in the cement industry	He and Lin (2019)

(continued)

Table 4.2 (continued)

Post-process	Specific process	Materials	Methodology	Advantages	References
	Fluidised bed combustion process	Glass and carbon fibres	– The process is useful for recovery of reinforcement fibres and to produce secondary fuels via depolymerisation process – The scrap was broken apart into 25 mm size before fed inside fluidised bed reactor – The reactor was operated at 450 °C for polyester resin composites and up to 550 °C for epoxy resin composites – The recovered fibres were clean and have a mean length of 6–10 mm – The whole process is summarised in Fig. 4.8	– The process is most efficient to recover recyclable materials and energy – It aided to recycle and recover fibres, while organic resin was used as energy sources	Nowak and Mirek (2013)

(continued)

Table 4.2 (continued)

Post-process	Specific process	Materials	Methodology	Advantages	References
	Pyrolysis recycling process	Long and high modulus fibre	– It involved the depolymerisation at 300–800 °C in absence of oxygen – The whole process is summarised in Fig. 4.9	– Both reinforcement fibre and the matrix materials were recovered	Meier et al. (2013)
Landfilling	– Dump the scrap inside a land hole – Permanent cross-links-matrix composite		– Large hole is dug in an open field for waste disposal	– Easiest to dispose, however, exhibits negative effects on the environment	Ziegler-Rodriguez et al. (2019)

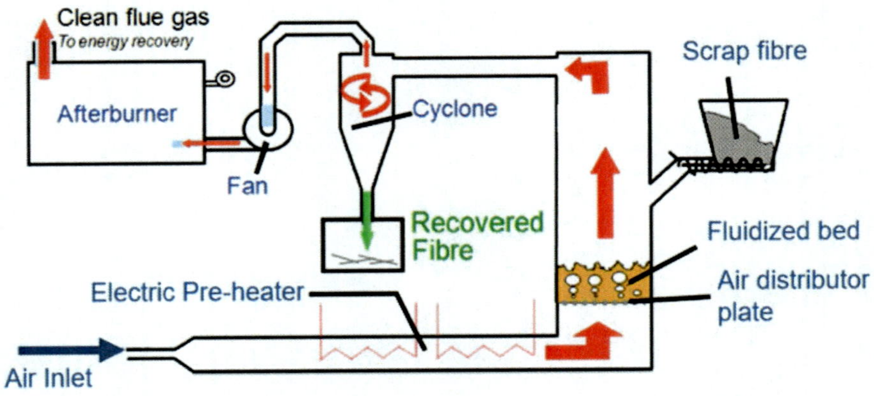

Fig. 4.8 Fluidised bed process (Vijay et al. 2016)

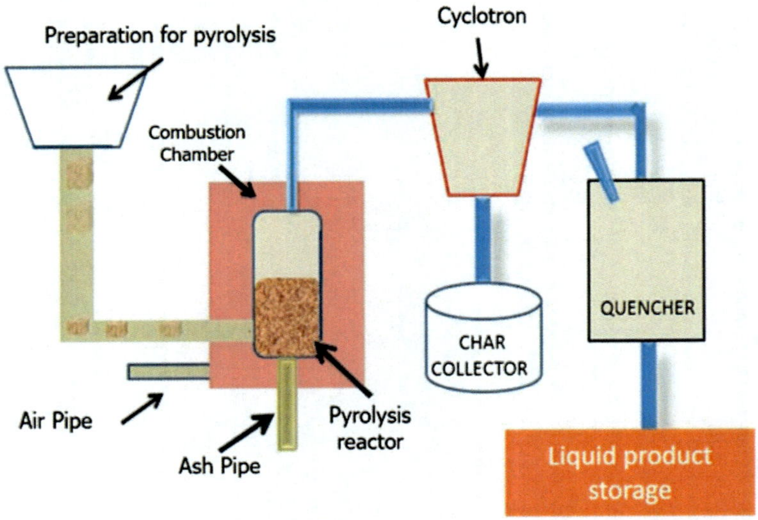

Fig. 4.9 Pyrolysis process (Zhao et al. 2011)

safety of the technician, all electrical equipments must be installed and maintained by the provisions, the devices must be de-energised and all capacitors discharged safely. Furthermore, discharged and de-energised conditions must be verified before proceeding. Another precaution that could be done is to ensure all new electrical equipments should be examined their safety certification mark. The safety certification mark covers the safety of the electrical appliances by the ability to protect from overload, use of a standard fuse, and following specific standards set by the electrical authorised body.

To summarise, specific precaution steps for electrical appliances usage are listed below (Office of the Vice President for Research 2020):

1. Ensure all electrical devices connection are properly insulated.
2. The damaged cords of any electrical devices are properly replaced.
3. Electrical equipment and power supplies are completely isolated.
4. The safety sign for electrical equipment is placed at a visible place with a mandate for non-sparking explosions and electrocuted signs.
5. The electrical appliances are placed in a secured area from the possibility of spills onto the equipment or flammable vapours carried into it.
6. The electrical equipment should be mounted on a wall or vertical panel.
7. These equipment are carefully grounded using a suitable flooring material to reduce the probability of electrical shock.

4.13.2 Equipment with Compressed Gases

Equipment with high-pressure operations can only be used by trained personnel. In this case, the apparatus with high-pressure operations should not carry out any chemical reactions or heat should not be applied to the apparatus to avoid from the explosive event unless it has been verified to withstand pressure. Moreover, equipment with high-pressure vessels should be ensured installed with pressure-relief devices, reading gauges, and safety systems. Relief devices used on pressure regulators do not require these seals or numbers (Compressed Gas Association 1990).

Apart from that, the equipment should be practiced with inspection periodically to ensure the device is safeguarded from danger to the operator. In terms of inspection of frequency, it is depending on the equipment type, nature of its usage, and the rate of use of these devices. During the inspection, the whole assembly equipment should be tested to the weakest points (threaded joining area, valves, and packing) in the assembled apparatus to detect leaking. Alternatively, the apparatus might be pressurised and monitored for pressure drop over time (Chugaev et al. 2017). On the other hand, during the repairing, storing, shipping, inspection, and maintenance operations, the equipment also has to remove all toxic, flammable, or other hazardous materials for safe handling. After the inspection, stamp inspection data should recorded (Pinney 1965).

4.13.3 Working with Extreme Pressures and Temperatures Equipment

Planning and special precaution steps are those essential works when dealing with hazardous chemicals at extreme pressures and temperatures. Various experimental works involve extremes pressures and temperatures with cryogenic liquids and high vacuum have to be managed simultaneously. Moreover, during the procedures related to low or high pressure, protection must be incorporated with appropriate equipment selection and the use of safety shields. In addition, the equipment should be included with appropriate temperature control and interlocks. Subsequently, this effort aids the equipment to control the temperature to not exceeding the desired limits even if the equipment fails. Lastly, another effort could be done by selecting and using glass apparatuses that can safely withstand thermal expansion or contraction at the designated pressure and temperature extremes (Golwalkar 2019).

4.14 Emergency Equipment and Response Procedures

As mentioned in the previous section, every composite laboratory should strictly develop and nurture a safety culture among laboratory users and members to have a safe working environment. This culture could be started by wearing personal protective equipment (PPE) during working with hazardous materials. Emergency responses with appropriate equipment are required in some circumstances such as in the fire, explosion, spill, and other laboratory accidents. Thus, all composite laboratory needs to have appropriate safety and emergency equipment in reachable places for the use of the laboratory personnel.

The general emergency procedures are intended to limit injuries and minimise damage during the occurrence of an accident. These included procedures are as follows (Richmond and Nesby-O'Dell 2002):

1. The sign of emergency help such as the Occupational Health and Safety phone number should be labelled accordingly at the visible area.
2. Fire-alarm pull stations and telephones with emergency contact numbers must be readily accessible.
3. Ascertain the situation by not entering and reentering the unsafe area.
4. Every laboratory should be provided with a sign to warn personnel in adjacent areas of any potential safety risks.
5. Safety equipment, including spill control kits, safety shields, fire safety equipment, respirators, safety showers and eyewash units, and emergency equipment should be available in well-marked highly visible locations in all chemical laboratories.
6. The first aid kit should be located in the reachable and visible areas.
7. The portable fire extinguisher and safety shower have to be placed and provided nearby to adjacent areas of any potential safety risks.
8. If clothing is on fire and a safety shower is immediately available, the person should be doused with water or rolled on the floor to smother the flames.
9. If harmful chemicals are spilled on the body, the chemicals should be removed, usually by flooding the exposed area with the safety shower.
10. If a chemical has splashed into the eye, the eyeball and the inner surface of the eyelid should immediately washed with water for 20 min. An eyewash unit should be used if available. Forcibly hold the eye open to wash thoroughly behind the eyelid.

11. Any medical helps and alerts should be called immediately when an accident happens.
12. The emergency equipment should be provided near to the hazards and the equipment has to provide with all information related to potential hazards and steps to manage hazards.

4.14.1 Fire Safety Equipment

For every buildings and room, a fire extinguisher is essential to be placed within a visible and reachable spot. The mass of a portable fire extinguisher is typically around 12 kg for a standard room size of 25 m^2 (2017). Generally, carbon dioxide and dry chemical fire extinguishers are used in a composite laboratory. Other types of extinguishers should be available if required for the works that will be performed in the laboratory (Asyraf et al. 2020b). In this case, there are four common types of portable fire extinguishers as listed in Table 4.3 with their classification by the fire types.

In general, every fire extinguisher should be marked with their classes of fires with the last inspected date. Moreover, the laboratory staff should also be trained on the fire accident operation and responsible for knowing the location, operation, and limitations of the fire extinguishers in the work area. To be specific, laboratory supervisors have to be responsible to guarantee and confirm that all laboratory users and members are aware of the fire extinguishers' locations with specific visible signs. After an extinguisher is used, personnel in-charge should promptly recharge or replace it.

4.14.2 Safety Showers and Eyewash Units

The safety shower is necessary emergency equipment to remove accidental chemical spills/splashes. The standard for the installation and maintenance of safety shower should follow ANSI Z358.1, Emergency Eyewash and Shower Equipment (2004). The application of a safety shower is for the mediate first-aid treatment of chemical splashes and for extinguishing clothing fires. This equipment must be used by trained laboratory personnel. The safety showers should be routinely tested to ensure the valve is operable and to remove any debris in the system.

Figure 4.10 shows the components of the safety shower in a laboratory. The safety shower operates by drenching out the chemical substances straightaway with a large amount of water during the chemical spill accident. It is functioned with a quick-opening valve at the top and connected with a shower switch handle (downward-pull delta bar). Besides, the shower is also commonly installed with drains to reduce the slip and fall risks and facility damage associated with flooding in a laboratory. On the other hand, the eyewash units are placed with a safety shower assembly. This

Table 4.3 Types of fire extinguishers with their respective fire classes (2020e)

Fire class	Symbols	Material classification	Examples	Extinguisher agents
A		Combustible materials	Paper and wood	Water, foam, dry powder, and wet chemical
B		Flammable liquids	Paint and petrol	Foam, dry powder, and carbon dioxide
C		Flammable gases	Butane and methane	Dry powder
D		Flammable metals	Potassium and lithium	Dry powder
E		Electrical equipment	Computers and generators	Dry powder and carbon dioxide
F		Deep fat fryers	Cooking oils, animal fats, and vegetable oils	Wet chemical

product is designed together with a safety shower to ease washing the eyes while showering. It could provide a soft stream or spray of aerated water for an extended period (20 min).

4.15 Conclusions

All in all, the safety of composite processing in a laboratory is a delicate and vital issue to govern the uncertainties around the work environment. Composite laboratory users are usually exposed to volatile toxic chemical compounds and very fine particles when dealing with the composite materials. This could negatively affect and harm the health and safety of the users in a short- and long-term duration. Thus, the best way to protect and reduce the risk of the users is by conducting training courses on safety for every laboratory personnel, provide personal protective equipment,

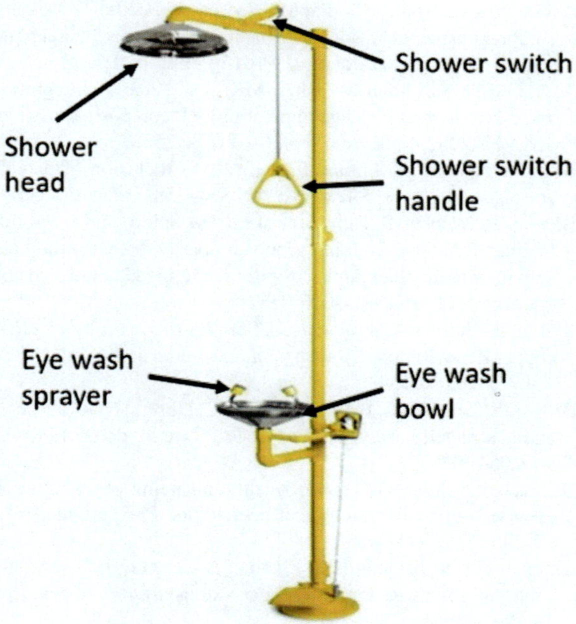

Fig. 4.10 Components of safety shower (Pate and Wilke 2020)

establish proper guidelines, visible signs, and MSDS for safety in the laboratory area. Moreover, emergency response actions with the availability of emergency equipment are provided and assisted by the trained staff when any accident occurred. Also, the laboratory should provide emergency contacts with their details for every equipment. Possible engineering controls should be identified and make available for the users by the laboratory supervisors. Thus, all laboratory users from the supervisors to the users should be responsible, conscious, and highly aware during conducting any works and dealing with hazardous substances and machines in the laboratory. This is because by following these safety guidelines, the risk of accidents and injuries among the students/customers in the laboratory can be mitigated and minimized.

Acknowledgements This chapter was written with Mohd Hafizie Manap.

References

Adane L, Abeje A (2012) Assessment of familiarity and understanding of chemical hazard warning signs among university students majoring chemistry and biology: A case study at Jimma University, Southwestern Ethiopia. World Appl Sci J 16:290–299

Al-Zyoud W, Qunies AM, Walters AUC, Jalsa NK (2019) Perceptions of chemical safety in laboratories. Safety 5. https://doi.org/10.3390/safety5020021

Alaimo PJ, Langenhan JM, Tanner MJ (2010) Safety teams: An approach to engage students in laboratory safety. J Chem Educ 87:856–861. https://doi.org/10.1021/ed100207d

Asyraf MRM, Ishak MR, Sapuan SM et al (2020a) Woods and composites cantilever beam: A comprehensive review of experimental and numerical creep methodologies. J Mater Res Technol 9:6759–6776. https://doi.org/10.1016/j.jmrt.2020.01.013

Asyraf MRM, Rafidah M, Ishak MR et al (2020b) Integration of TRIZ, Morphological Chart and ANP method for development of FRP composite portable fire extinguisher. Polym Compos 41:2917–2932. https://doi.org/10.1002/pc.25587

Awais M, Tausif M, Ahmad F et al (2015) Inclusion of recycled PPTA fibre in development of cut-resistant gloves. J Text Inst 106:354–358. https://doi.org/10.1080/00405000.2014.922246

Barclay R, Webber S, Ripat J et al (2018) Safety and feasibility of an interactive workshop and facilitated outdoor walking group compared to a workshop alone in increasing outdoor walking activity among older adults: A pilot randomized controlled trial. Pilot Feasibility Stud 4. https://doi.org/10.1186/s40814-018-0367-4

Bellini E, Gambassi G, Nucci G et al (2016) Death by electrocution: Histological technique for copper detection on the electric mark. Forensic Sci Int 264:24–27. https://doi.org/10.1016/j.forsciint.2016.03.013

Brown JS, Zeman KL, Bennett WD (2002) Ultrafine particle deposition and clearance in the healthy and obstructed lung. Am J Respir Crit Care Med 166:1240–1247. https://doi.org/10.1164/rccm.200205-399OC

Bucksch J, Schlicht W (2006) Health-enhancing physical activity and the prevention of chronic diseases—An epidemiological review. Soz Praventivmed 51:281–301. https://doi.org/10.1007/s00038-006-5043-4

Chugaev SS, Strizhenov EM, Zherdev AA et al (2017) Fire- and explosion-safe low-temperature filling of an adsorption natural gas storage system. Chem Pet Eng 52:846–854. https://doi.org/10.1007/s10556-017-0281-2

Compressed Gas Association (1990) Handbook of compressed gases, 3rd edn. Chapman and Hall, Arlington, Virginia, USA

Dave J, Wilcox MH, Kellett M (1999) Glove powder: Implications for infection control. J Hosp Infect 42:283–285. https://doi.org/10.1053/jhin.1998.0592

de Hartog JJ, Boogaard H, Nijland H, Hoek G (2010) Do the health benefits of cycling outweigh the risks? Environ Health Perspect 118:1109–1116. https://doi.org/10.1289/ehp.0901747

EHS Today (2000) Protecting hands from chemicals. In: Endeavor Bus. Media. https://www.ehstoday.com/ppe/hand-protection/article/21908760/protecting-hands-against-chemical-exposures#:~:text. While there are dozens of, research chemist for Best Manufacturing. Accessed 9 June 2020

Faller G, Mikolajczyk RT, Akmatov MK et al (2010) Accidents in the context of study among university students-A multicentre cross-sectional study in North Rhine-Westphalia, Germany. Accid Anal Prev 42:487–491. https://doi.org/10.1016/j.aap.2009.09.012

Golwalkar KR (2019) Safety precautions during maintenance and energy recovery. In: Golwalkar KR (ed) Integrated maintenance and energy management in the chemical industries. Springer Nature, Gewerbestrasse, Switzerland, pp 313–325

Gupta M (2009) Combustion of composite materials : assessment and quantification of hazards. DRDO Sci Spectr 27–30

Hanton SD (2017) lab safety rules and guidelines. Lab Manag

He J, Lin B (2019) Assessment of waste incineration power with considerations of subsidies and emissions in China. Energy Policy 126:190–199. https://doi.org/10.1016/j.enpol.2018.11.025

Heragu SS (2018) Facilities design, 4th edn. CRC Press, Boca Raton, USA

Ilyas RA, Sapuan SM, Atiqah A et al (2020a) Sugar palm (Arenga pinnata [Wurmb.] Merr) starch films containing sugar palm nanofibrillated cellulose as reinforcement: Water barrier properties. Polym Compos 41:459–467. https://doi.org/10.1002/pc.25379

Ilyas RA, Sapuan SM, Norrrahim MNF et al (2020b) Nanocellulose/starch biopolymer nanocomposites: Processing, manufacturing, and applications. In: Al-Oqla FM, Sapuan SM (eds) Advanced processing, properties, and applications of starch and other bio-based polymers, 1st edn. Elsevier Inc., Amsterdam, Netherland, pp 65–88

Jaafar CNA, Rizal MAM, Zainol I (2018a) Effect of kenaf alkalization treatment on morphological and mechanical properties of epoxy/silica/kenaf composite. Int J Eng Technol 7:258–263. https://doi.org/10.14419/ijet.v7i4.35.22743

Jaafar CNA, Zainol I, Rizal MAM (2018b) Preparation and characterisation of epoxy/silica/kenaf composite using hand lay-up method. 27th Scientific Conference of the Microscopy Society Malaysia (27th SCMSM 2018). Melaka, Malaysia, pp 2–6

Johari AN, Ishak MR, Leman Z et al (2019) Fabrication and cut-in speed enhancement of savonius vertical axis wind turbine (SVAWT) with hinged blade using fiberglass composites. In: Seminar Enau Kebangsaan. Bahau, Negeri Sembilan, Malaysia, pp 978–983

Ladewski BJ, Al-Bayati AJ (2019) Quality and safety management practices: The theory of quality management approach. J Safety Res 69:193–200. https://doi.org/10.1016/j.jsr.2019.03.004

Loibner M, Hagauer S, Schwantzer G et al (2019) Limiting factors for wearing personal protective equipment (PPE) in a health care environment evaluated in a randomised study. PLoS One 14. https://doi.org/10.1371/journal.pone.0210775

Lunar BC, Rhea V, Padura S, et al (2014) Familiarity and understanding of chemical hazard warning signs among select college students of De La Salle Lipa. Asia Pacific J Multidiscip Res P 2:2350–7756

Maynard AD, Kuempel ED (2005) Airborne nanostructured particles and occupational health. J Nanoparticle Res 7:587–614. https://doi.org/10.1007/s11051-005-6770-9

Meier D, Van De Beld B, Bridgwater AV et al (2013) State-of-the-art of fast pyrolysis in IEA bioenergy member countries. Renew Sustain Energy Rev 20:619–641. https://doi.org/10.1016/j.rser.2012.11.061

Ministry of Manpower (2020) Workplace safety and health report 2019. Singapore

MSG (2020) Material handling. Manag. Study Guid. https://www.managementstudyguide.com/material-handling.htm. Accessed 7 June 2020

Newill CA, Koegel AE, Prenger VL et al (1989) Utilization of personal protective equipment by laboratory personnel at a large medical research institution. Appl Ind Hyg 4:205–209. https://doi.org/10.1080/08828032.1989.10390433

Niu Y, Lu W, Xue F et al (2019) Towards the "third wave": An SCO-enabled occupational health and safety management system for construction. Saf Sci 111:213–223. https://doi.org/10.1016/j.ssci.2018.07.013

Nowak W, Mirek P (2013) Circulating fluidized bed combustion (CFBC). In: Scala F (ed) Fluidized bed technologies for near-zero emission combustion and gasification. Woodhead Publishing Limited, Philadelphia (PA), USA., pp 701–764

Occupational Safety and Health Act (1994) Guidelines on Occupational Safety and Health Act 1994. Minist. Hum. Resour

Office of the Vice President for Research (2020) Electrical safety in the laboratory. In: Univ. Iowa. https://ehs.research.uiowa.edu/electrical-safety-laboratory. Accessed 8 June 2020

Pate WJ, Wilke J (2020) Development and implementation of a clinical eyewash and safety shower risk assessment process in an academic medical center. ACS Chem Heal Saf 27:15–19. https://doi.org/10.1021/acs.chas.9b00003

Pinney G (1965) Compressed gas cylinders and cylinder regulators used in laboratories. J Chem Educ 42:A976. https://doi.org/10.1021/ed042pa976

Richmond JY, Nesby-O'Dell SL (2002) Laboratory security and emergency response guidance for laboratories working with select agents. MMWR Morb Mortal Wkly Rep 51:1–6

Roberge RJ (2016) Face shields for infection control: A review. J Occup Environ Hyg 13:235–242. https://doi.org/10.1080/15459624.2015.1095302

Rostykus W, Baker R (2018) ISO 45001: A model for managing workplace ergonomics. In: EHSToday. https://www.ehstoday.com/health/iso-45001-modelmanaging-workplace-ergonomics. Accessed 5 June 2020

Shahroze RM, Chandrasekar M, Senthilkumar K et al (2019) A review on the various fibre treatment techniques used for the fibre surface modification of the sugar palm fibres. In: Seminar Enau Kebangsaan. Bahau, Negeri Sembilan, Malaysia, pp 48–52

Shrivastava SK (2017) Safety procedures in science laboratory. Int J Sci Eng Res 5:54–64

Shvedova AA, Kisin ER, Mercer R et al (2005) Unusual inflammatory and fibrogenic pulmonary responses to single-walled carbon nanotubes in mice. Am J Physiol Lung Cell Mol Physiol 289. https://doi.org/10.1152/ajplung.00084.2005

Simplify Compliance (2020) Laboratory safety training. Simpl. Train. https://simplifytraining.com/course/laboratory-safety-training/. Accessed 6 June 2020

Tate JS, Cook R (2020) Advanced composites lab: rules and policies. Ingram School of Engineering, Texas State University, San Marcos

USDOL (2018) Employer-reported workplace injuries and illnesses—2016. U.S. Bureau of Labor—News Release. https://www.bls.gov/news.release/archives/osh_11092017.pdf. Accessed 5 June 2020

Van Kets K, Delva L, Ragaert K (2019) Structural stabilizing effect of SEBSgMAH on a PP-PET blend for multiple mechanical recycling. Polym Degrad Stab 166:60–72. https://doi.org/10.1016/j.polymdegradstab.2019.05.012

Vijay N, Rajkumara V, Bhattacharjee P (2016) Assessment of composite waste disposal in aerospace industries. Procedia Environ Sci 35:563–570. https://doi.org/10.1016/j.proenv.2016.07.041

Yang Y, Boom R, Irion B et al (2012) Chemical engineering and processing: process intensification recycling of composite materials. Chem Eng Process Process Intensif 51:53–68. https://doi.org/10.1016/j.cep.2011.09.007

Yarahmadi R, Moridi P, Roumiani YA (2016) Health, safety and environmental risk management in laboratory fields. Med J Islam Repub Iran 30

Zhang Z, Wu J, Hao L et al (2020) Development of biosafety equipment for high containment laboratory and for personal protection in China. Biosaf Heal 2:12–17. https://doi.org/10.1016/j.bsheal.2019.12.008

Zhao D, Zhang C, Hu H, Zhang Y (2011) Preparation and characterization of three-dimensional carbon fiber reinforced zirconium carbide composite by precursor infiltration and pyrolysis process. Ceram Int 37:2089–2093. https://doi.org/10.1016/j.ceramint.2011.02.024

Ziegler-Rodriguez K, Margallo M, Aldaco R et al (2004) ANSI Z358.1—Emergency eyewash and shower equipment. Chicago, USA

Ziegler-Rodriguez K, Margallo M, Aldaco R et al (2019) Transitioning from open dumpsters to landfilling in Peru: Environmental benefits and challenges from a life-cycle perspective. J Clean Prod 229:989–1003. https://doi.org/10.1016/j.jclepro.2019.05.015

Ziegler-Rodriguez K, Margallo M, Aldaco R et al. Safety glass goggle. Pro Saf. https://prosafety.com/en/mask-glasses/2692-uvex-safety-goggles--carbonvision.html. Accessed 20 June 2020

Ziegler-Rodriguez K, Margallo M, Aldaco R et al (2020a) Training for laboratory personnel. Environ Heal Safety (Princeton University). https://ehs.princeton.edu/laboratory-research/laboratory-safety/training-laboratory-personnel

Ziegler-Rodriguez K, Margallo M, Aldaco R, et al (2020b) Personal Protective Equipment. Georg Inst Technol. https://www.ehs.gatech.edu/chemical/lsm/7-6. Accessed 20 June 2020

Ziegler-Rodriguez K, Margallo M, Aldaco R et al (2020c) Safety jacket. Shaoma NIG Ltd. http://shaomanig.com/portfolio-items/engineering-safety-jackets/. Accessed 20 June 2020

Ziegler-Rodriguez K, Margallo M, Aldaco R et al (2020d) PPE: Respiratory protection. University of California. https://ehs.ucmerced.edu/researchers-labs/ppe/selection. Accessed 20 June 2020

Ziegler-Rodriguez K, Margallo M, Aldaco R et al (2020e) Types of fire extinguisher. Surrey Fire & Safety Ltd. https://surreyfire.co.uk/types-of-fire-extinguisher/. Accessed 19 Feb 2020

Chapter 5
Design for Safety in Composites

Abstract This chapter introduces the basic fundamental and concept for design for safety. It reveals the technical gaps between design for safety and current state of the art to produce a new products. This chapter introduces the background of design for safety which covers those ten paradigms for safe designs to guide the designer to do right things at the right times. Moreover, the several concurrent engineering approaches have been described and used to be integrate with the design for safety in order to produce a new product. Detail discussions has been expanded with its examples and current progresses.

Keywords Composite · Safety and health · Design for safety · Concurrent engineering · Product development

5.1 Introduction

Safety is one of the vital aspects needed to be covered in designing many products and systems in fulfilling their design intentions. Thus, by incorporating the safety aspect in design development, design for safety has been introduced and applied in the design process. It is one of the elements of design for manufacturability (DFM) that is principally explained as the design development process which incorporates any actual potential risks and hazards. It is considered the most effective risk control measure which is achieved by eliminating the hazards at the source during the design process. To be specific, the design for safety usually takes place by measuring the risk. It is a framework of design with its specification to avoid such hazards or to mitigate their effects. In this context, there are numerous parties involved in the design for the safety process, such as the environment, community, customers, maintenance personnel, and operators (Moriarty 2012).

The application of the design for safety method emphasised the reliability of the product and system including the main system, subsystems, and components of all areas of the product. The process involves the exploitation of the current information of the product in the course of examining a design to insert the reliability data. Later, risk assessment codes for each hazard found in the design can be obtained

S. M. Sapuan et al., *Safety and Health in Composite Industry*,
Composites Science and Technology ,
https://doi.org/10.1007/978-981-16-6136-5_5

and achieved (Hadikusumo and Rowlinson 2004). Design for safety approach also implements directly compatible use of the information, to conclude that the design can be used reliably with the hazards defined as acceptable hazards. Subsequently, it aids to assure that all known hazards have been considered for a product being designed (Real Estate Developers' Association of Singapore 2016). Usually, the project outcome is oriented to be an alternative to the current product in the market, as well as demonstrated the application of safety to the users and manufacturers in the composite market industry.

As mentioned before, design for safety requires a reliable enough system to perform a set of functions accordingly with acceptable risk during their operations. In this case, system safety is usually defined as the integrating process of engineering with management philosophies, criteria, and approaches in order to attain a satisfactory value of risk within the constraints of operational efficacy in entire stages of the system life cycles (Dixon 2012). The philosophy and the concept of a safety system were started in the 1960s. This discipline has been widely used by various organizations which integrates the mainstream of systems engineering along with supported by management values that give a value-added along with product development (Roland and Moriarty 1990; Ericson 2005). Numerous examination methods have been produced and regenerated several times to ensure the system safety is efficiently and effectively operated. The application of system safety in product design and development has proven valuable in reducing accidents and product liability (Bahr 2014).

Bralla (1996) elaborated a proper guideline used in the design for safety, including good ergonomic shape and product surface finishing. To ensure this issue can be catered accordingly, the materials specifications have to follow occupational safety and health (OSH) regulations, rules on safety during disposal, and actions to inhibit the toxicity causing health problems to users and manufacturers. Other regulations and guidelines in design for safety include designing a product with the easiness of safe use. During the testing of the prototype, it should reveal any hazards that were overlooked during the initial designing stage and improvement could be made before the product gets to the consumer.

Design and product designers usually implement the following design traditional approach to manufacture composite products in various sectors. Up to this date, there is no state-of-art review exclusively for design procedures on composite products related to design for safety. Thus, this article is focused and narrowed on design for safety (DfSa) in composite product development. In this case, the designers should consider risk evaluation related to manufacturing at the beginning of the product development (Sapuan 2017). The design for safety is also one of the branches in the field of concurrent/simultaneous engineering to solve the issues related to consumers. Figure 5.1 displays three vital components of design for safety (Dieter and Schmidt 2009).

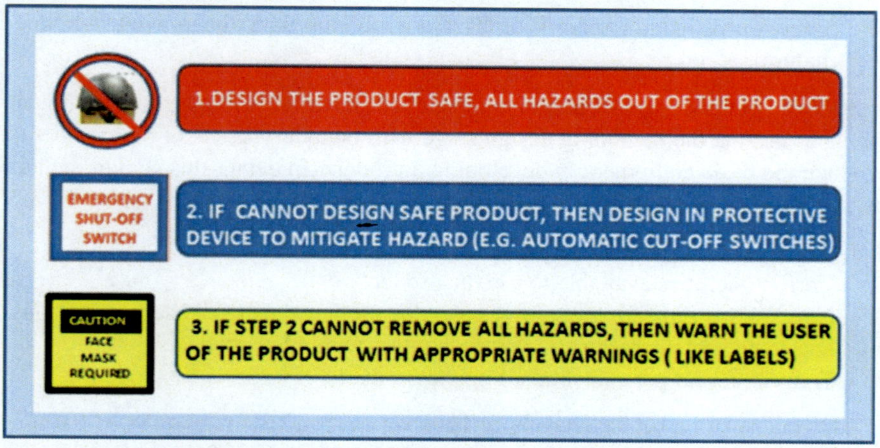

Fig. 5.1 Three vital aspects of design for safety

5.2 Background of Design for Safety

Design for safety is implemented in system safety in order to complete information of a system from its original design to final design, which also covers the analysis of expected risks. These risks might induce hazards including human errors that might lead to unsafe consequences (Ladewski and Al-Bayati 2019). Among the activities involved in this approach include classifying the significant safety attributes of each design analysis. In this works, a hierarchy system is applied to revise and examine the previous product safety status (Roland and Moriarty 1990). A comprehensive study on the previous product is carried out to produce and set a base understanding for the upcoming improvements, which serve as a revision update to the hierarchy system. This process would endure to the final operational field use for the expected future product.

The design for safety also implies the investigation on boundaries in previous product design. Initially, the boundaries are mentioned before including the definition of their functional system with their operations. Next, a general view on the specification of safety hazard analysis should be covered, which being a part of total hazard analysis work. At this point, these boundaries must be included in the requirements for the design for reliability for safety (Raheja 1995). To achieve a good design for safety during product development, these information should be provided:

1. System design of the previous product (datum).
2. System design of the revised or improved product.
3. Specifications of the previous product (datum).
4. Functional and operational specifications of a revised or improved product.
5. Hazard analysis data on the safety of the previous product (datum).
6. Any recorded safety concerns of the previous product (datum).

7. Safety investigation reports of hazard problems with the previous product (datum).

From the generic list of information mentioned above, these information are fundamentally used at the beginning design stage with boundary levels of the product to produce the basic equipment in developing a product. In detail, this should include the:

1. Potential conditions of the product after fabrications.
2. Current facilities available to use the products.
3. Necessary actions for the safety task of the product during all usage conditions.
4. Differences of characteristics of the product with other products.
5. Training for all personnel handling the product.

The interrelation and connectivity between operations, maintenance, training procedures, environment and personnel, and functionalities of the equipment are essential in developing a product. For instance, if one factor in this relationship is diminished such as elevated temperature above the specification, it could jeopardize the other parts of the design. A proper understanding of the relationship of each of these factors is considered as a fragment of safety evaluation required as the product is developed (Hadikusumo and Rowlinson 2004; Raheja et al. 2017).

Fundamentally, design for safety has its paradigm or "rule of thumb" describing their framework and model in designing the process. These hazards and failure prevention actions include doing the right safety analyses, using the ten paradigms, and incorporating robust design risk mitigations early. In this case, there are ten paradigms on design for safety, which highlight the important criteria of safety, as mentioned in Raheja et al. (2017). The paradigms as mentioned in the literature are:

1. Always aim for zero accidents.
2. Be courageous and "just say no."
3. Spend significant effort on systems requirements analysis.
4. Prevent accidents from a single as well as multiple causes.
5. If the solution costs too much money, develop a cheaper solution.
6. Design for Prognostics and Health Monitoring (PHM) to minimise the number of surprise disastrous events or preventable mishaps.
7. Always analyse the structure and shape for the safety of complex systems.
8. Develop a comprehensive safety training program to include handling of systems by operators and maintainers.
9. Taking no action is usually not an acceptable option.
10. If you stop using the wrong practices, you are likely to discover the right practices.

5.3 Total Design Model in Composite Development Under the Scope of Design for Safety

To develop higher quality and better performance products with lesser cost and production time, concurrent engineering approaches such as design for safety should be applied. Shorter time and total cost production could be attained via life-cycle management without compromising the safety issue of the product. Various tools and techniques have been introduced by several designers to create and develop a product in a shorter time with less costly production. In this case, Prasad (1995) suggested several computer tools could be executed to reduce cycle time without conceding safety attributes in developing a product if the tools and the techniques used are right. Among the current tools implemented are the Pugh concept selection matrix and Pugh total design approach (Pugh 1996).

To develop composite products, it is vital to conduct a design process starting from identifying current problems (market analysis), documenting product design specification, generating concept designs, design selections, and fabrication of the final product (Asyraf et al. 2019b, 2020b). To achieve this process, the total design method created by Pugh (1991) is one of the tools that integrate the identification of problems via current market analysis, generating product design specification (PDS), concept development and evaluation, design detail, and fabrication of the prototype. The method requires simultaneous consideration of the manufacturing process so that the component is designed to be compatible with the manufacturing method. To be specific, any components of the product should be facilitated with ease of fabrication in order to increase the production quantity (Pahl and Beitz 1996). Moreover, the total design method should be also embedded with the potential risks and hazards in all relevant stages of the design process to ensure their functional system in accordance with good safety. Thus, the component development for composite material requires the use of a thorough design process.

5.4 Design for Safety in Composite Product Development

In the design of a composite product, there are several guidelines to follow in accordance to design for safety. According to the guidelines, the ergonomic design should be incorporated in developing the composite product design. The design aids the product or prototype to be developed for efficiency and comfort during operating their functions. Moreover, the application of design for safety should use painting and surface finishing following the occupational health and safety regulations. The surface finish might involve large varieties of coatings, either solvent-based or water-based. The coatings are usually applied after the product assembly or in the flat line assembly stage. In general, the coating can be formed of stains, glazes, varnishes, fillers, paints, sealers, lacquers, etc. They are typically applied by brush, flow-coating machine, dip, spray, and roller (Matthews 1984). Apart from that, the design of the

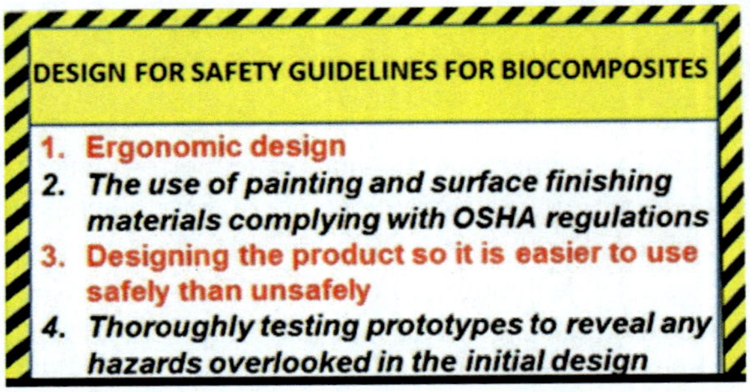

Fig. 5.2 Design for safety guidelines for composite products

composite products should weigh more on safety. Lastly, the fabricated prototype should reveal any possible risks or hazards, which might be overlooked in the initial design stage. This article discusses four rules of design for safety given by Bralla (1996) as shown in Fig. 5.2.

5.4.1 Ergonomic Design: Sharp Edge and Corners

The ergonomic design has a strong relationship with safe design in design for safety technique. One of the main efforts that could be incorporated on ergonomic design elements is to remove any sharp edges and corners to minimise the risks of cut and bleeding to the users, especially children. Moreover, it is not only for safety reasons, as far as the fabrication process is concerned, sharp edges and corners/radii also are not desirable. This is due to the rule that the products should be made as generous as possible as sharp edges can concentrate stress during the composite fabrication process (Murphy 1994).

For instance, one of the attempts taken by Universiti Putra Malaysia researchers was to develop PUTRAFrame from kenaf fibre reinforced polymer composite without any sharp ends (Zakaria 2019). The PUTRAFrame was developed by a team of researchers from the Institute of Tropical Forestry and Forest Products (INTROP), Universiti Putra Malaysia (UPM) with the concept of minimalist, sleek, ergonomic, and versatile. The materials used to develop the frame was kenaf core fibre reinforced high-density polyethylene (HDPE) composites without making any painting or coating on its surface. This is to retain the original natural brown color of kenaf fibre since the fibre was used in the formulation of the pallets. Figure 5.3 illustrates the final PUTRAframe product without any sharp edges and corners.

Fig. 5.3 PUTRAFrame: sharp edges and corners avoided

5.4.2 Painting and Finishing Following Occupational Safety and Health (OSH) Regulations

Painting and finishing in accordance with occupational safety and health (OSH) regulations is another vital component in design for safety in developing composite products. Recently, UPM researchers have designed and fabricated composite safety helmet and multi-purpose table which involved painting and finishing processes as shown in Fig. 5.4. To ensure the safety of the product to the user, the finishing processes should follow the OSH regulations and rules (Legal Research Board 2019). This was due to composite products that generally involve finishing processes involving chemical compounds that can emit volatile toxic substances to the user. The chemical compounds used in the finishing processes are gel coat and sprays. Typically, there are four types of spray used for a good surface finish of plastic products such as plastic primer, putty primer, clear spray, and black colour spray. Sandpaper is also used to smoothen the surface of the product by removing gel coat excess. To guarantee the surface finish of the product is in a good manner, a surface roughness

Fig. 5.4 Painting and finishing processes of composite products

test has to be conducted either by using scanning electron microscopy (SEM) or optical microscopy. This would aid the engineers and designers to detect any surface defects before getting the end products to consumers (Nurul Nathirah 2019). Apart from that, painting composite products is essential to accomplish an aesthetically pleasing finish (Fig. 5.5). It was ensured that all the activities were conforming to OSH regulations (Legal Research Board 2019).

5.4.3 Designing the Product So It Is Easier to Use Safely Than Unsafely

In general, it is required for designers to design composite products to study and embed the design in proper safety requisite. Usually, the designing of composite products involved proper material selection to evaluate and match optimally high-performance material with lesser hazards by-product when in use. The material selection usually covers the optimum value of mechanical, thermal, physical, and chemical properties of the combination of natural fibres and their polymers (Johari et al.

Fig. 5.5 Composite tables painted in different colours

2020a, b; Ilyas et al. 2020, 2021; Asyraf et al. 2021a, b; Omran et al. 2021). Thus, the application of biodegradable materials either in the form of fibres or polymeric resin would provide a safer use of products. In some cases, several non-biodegradable materials might be unsafe for marine creatures in developing marine-based products (Sapuan 2019). Other than that, it is safer to implement composite materials rather than conventional composites (glass or carbon fibre) as automotive interior components. The use of composite, which did not form sharp-edge fracture in these automotive components would provide safety to passengers during crashes (Wood 2020).

5.4.4 Thoroughly Testing Prototypes to Reveal Any Hazards Overlooked in the Initial Design

In the design for safety principle, it is required for the designer to thoroughly test prototypes to reveal any hazards overlooked in the initial design. In this stage, multiple preproduction should be carried out on the prototype of composite products during fabrication for testing and refinement. For instance, a group of researchers has designed the composite safety helmet made up for oil palm plantation workers (Shamsul Bahri et al. 2019). The early version of prototypes called alpha prototypes were fabricated with production-intent parts as depicted in Fig. 5.6a and b. The alpha prototypes also have the same size and geometrical shapes with similar material behaviors proposed for the final product design. Nevertheless, it is not supposedly be manufactured using the actual processes to be used in production (Ulrich and Eppinger 2004). For the alpha prototype of composite safety helmet, it is fabricated using hand lay-up method instead of resin transfer moulding (RTM) or compression moulding process in actual production. Figure 5.6c displayed the testing of the alpha prototype of the composite helmet. An effective product refinement and performance

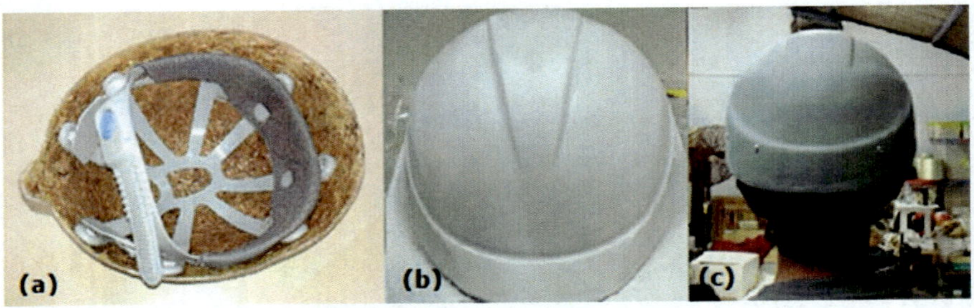

Fig. 5.6 Alpha prototype of a composite helmet and its testing

improvement could be done via a series of testing using the actual product. Thus, the testing and refinement process in this stage is to expose any hazards which might be overlooked during the initial design process.

5.5 Integration of Other Concurrent Engineering Tools with Design for Safety in Composite Product Development

The application of concurrent engineering techniques is essential in the development of composite goods since they are custom-made products requiring a specific material selection and their own fabrication process (Mansor et al. 2015; Asyraf et al. 2019a; Alsubari et al. 2021). In this case, those main issues have to be studied in the early stage of the designing process, which are covered with their safety attributes. For composite products, they cannot be treated as metal goods because composites entail not only knowledge of the material properties but also the knowledge of their specific manufacturing methods. Commonly, the properties of composite products are highly dependable on the fabricating process of the final product along with to fulfill product safety requirements (Mayer 1993; Sapuan et al. 2005). Such consideration is the essence of concurrent engineering.

The approach of concurrent engineering is very crucial in the manufacturing of composite products to comprehend costly modifications during the end-stage in product development (Hambali et al. 2009). At this point, a composite product development process usually covers the elements of the component life cycle based on the market analysis (sales trend) such as cost, performance, and consumer demands (Sapuan 2015; Mazani et al. 2019). These elements are used and integrated with various computer software tools such as material selection tool (Ashby 2005), computer-aided drawing (CAD) (Azammi et al. 2018), and finite element analysis (FEA) (Asyraf et al. 2020a) to pre-evaluate the initial design. Moreover, integration and combination of several concurrent engineering tools such as problem identification tools (Ghani et al. 2016), refining attributes techniques (Sapuan 2006), and selection design tools (Ertay et al. 2005) would provide optimum quality and values in the product development process. Subsequently, it would result in the reduction of manufacturing cost and better product performance compared to the previous product design. Thus, a design is considered in total including market investigation, development of product design specification (PDS), concept development and evaluation, detailed design, and manufacturing.

In the early stage of product development, it necessary to conceptualize the idea of product design with any possible solutions to satisfy the intended design. Overall, the conceptual design stage contains four (4) crucial sub-areas; concept generation, concept clarification, concept selection, and concept development (Yeh et al. 2011). From these sub-areas, they could aid by forming numerous techniques that can be used systematically by designers to produce ideas for design concepts. Moreover, they are very useful in conducting a design selection process after a set of generated ideas and concept designs were formed. For the concept development process, various tools and techniques could be implemented to attain this objective (Sapuan and Willmot 1996). In general, six main concept design methods have been widely used in product development, such as Systematic Exploitation of Proven Ideas or Experience, Extending the Search Space, Gallery method, Morphological Method, Voice of Customer, and Theory of Inventive Problem Solving (TRIZ).

5.5.1 *Systematic Exploitation of Proven Ideas or Experience*

The analysis of the current systems or products is one of the methods used by many designers and researchers to initiate new models or prototypes with better solutions (Pahl and Beitz 1984). This type of method, also called brainstorming discussions covers the physical analysis of current products. The discussion might produce a clearer picture by mind mapping of problems, ideas generation, concept design production, fabrication processes, and finalised prototypes (Mazani et al. 2019). Usually, the analysis of the existing product would be the competitors' product, older products of one's own company, and similar products which have several sub-functions of function structures.

5.5.2 *Extending the Search Space*

This method is also called the "Why? Why? Why?" technique that extends the search option by asking the factors that contribute to the problems (Cross 2005). For instance, "why do we need safety in composite products?" Each answer should be followed by another why question until a conclusion is reached or an unexpected answer prompts a solution. This technique also uses a conventional brain-storming innovation approach which is highly dependent on luck. Brainstormings are often conducted ad hoc to generate a solution to a particular problem.

5.5.3 Morphological Method

The objective of a morphological chart is to create different arrangements and enable designers to choose a new combination of elements. The term "Morphology" means the study of shape or form, whereby "Morphological Chart" led to a meaning of summary of systematic attempts to analyse the form that a product might choose (Ulrich and Eppinger 1995; Ullman 2003). In this case, the chart provides a range of options for each element and component that can be integrated to form a solution. A study conducted by Sapuan (Salit et al. 2005) demonstrated that the implementation of the morphological chart in creating new automotive pedal designs by using polymer composites. The method was based on function analysis where new design concepts were generated through the combination of multiple design features based on the product functions.

5.5.4 Gallery Method

The method is a way to display a large number of concepts concurrently for discussion (McKoy et al. 2001). Sketches, commonly one concept to a sheet are taped to the wall of the designer's room. Designers would circulate and look at each point and propose the improvements of the concept or suddenly generate related ideas.

5.5.5 Voice of Customer

This is an essential technique of quality function deployment (QFD) (Yeh et al. 2011), whereby it hears the voice of the customers to generate ideas for design solutions. It establishes information to other multidisciplinary experts to determine the satisfaction of customer desires and demands.

5.5.6 Theory of Inventive Problem Solving (TRIZ)

TRIZ method is another concurrent engineering method to generate solution ideas from problems that occurred by using inventive principles (Dali et al. 2015). The TRIZ is functioned to remove any trade-off that might arise from the solution created, which is focused on the root cause of the problem (Rosli et al. 2013; Ahmad et al. 2015; Li et al. 2015). It is also used to identify the purpose subject of the design, which is also known as design intends. TRIZ principles are divided into four crucial approaches to find the solution based on their complexity levels. Those main approaches are Su-field modeling and 76 Standard Inventive Solutions, Prediction of Technology Trends, contradiction engineering with 40 inventive principles, and Algorithms of Inventive Problem Solving (ARIZ) (San et al. 2011). The tools are very useful in order to solve a problem systematically by identification of opportunity and innovation techniques to address the problem (Li 2010). For instance, Cascini et al. (2011) documented a new concept of sheet metal snips using the TRIZ method focusing on the contradiction matrix. They found the solution by comparing the improving and worsening parameters, which were later matched using a contradiction matrix to find 40 inventive solution principles. In the end, they refined and visualised their design concept using a CAD-based design optimization tool.

5.5.7 Other Concurrent Engineering Tools and Techniques

In some cases, a few researchers executed the use of other concurrent engineering techniques to refine their concept development process. Typically, those researchers used available methods called multi-criteria decision making (MCDM) to choose the best concept design using numerical evaluation. These MCDM methods such as Analytic Hierarchy Process (AHP) (Shaharuzaman et al. 2018), Analytic Network Process (ANP) (Hambali and Amira Farhana 2018), Technique for Order Preference by Similarity to Ideal Solution (TOPSIS) (Abidin et al. 2016), and *VIseKriterijumska Optimizacija I Kompromisno Resenje* (VIKOR) (Tiwari et al. 2016) would establish crucial solution in the decision-making process where multiple attributes and design alternatives have to be analyzed simultaneously. In the end, the alternative proposed designs will be suited to the intended design specification and also able to be applied in the group decision making process. For example, Hambali (Hambali et al. 2009) carried out a study on developing an automotive bumper beam from polymer composites using the AHP approach. He found out that the proposed selection frameworks, AHP together with other concurrent engineering approaches could aid the designer to govern the best materials, conceptual design, and fabrication process for the automotive bumper beam from composite materials.

5.6 Recent Developments and Advancements of Design for Safety

Several research works conducted the designing of a composite product using design for safety to ensure the safety aspect is covered. Table 5.1 depicts recent development and advancement in design for safety related to the composite industry.

Table 5.1 Recent development and advancement of design for safety (DfSa) related to the composite industry

Product	Final concept design	Concurrent engineering techniques			No. of concept design produce	Application	References
		Problem identifier tool	Refine problem identifier tool	Selection concept design tool			
Auto-motive bumper beam	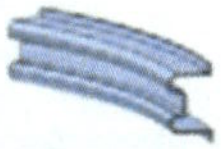	Voice of customer	Finite element analysis	TOPSIS	8	Automotive	Davoodi et al. (2008)
Fire extinguisher		TRIZ (Contra-diction matrix)	Morphological chart	ANP	4	Safety equipments	Asyraf et al. (2020c)
Safety helmet for oil palm worker		Bench-marking and trend analysis + Creativity methods	Requirement analysis	Egro-nomic principle + evaluation and selection methods	1	Safety equipments	Shamsul Bahri et al. (2019)
Anti-roll bar		TRIZ-Blue Ocean Strategy (BOS) four-action frame-work	Morphological chart	AHP	42	Automotive	Mastura et al. (2017)
Auto-motive crash box		Function analysis diagram (FAD) + Free Body Diagram (FBD)	TRIZ + morphological chart	AHP	5	Automotive	Yusof et al. (2020)
Automotive bumper		Brain-storming + Bench-marking method	–	Finite element analysis	1	Automotive	Pathania et al. (2018)

(continued)

Table 5.1 (continued)

Product	Final concept design	Concurrent engineering techniques			No. of concept design produce	Application	References
		Problem identifier tool	Refine problem identifier tool	Selection concept design tool			
Side door impact beam		TRIZ + Biomi-metics	Finite element analysis	VIKOR	8	Automotive	Shaharuzaman et al. (2020)

5.7 Conclusions

Design for safety is one of the designs for the manufacturing branches needed in developing sustainable composite products. However, there is no design for sustainability route exclusively developed for composite products so far. The design for safety in composite products should be conducted by designers and engineers using its general guidelines and steps proposed by the experts, e.g. ergonomic design, painting and finishing complying with OSH regulations, designing products for easier use safely, and testing and refining of alpha prototypes to reveal any hazards overlooked in the initial design. These concurrent engineering also could be integrated and hybridised with other concurrent engineering methods to cater to the current problems holistically. This chapter also highlights and summarizes the current composite products manufactured via this concurrent engineering technique, Design of Safety.

Acknowledgements This chapter was written with Nurul Huda Mohamad Isham, Eugenia Paul, Nur Sabrina Alauddin, Patricia Patinadan, and Ting Li Fung.

References

Abidin MZ, Rusli R, Shariff AM (2016) Technique for order performance by similarity to ideal solution (TOPSIS)-entropy methodology for inherent safety design decision making tool. Procedia Eng 148:1043–1050. https://doi.org/10.1016/j.proeng.2016.06.587

Ahmad SA, Ang MC, Ng KW, Abdul Wahab AN (2015) Reducing home energy usage based on TRIZ concept. Adv Environ Biol 9:6–11

Alsubari S, Zuhri MYM, Sapuan SM et al (2021) Potential of natural fiber reinforced polymer composites in sandwich structures: a review on its mechanical properties. Polymers (basel) 13:423. https://doi.org/10.3390/polym13030423

Ashby MF (2005) Materials selection in mechanical design, third. Elsevier Butterworth Heinemann, Oxford

Asyraf MRM, Ishak MR, Razman MR, Chandrasekar M (2019a) Fundamentals of creep, testing methods and development of test rig for the full-scale crossarm: a review. J Teknol 81:155–164. https://doi.org/10.11113/jt.v81.13402

Asyraf MRM, Ishak MR, Sapuan SM, Yidris N (2019b) Conceptual design of creep testing rig for full-scale cross arm using TRIZ-Morphological chart-analytic network process technique. J Mater Res Technol 8:5647–5658. https://doi.org/10.1016/j.jmrt.2019.09.033

Asyraf MRM, Ishak MR, Sapuan SM, et al (2020a) Evaluation of design and simulation of creep test rig for full-scale cross arm structure. Adv Civ Eng 6980918. https://doi.org/10.1155/2019/6980918

Asyraf MRM, Ishak MR, Sapuan SM, Yidris N (2020b) Conceptual design of multi-operation outdoor flexural creep test rig using hybrid concurrent engineering approach. J Mater Res Technol 9:2357–2368. https://doi.org/10.1016/j.jmrt.2019.12.067

Asyraf MRM, Rafidah M, Ishak MR et al (2020c) Integration of TRIZ, morphological chart and ANP method for development of FRP composite portable fire extinguisher. Polym Compos 41:2917–2932. https://doi.org/10.1002/pc.25587

Asyraf MRM, Ishak MR, Sapuan SM, Yidris N (2021a) Comparison of static and long-term creep behaviors between Balau wood and glass fiber reinforced polymer composite for cross-arm application. Fibers Polym 22

Asyraf MRM, Rafidah M, Azrina A, Razman MR (2021b) Dynamic mechanical behaviour of kenaf cellulosic fibre biocomposites: a comprehensive review on chemical treatments. Cellulose. https://doi.org/10.1007/s10570-021-03710-3

Azammi AMN, Sapuan SM, Ishak MR, Sultan MTH (2018) Conceptual design of automobile engine rubber mounting composite using TRIZ-morphological chart-analytic network process technique. Def Technol 14:268–277. https://doi.org/10.1016/j.dt.2018.05.009

Bahr NJ (2014) System safety engineering and risk assessment: a practical approach, second edition, 2nd edn. Taylor & Francis, Boca Raton, USA

Bralla JL (1996) Design for excellence. McGraw-Hill Education, New York, USA

Cascini G, Rissone P, Rotini F, Russo D (2011) Systematic design through the integration of TRIZ and optimization tools. Procedia Eng 9:674–679. https://doi.org/10.1016/j.proeng.2011.03.154

Cross N (2005) Engineering design methods. Des Stud 11:210. https://doi.org/10.1016/0142-694x(90)90015-5

Dali MNABM, Ghani JA, Haron CHC (2015) Triz approach for analyzing method of dimple structure fabrication. ARPN J Eng Appl Sci 10:6257–6262

Davoodi MM, Sapuan SM, Yunus R (2008) Conceptual design of a polymer composite automotive bumper energy absorber. Mater Des 29:1447–1452. https://doi.org/10.1016/j.matdes.2007.07.011

Dieter GE, Schmidt LC (2009) Engineering design, 4th edn. McGraw-Hill Education, Boston, United States America

Dixon J (2012) risk management, exception handling, and change management. In: Raheja D, Gullo LJ (eds) Design for reliability, 16th edn. Wiley, Hoboken, New Jersey, USA, pp 235–251

Ericson CA (2005) Hazard analysis techniques for system safety, 2nd edn. Wiley, Hoboken, New Jersey, USA

Ertay T, Büyüközkan G, Kahraman C, Ruan D (2005) Quality function deployment implementation based on analytic network process with linguistic data: an application in automotive industry. J Intell Fuzzy Syst 16:221–232

Ghani JA, Natasha AR, Che Hassan CH, Syarif J (2016) TRIZ approach for machining process innovation in cryogenic environment. Int J Mater Prod Technol 53:286–297. https://doi.org/10.1504/IJMPT.2016.079200

Hadikusumo BHW, Rowlinson S (2004) Capturing safety knowledge using design-for-safety-process tool. J Constr Eng Manag 130:281–289. https://doi.org/10.1061/(ASCE)0733-9364(2004)130:2(281)

Hambali A, Sapuan SM, Ismail N, Nukman Y (2009) Application of analytical hierarchy process in the design concept selection of automotive composite bumper beam during the conceptual design stage. Sci Res Essays 4:198–211

Hambali A, Amira Farhana MT (2018) Development of integrated analytic network process (Anp) and theory of inventive problem solving (Triz) in the conceptual design selection. J Eng Sci Technol 13:2716–2733

Ilyas RA, Sapuan SM, Atiqah A et al (2020) Sugar palm (Arenga pinnata [Wurmb.] Merr) starch films containing sugar palm nanofibrillated cellulose as reinforcement: water barrier properties. Polym Compos 41:459–467. https://doi.org/10.1002/pc.25379

Ilyas R, Sapuan S, Atikah M et al (2021) Effect of hydrolysis time on the morphological, physical, chemical, and thermal behavior of sugar palm nanocrystalline cellulose (Arenga pinnata (Wurmb.) Merr). Text Res J 91:152–167. https://doi.org/10.1177/0040517520932393

Johari AN, Ishak MR, Leman Z et al (2020a) Influence of $CaCO_3$ in pultruded glass fibre/unsaturated polyester composite on flexural creep behaviour using conventional and TTSP methods. Polimery 65:46–54. https://doi.org/10.14314/polimery.2020.11.6

Johari AN, Ishak MR, Leman Z et al (2020b) Creep behaviour monitoring of short-term duration for fiber-glass reinforced composite cross-arms with unsaturated polyester resin samples using conventional analysis. J Mech Eng Sci 14:7361–7368. https://doi.org/10.15282/jmes.14.4.2020.04.0578

Ladewski BJ, Al-Bayati AJ (2019) Quality and safety management practices: the theory of quality management approach. J Safety Res 69:193–200. https://doi.org/10.1016/j.jsr.2019.03.004

Legal Research Board (2019) Occupational Safety and Health Act 1994 (Act 514). Regulation & Orders. International Law Book Services, Petaling Jaya, Malaysia

Li M, Ming X, He L et al (2015) A TRIZ-based trimming method for patent design around. CAD Comput Aided Des 62:20–30. https://doi.org/10.1016/j.cad.2014.10.005

Li T (2010) Applying TRIZ and AHP to develop innovative design for automated assembly systems. Int J Adv Manuf Technol 46:301–313. https://doi.org/10.1007/s00170-009-2061-4

Mansor MR, Sapuan SM, Hambali A (2015) Conceptual Design of Kenaf Polymer Composites Automotive Spoiler Using TRIZ and Morphology Chart Methods 761:63–67. https://doi.org/10.4028/www.scientific.net/AMM.761.63

Mastura MT, Sapuan SM, Mansor MR, Nuraini AA (2017) Conceptual design of a natural fibre-reinforced composite automotive anti-roll bar using a hybrid approach. Int J Adv Manuf Technol 91:2031–2048. https://doi.org/10.1007/s00170-016-9882-8

Matthews I (1984) Encyclopaedia of occupational health and safety. J Epidemiol Community Heal 38:180–180. https://doi.org/10.1136/jech.38.2.180

Mayer RM (1993) Design with reinforced plastics: a guide for engineers and designers. The Design Council, London, UK

Mazani N, Sapuan SM, Sanyang ML et al (2019) Design and fabrication of a shoe shelf from kenaf fiber reinforced unsaturated polyester composites. In: Ariffin H, Sapuan SM, Hassan MA (eds) Lignocellulose for future bioeconomy, 1st edn. Elsevier Inc., Amsterdam, Netherland, pp 315–332

McKoy FL, Vargas-Hernández N, Summers JD, Shah JJ (2001) Influence of design representation on effectiveness of idea generation. Proc ASME Des Eng Tech Conf 4:39–48

Moriarty B (2012) Integrating design for reliability with design for safety. In: Raheja D, Gullo LJ (eds) Design for reliability. Wiley, Hoboken, New Jersey, USA., pp 253–266

Murphy J (1994) Reinforced plastics handbook. Elsevier Advanced Technology, Oxford, UK

Nurul Nathirah BK (2019) Finishing and joining of sugar palm fibre reinforced unsaturated polyester composites. Universiti Putra Malaysia

Omran AAB, Mohammed AABA, Sapuan SM et al (2021) Micro- and Nanocellulose in polymer composite materials: a review. Polymers (basel) 13:231. https://doi.org/10.3390/polym13020231

Pahl G, Beitz W (1996) Engineering design: a systematic approach. Springer, London, UK

Pahl G, Beitz W (1984) Engineering design. Design Council, London, UK

Pathania MS, Tomar A, Chaturvedi A (2018) Design and analysis of automotive bumper using composite materials. Int Res J Eng Technol 5:266–269

Prasad B (1995) Concurrent engineering fundamentals: integrated product and process organization, volume I, 1st edn. International series in industrial and systems engineering, Upper Saddle River, New Jersey, USA

Pugh S (1991) Total design: Integrated methods for successful product engineering. Qual Reliab Eng Int 7:119–119. https://doi.org/10.1002/qre.4680070210

Pugh S (1996) Concept selection—a method that works. In: Clausing D, Andrade R (eds) creating innovative products using total design. Addison-Wesley Publishing Company, Boston, United States America

Raheja D (1995) Product assurance technologies: principles and practices. Wiley, New York, USA

Raheja D, Gullo LJ, Dixon J (2017) Design for safety paradigms. In: Gullo LJ, Dixon J (eds) Design for safety. Wiley, Chichester, West Sussex, UK, pp 1–16

Real Estate Developers' Association of Singapore (2016) DfS & WSH good practice and guide. Singapore

Roland HE, Moriarty B (1990) System safety engineering and management. Wiley, Hoboken, New Jersey, USA

Rosli MU, Ariffin MKA, Sapuan SM, Sulaiman S (2013) Integrated AHP-TRIZ innovation method for automotive door panel design. Int J Eng Technol 5:3158–3167

Salit MS, Syed M, Molla A, Ali L (2005) Conceptual design of an automotive composite brake pedal. Suranaree J Sci Technol 12:173–177

San YT, Jin YT, Li SC (2011) TRIZ : systematic innovation in manufacturing. Firstfruit Sdn. Bhd., Selangor

Sapuan SM (2006) Using the morphological chart technique for the design of polymeric-based composite automotive pedals. Discov Innov 18:311–317

Sapuan SM (2015) Concurrent engineering in natural fibre composite product development. Appl Mech Mater 761:59–62. https://doi.org/10.4028/www.scientific.net/amm.761.59

Sapuan SM (2017) Composite materials: concurrent engineering approach. Butterworth-Heinemann, Oxford, UK

Sapuan SM (2019) (Keynote Adress), Sustainable bio-packaging from agricultural products toward environmentally benign society. In: second international conference on chemistry, industry and environment. Aligarh Muslim University, Aligarh, India

Sapuan SM, Willmot P (1996) The application of engineering design methodologies in the design of a weather protection system. Asean J Sci Technol Dev 13:67–86

Sapuan SM, Maleque MA, Hameedullah M et al (2005) A note on the conceptual design of polymeric composite automotive bumper system. J Mater Process Technol 159:145–151. https://doi.org/10.1016/j.jmatprotec.2004.01.063

Shaharuzaman MA, Sapuan SM, Mansor MR (2018) Prioritizing the product design specification of side-door impact beam using analytic hierarchy process method. Mech Eng Res Day 2018:34–35

Shaharuzaman MA, Sapuan SM, Mansor MR, Zuhri MYM (2020) Conceptual design of natural fiber composites as a side-door impact beam using hybrid approach. J Renew Mater 8:549–563. https://doi.org/10.32604/jrm.2020.08769

Shamsul Bahri MT, Guan NY, Abd Wahib KN, et al (2019) Creativity in design of safety helmet for oil palm workers. In: Advances in intelligent systems and computing, pp 1044–1047

Tiwari V, Jain PK, Tandon P (2016) Product design concept evaluation using rough sets and VIKOR method. Adv Eng Informatics 30:16–25. https://doi.org/10.1016/j.aei.2015.11.005

Ullman D (2003) The mechanical design process. Des Stud 15:448. https://doi.org/10.1016/0142-694x(94)90041-8

Ulrich K, Eppinger S (1995) Product design and development. McGraw Hill, New York, USA

Ulrich KT, Eppinger SD (2004) Product design and development. McGraw-Hill, Boston, United States America

Wood L (2020) Biocomposites market by fiber, polymer, product, end-use industry and region—global forecast to 2022—research and markets. Bus. Wire. https://www.businesswire.com/news/home/20171002005936/en/. Accessed 28 June 2020

Yeh CH, Huang JCY, Yu CK (2011) Integration of four-phase QFD and TRIZ in product R&D: a notebook case study. Res Eng Des 22:125–141. https://doi.org/10.1007/s00163-010-0099-9

Yusof NSB, Sapuan SM, Sultan MTH, Jawaid M (2020) Conceptual design of oil palm fibre reinforced polymer hybrid composite automotive crash box using integrated approach. J Cent South Univ 27:64–75. https://doi.org/10.1007/s11771-020-4278-1

Zakaria A (2019) PUTRA frame, a frame made of kenaf core fibre and industrial plastic. Universiti Putra Malaysia. https://upm.edu.my/news/putra_frame_a_frame_made_of_kenaf_core_fibre_and_industrial_plastic-48177. Accessed 27 June 2020

Chapter 6
Carbon Footprint in Healthcare

Abstract The phrase "carbon footprint" has been increasingly used over the last decade and is now all over the media. This can be credited to the strong correlation between carbon emissions and climate change, not to mention the ever-worsening effect on the earth and its inhabitants. Carbon footprint calculations are in demand due to climate change being high up on the political and corporate agenda. Despite its popularity, there is a relative lack of academic papers expressing the relationships between carbon footprint and the most important sector in the human world, healthcare. This review intends to explore and summarize research articles related to carbon footprint and healthcare; the carbon emissions by healthcare, the ways to reduce carbon emissions by the healthcare sector, and the effects of the carbon footprint on healthcare. Here, we have found that the most used method of analysis to be the Life-cycle Assessment methodology (LCA cycle): process-based LCAs and economic input–output LCAs. Though there is a lack of research on the carbon emissions by the healthcare sector, we found that the healthcare sector, especially the hospitals produce the most carbon emissions compared to other non-industrial buildings as they operate all day round, 365 days a year. Hospitals contain several high-energy consuming activities, that includes advanced temperature-adjustment and ventilation systems, computer use, laboratory equipment, sterilization, refrigeration, laundry, as well as food preparation. Medical activities such as conferences and operations were also found to emit relatively high carbon dioxide. This study have concluded a few ways to reduce these emissions and that recycling is not enough to significantly reduce the emissions. The emissions from the healthcare sector are ironically affecting the healthcare sector itself. Studies have shown that climate change and air pollution increase healthcare expenses. When more people require health care services, healthcare, in turn, needs to use more energy and emit more greenhouse gasses. This continues to be a vicious cycle that has to be stopped.

Keywords Healthcare · Health services · Carbon footprint · Carbon emissions · Energy

S. M. Sapuan et al., *Safety and Health in Composite Industry*,
Composites Science and Technology ,
https://doi.org/10.1007/978-981-16-6136-5_6

6.1 Introduction

The carbon footprint comes from the conception of ecological footprint, meaning a quantity of the human's demand from Earth's resources. This may also be measured by the planets's ability to regenerate resources, compared to a uniform measure of demand from earth's ecosystem. It embodies the quantity of biologically productive land and ocean space needed to provide the capitals to fulfill people's demand at a certain place.

Nevertheless, a recognized and standardised definition of a carbon footprint does not exist as of now. What does exist is the notion of a footprint. A mostly accepted concept presented by Wiedmann et al. (2006): "the carbon footprint is a measure of the total amount of carbon dioxide emissions directly and indirectly caused by an activity or accumulation from the life stages of a product. Meanwhile, the carbon footprint is a measure of carbon dioxide emissions" (Lenzen 2000).

Climate change is now a global threat for the healthcare sector, and it requires a lot of planning and mitigation efforts to be integrated into the current healthcare system (Costello et al. 2009a, b; Pandey 2017). In Scotland, the National Health Service (NHS) is required to measure their carbon emissions (Scottish Government Climate change (Scotland) Act 2009a, b) in order to achieve energy consumption and GHG reduction goals. In the United States, the healthcare system generates an estimated of 8–10% total emissions of greenhouse gasses (Chung and Weltzer 2009) and in the UK, their national healthcare system, the National Health Service (NHS), emits a total of 25% of public sector emissions (NHS Sustainable Development Unit 2016).

The relationship between climate change and increasing atmospheric concentrations of greenhouse gases, like carbon dioxide, can be seen clearly. Climate change represents one of the greatest health threats of the twenty-first century (Costello et al. 2009a, b; Watts et al. 2015) that resulted in the rising of temperature-related morbidity and mortality, including diseases that are transmitted through food and vector (Ebi et al. 2006; Campbell-Lendrum et al. 2015). Individuals whose health is adversely affected are more likely to seek for treatments at healthcare facilities, continuing the carbon emission cycle. Sectors with significant carbon footprints are as discussed in the Article "The Lancet Planetary Health" (Malik et al. 2018).

6.2 Methods of Analysis

Every carbon emission source varies from one another, and in order to control the emission to be within the acceptable range, standards have been developed by the governments and international organisations. Establishing these standards requires a series of research works. Among organisations that participate in introducing the different types of carbon footprint assessment standards are the World Business Council For Sustainable Development (WBCSD) International Organisation For

Standardisation (ISO) and the British Standards Institution (BSI). The standards are mostly focusing on the organisations and their products.

Years pass by, and many improvements have been done to these assessment standards in order to create a higher awareness about the carbon footprint. Some of the improvements can be seen through the development of ISO14064, GHG Protocol, and PAS2050. The enforcement of these standards in every organisation reduced carbon emissions. Unfortunately, the consequence of these standards is those counting methods that are not uniform. However, the issue can be refuted with further studies and analysis, especially in organisation and product fields.

The life–cycle assessment (LCA) methodology as used in Malik et al. (2018) study stated that the carbon footprint of Australian healthcare is one of the methods being used by the organisations as in these process-based LCA, all materials inputs are taken into account to measure all emissions into the environment through individual processes or multiple processes combined to create a final service or product. There are two basic elements of environmental footprinting or LCA which are process-based LCA and economic input–output LCA. The process–based LCA is a suitable approach to be used in examining small amounts of specific data but impractical to be used when examining the whole supply chain in health care, especially in environmental effects context as this process requires selection of a system boundary.

This statement is supported in Junnila (2006) study, where the supply chain that falls under this framework or system boundary is being considered with the use of LCA whereas the others are deemed negligible. Unfortunately, when choosing the system boundary, an error called the truncation error might occur that can be fixed through scanning the whole upstream supply chains of goods, process, or even a whole nation by using the input–output analysis, as written in Lenzen (2000) article about errors in conventional and input-output based life-cycle inventories. Economic input–output LCA is a macroeconomic technique that explains the complex interdependencies among the various division of the economy.

All bodies or organisations for Economic Co-operation and Development (OECD) nations and 27 non-member economies create input–output tables to enable this economic analysis to happen. Economic input–output LCA takes into account all infinite upstream supply chains, with the absence of choosing a system boundary, hence providing a comprehensive figure by making sure that both direct (on-site) and total (direct plus indirect) effects are detected. As an example, the emission of CO_2 might come from the burning of fuel and gas in a hospital where the indirect result would be the hospital's electricity use as the electricity was routinely generated elsewhere. Referring to Fenner et al. (2018) the LCA cycle is considered as a "cradle-to-cradle" approach where the entire process of consumption in each product or process is being assessed (product use and disposal). In environmental aspects, e.g. environmental impact assessment, integrated waste management, and pollution studies, LCA do play important roles. According to ISO 14000, the environmental management standard has implemented the LCA methodologies consisting of four-stage standard major characteristics:

I. Scope definitions (identifies goals and boundaries, functional unit, and main definitions)
II. Inventory analysis (data collection about energy and material flows for each stage of a product life-span, include inputs of water, energy, and raw materials releases to land, water, and air)
III. Impact assessment (classify, cumulate, and characterise several midpoints and endpoints of environmental impacts by weighting and normalisation methodologies
IV. Interpretations (interpreting analysis to assist in the selection of environmentally- friendly products and to provide recommendations).

Meanwhile, in the study by Gao et al. (2014) a comparative study of carbon footprint and assessment standards25 was conducted that explained the analysis method based on IO analysis for assessment of organisational carbon footprint. Organisational carbon footprint refers to direct and indirect carbon dioxide emissions generated within an organisation. The steps are as follows:

I. Defining organisational boundaries: It is an essential strategy to set clear, explicit limit on which section of an organisation that is associated with the organisational carbon footprint. In an organisation, it might build up from one or more amenity, which applies controls and sharing value approaches to deal with greenhouse gas (GHG) emission levels and expulsions at the organisation level.
II. Establishing operational boundaries: The operational division identifies sources that release the carbon and these sources will be counted or quantify. The emission that comes from the activities under operational control should be measured in a full range. All by-products produced under scope 1 and 2 should be included, while scope 3 can be chosen to be included or not, as shown in Figs. 6.1 and 6.2
III. Calculating carbon footprint: The validity of the footprint depends on the data consumption collected for all of the emission sources that lie within the accepted limit or boundary. It is necessary to identify any gaps in the data and any assumptions made in estimating the carbon footprint. The carbon footprint is usually measured by using data collated multiplied by standard

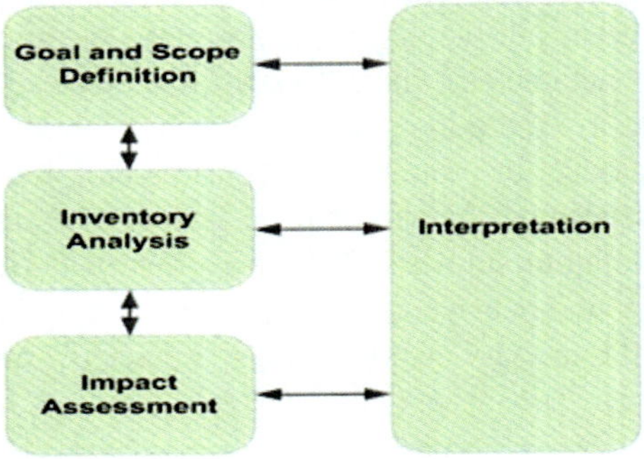

Fig. 6.1 Illustration of LCA phases

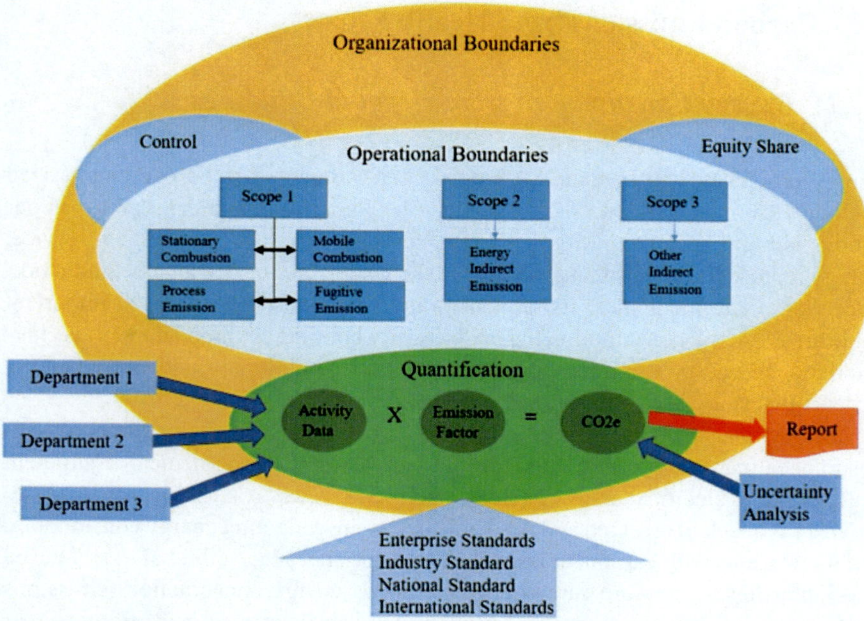

Fig. 6.2 Assessment procedures of the organisational carbon footprint (Gao et al. 2014)

emissions factors. However, there are other several ways of measuring the carbon footprint, such as the use of models.

IV. Reporting and verifying: Organisation should produce a report to assist inventory identification, participation in a GHG program, or to notify external and internal users. On the other hand, the third-party identification of carbon footprint is recommended to be performed in order to increase the credibility and confidence to report carbon emission for public disclosure.

As for assessment standards, both GHG Protocol and ISO14064 provide requirements for quantifying the GHG impact of an organisation, while harmonisation on all qualification methodologies was sought during the development of both standards, with some minor differences remain. Assessment standards of an organisational carbon footprint include:

I. The GHG protocol
Provides sector specifics and general calculation tools and deals with the quantification of greenhouse effects.

II. ISO 14064
International standards for determination of boundaries, quantification, mitigation, and removal are used to guide governments and companies to measure and control greenhouse gas and carbon emissions.

6.3 Carbon Emissions from Health Sector

6.3.1 Energy Consumption from Hospital Buildings

Energy efficiency will become a necessity that is enabled to be overemphasized for long-term management. The demand for healthcare and clinical services in our nations is high, hence, contributing to the cost of healthcare and medical service. Energy represents the third largest value in the healthcare service area behind workforce wages and medicines, subsequently, and is diagnosed as the most important value area. Energy consumption in health facility buildings exhibits numerous characteristics in energy use, including the central air conditioning and water heating systems that function 24 h a day, all year round.

Back-up machines are required for multi-function services, e.g. surgery, diagnostic, healing, monitoring, food production, laundry, and medical equipment consume large electrical energy such as MRT, X-ray, etc.

There is a lack of research and academic papers investigating energy consumption and its relations with carbon emissions, except for one. MacNeill et al. (2017) used grid intensities to measure carbon emission using energy consumption (emissions produced per kWh of electricity generated) provided by local electrical utilities. Their study revealed that their studied hospitals produced from 534,194 to 4,344,150 kg CO_2 emission per year. Though their grid intensity formulas cannot be a point of reference for other studies as there is a reasonable variableness in carbon potency of electricity reflecting the use of electricity generated from water or coal-dominated electrical grids.

Chung and Weltzer (2009) stated in their research on the healthcare sector, counting upstream supply-chain activities contributed to a measured amount of up to 546 $MMTCO^2$, which 254 $MMTCO^2$ (46%) from the total amount was inferable to coordinate activities. The biggest donors were the hospital and sedative medicine (39% and 14%, separately). Roughly, 80% of the total global warming potential is due to carbon dioxide outflows.

The annual average total energy consumption for healthcare centres are 407gandg274.7 kWh/m^2, respectively, which are way higher in comparison to 187 kWh/m^2 in office buildings, 152 kWh/m^2 in commercial buildings, and 92 kWh/m^2 in school buildings. The nonstop use of heating systems and air-conditioning to preserve satisfactory temperature and indoor air quality levels for the patients, artificial light in a few zones of the hospitals, and operation of a few medical equipments result in moderately higher energy consumption when compared to other buildings. Table 6.1 summarized the major energy consumption in Tapei City Hospital. The yearly normal thermal energy utilisation is 332 kWh/m^2 in hospitals and 260 kWh/m^2 in clinics, whereas the yearly normal electricity utilisation in clinics is 133 kWh/m^2 and in a few clinics is 102 kWh/m^2. Monthly electricity consumption from hospital buildings are shown in Fig. 6.3.

Table 6.1 The major energy consumption in Tapei City Hospital

Items	Description	Max load (kW)
1	600 RT chiller × 3 (420 kW × 3), chilled water pump × 3 (30 kW × 3), cooling tower fan × 3 (22 kW × 3), cooling water pump × 3 (95 kW × 3)	1701
2	350 RT absorption chiller (7.5 kW), chilled water pump × 3 (30 kW × 3), cooling tower fan (22 kW, back up), cooling water pump (75 kW)	135
3	Zone water pumps × 7 (55 * 3 + 45 * 2 + 19 * 2) kW = 293 kW	293
4	Air handling unit 90 sets	381
5	Fan coil unit 720 sets	95
6	Exhaust fan, 87 sets	170
7	Supply fan, ventilating fan, 33 sets	161
8	Packaged air-conditioning, 5 sets	78
9	Kitchen exhaust fan, 13 sets	41
10	Kitchen equipment	100
11	Lighting	708
12	Elevator, escalator	170
13	Drainage pump	194
14	Pressurizing pump	77
15	Gas	70
16	Water treatment	56
17	Computer	48
18	Medical equipment	854
19	CSR supplier	54
20	Other equipment	100

Eckelman and Sherman (2016) agreed and reported that hospitals are the first runner-up of the most energy-dependent industrial buildings in the nation, champion of which is the facility that houses food service. Hospitals are normally large buildings, which are open 24 h a day, 365 days a year, and include a number of high energy-requiring activities, consisting of advanced heating, cooling, ventilation systems, administration, laboratory use, laundry, as well as food preparing. Figure 6.4 shows the comparison of typical household and hospital energy consumption. Not only that, but the health care centres also use huge sums of high voltage goods and services, such as pharmaceuticals and clinical appliances, which require large energy inputs for their productions. Figure 6.5 displays characteristics and energy consumption in large hospital buildings in US.

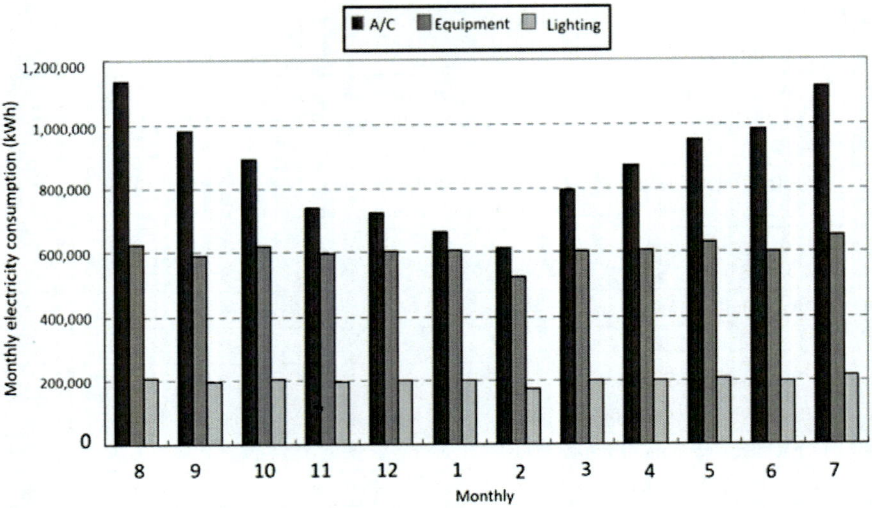

Fig. 6.3 Monthly electricity consumption from hospital buildings

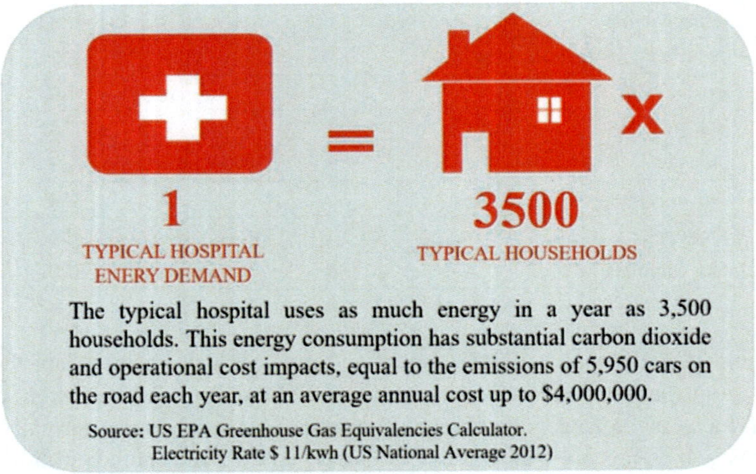

Fig. 6.4 Comparison of typical household and hospital energy consumption

6.3.2 Carbon Emission from Medical Activities

The danger to the health of the human race from the changes of weather patterns through lack of healthy sustenance, disease, and flooding is real. In few sections of the world, the changes are instant. The global change of health burden is mostly carried by children in emerging nations. Doctors are the ones who are held accountable for making sure health and wellbeing. Ironically, they are the contributing factors to

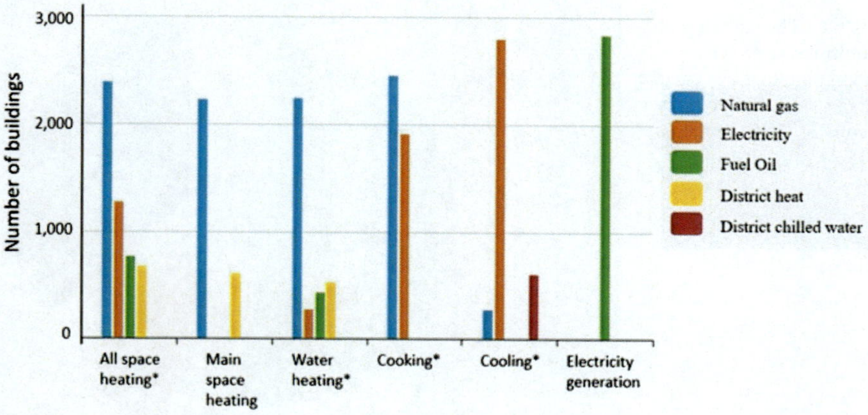

Fig. 6.5 Characteristics and energy consumption in large hospital buildings in the US (US Energy Information Administration 2007)

global warming through unnecessary attendances at international conferences. The physicist and past president of the Royal Society, William Thomson who is better known as Lord Kelvin said, "if you are not able to measure it, you will not be able to improve it." He also said, "heavier than air flying machines are beyond the bounds of possibility," but the doctors proved him wrong as they used such machines to attend the European Respiratory Society annual congress in Munich the past few years. Director of the Wellington Asthma Research Group, Julian Crane roughly calculated that the 17,000 representatives generated about 4000 tonnes of carbon dioxide from traveling alone. A meeting on American Thoracic Society in San Diego reported that it generated around 10,779 tonnes of carbon dioxide by the attendance of 15,000 representatives via air traveling. However, are we able to put these massive numbers in context? Researchers found that the yearly per capita of carbon dioxide released in the United States is about 20 tonnes. Thus, the 11,000 tonnes from the American Thoracic Society meeting is equivalent to that produced by approximately 550 US citizens in a year. However, the United States is not the best comparator although they are the most energy-hungry nations on earth. It was found that approximately 11,000 tonnes of carbon dioxide are equivalent to the production of carbon dioxide by 11,000 people in India and 110,000 people in Chad, in one year (Roberts and Godlee 2007).

Sherman et al. (2012) found that assessments of the health sector life cycle showed that 60% of carbon emissions come from procurement. Pharmaceuticals and medical equipment contribute to more than half of the total emissions. Understandably, the environmental footprint of perioperative services is among the largest in all of health-care. Besides that, a study conducted by Morris et al. (2013) found out the carbon footprint of specific cataract surgery in the University Hospital of Wales, Cardiff revealed that the carbon footprint of one patient going through first eye cataract surgery in Cardiff was 181.8 kg CO_2. The contributions of the primary sectors are shown in Fig. 6.6 and Table 6.1.

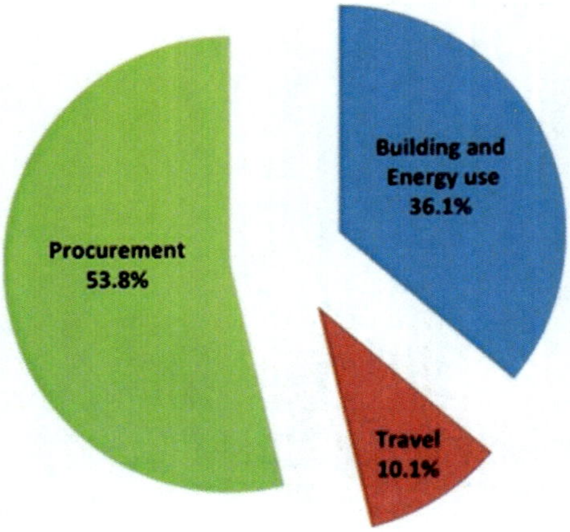

Fig. 6.6 The carbon footprint of a patient undergoing first eye cataract surgery divided into primary sectors of GHG emissions (Morris et al. 2013)

Figure 6.6 shows the contribution of the primary sectors, which are buildings construction and energy use, travel, and procurement contributed to 36.1%, 10.1%, and 53.8% respectively of the total GHG emissions. From Table 6.1, the amounts are 65.7 kg CO_2eq, 18.3 kg CO_2eq, and 97.8 kg CO_2eq, respectively. The outcome of this study found that 2230 patients had cataract surgeries during 2011 in University Hospital of Wales, Cardiff, and were affiliated to an annual carbon footprint of 405,392 kg CO_2eq, which was equivalent to 405 tonnes CO_2eq as shown in Table 6.2.

Another study by MacNeil et al. (2017) was carried out to estimate the carbon footprint of surgical suites in three hospitals from different healthcare systems. Surgery is a resource-intensive health care activity that includes sterilisation procedures, high priced equipment, obligatory life support systems, and advanced operative technologies. Quite an amount of energy was consumed and a large amount of waste volumes were produced from these activities. Whereas, the global impacts related to health care activities are necessary for the provision of quality care have not been strictly subjected to crucial analysis. Due to heating, ventilation, and air conditioning requirements, this study found that operating theatres were three to six times more energy-intensive than the hospital as a whole. The overall calculated carbon footprint of surgery in the three different countries is approximately around 9.7 million tonnes of CO_2 per year. Tables 6.3, 6.4 and 6.5 displays the annual operating theater energy requirements and GHG; annual waste volumes and GHG emissions due to surgical consumables; and annual operating theater GHG emissions respectively.

Last but not least, a study conducted by Thiel et al. (2015) had found over 70% of energy impacts in the operating theater were caused by heating, ventilation, and air conditioning (HVAC). On average, anesthetic gases used in medical activities contributed to one-third of the greenhouse gas emissions from robotic and laparoscopic hysterectomies and two-thirds of abdominal and vaginal hysterectomies. 98% of the ozone depletion potential for abdominal and vaginal hysterectomy is due to

Table 6.2 The GHG emissions attributed to an individual patient undergoing first eye cataract surgery (Morris et al. 2013)

Sector	Subsector	GHG emissions (kg CO_2eq)	Percentage of total GHG emissions (%)
Building energy use	Total building energy use	65.7	36.1
Travel	Staff commuting	5.5	3.0
	Patient travel	12.8	7.0
	Total travel	18.3	10.1
Procurement	Pharmaceuticals	32.7	18.0
	Medical equipment	59.2	32.6
	Paper and ink	0.9	0.5
	Food	0.1	0.0
	Laundry services	0.9	0.5
	Information technology	0.4	0.2
	Water	0.2	0.1
	Waste	3.4	1.9
	Total procurement	97.8	53.8
Total per patient		181.8 kg CO_2eq	100

Table 6.3 Annual operating theater energy requirements and GHG emissions (MacNeil et al. 2017)

	Energy (MWh/year)			CO_2e (kg/year)		
	VGH	UMMC	JRH	VGH	UMMC	JRH
Heating	2518	2204	6971	514,340	610,702	2,283,426
Cooling	66	357	1312	1523	195,629	787,149
Ventilation	449	1062	2045	10,317	581,938	1,104,386
Lighting[a]	236	177	313	5423	96,959	169,189
Plug-loads	113	56	–	2591	30,535	–
Total	3382	3856	10,641	534,194	1,515,763	4,344,150

CO_2e CO_2 equivalents. *VGH* Vancouver General Hospital. *UMMC* University of Minnesota Medical Center. *JRH* John Radcliffe Hospital. [a]At VGH and UMMC, theatre submetering included plug-loads and surgical spotlights, but not overhead lighting; overhead lighting is reported separately based on lighting audits; at JRH, all lighting was captured in theatre submetering, hence only one value is reported for both lighting and plug-loads

the usage of anesthetics. Sevoflurane or desflurane are the different types of inhalation anesthetics used that are used either with or without nitrous oxide (N_2O) as a carrier gas. Moreover, desflurane has a higher potential to cause global warming than sevoflurane. Intravenous propofol that is not a GHG was operated on four vaginal cases. As such, greenhouse gas emissions for vaginal hysterectomy from anesthetics

Table 6.4 Annual waste volumes and GHG emissions due to surgical consumables (MacNeil et al. 2017)

Activity	Waste (kg/year[a])			CO_2e (kg/year)		
	VGH	UMMC	JRH	VGH	UMMC	JRH
Municipal solid waste	111,255	105,975	83,060	438,167	423,060	327,122
Hazardous waste	21,933	9374	81,121	63,028	26,938	233,122
Reusable textiles	178,176	87,120	33,597	53,336	52,248	12,419
Fluid waste	15,526	…	15,525	194	…	194
Sharps	1793	1076	9698	4913	2980	44,229
Cytotoxic waste	902	598	…	4114	2728	…
Recycling+	30,991	10,154	4620	85,264	26,913	11,445
Domestic waste	…	…	993	…	…	2327
Transport	1855	1818	1404	1421	1393	1727
Total	360,576	214,297	228,615	650,436	523,260	632,574

CO_2e CO_2 equivalents. *VGH* Vancouver General Hospital. *UMMC* University of Minnesota Medical Center. *JRH* John Radcliffe Hospital. [a]Except transport where the units are km/year. Recycling includes cardboards, plastic, and surgical blue wrap (polypropylene) at UMMC, versus cardboard and plastic only at VGH and UMMC; production emissions factors used were 1038 kg CO_2e/tonne for cardboard, 3179 kg CO_2e/tonne for average plastics, and 3254 kg CO_2e/tonne for polypropylene; net emissions with recycling were-240 kg CO_2e/tonne for cardboard,-282 kg CO_2e/tonne for average plastics, and 12 kg CO_2e/tonne for polypropylene. Assuming 7.15 miles per gallon average fleet fuel efficiency (Natural Resources Canada)

Table 6.5 Total annual operating theater GHG emissions (kg CO_2e/year) (MacNeil et al. 2017)

Scope	VGH	UMMC	JRH
Scope 1	2,034,277	2,129,841	211,212
Scope 2	534,194	1,515,763	4,322,150
Scope 3	650,436	536,260	632,574
Total	3,218,907	4,181,864	5,187,936

CO_2e CO_2 equivalents. *VGH* Vancouver General Hospital. *UMMC* University of Minnesota Medical Center. *JRH* John Radcliffe Hospital

has widened drastically from case to case, from as low as 0.001 kg CO_2-eq/case to 505 kg CO_2-eq/case.

6.4 Reducing Carbon Emissions in Healthcare

The carbon footprint has been a global conscious and currently used by corporations to calculate their total carbon production and carry out suitable actions to cut down emissions, as well as to meet the demands from green consumers and government bodies. It also gives out chances to motivate enterprises to enhance overall production efficiency and reduce resource expenditure whilst promoting the growth of innovation and technology by achieving sustainable development. Healthcare services are one of the organisations that must provide secured, affordable clinical care while gathering efficient, environmental, and social targets. These targets include achieving reduced GHG emissions.

6.4.1 Operating Theaters

The largest carbon footprint savings came from selecting precise anesthetic gases and minimising substances used in surgeries (Thiel et al. 2015). Power-related interventions resulted in a low reduction in carbon footprint value per case but would create much bigger savings for the entire facility. To reduce the GHG emissions from surgeries, health care providers need to apply a combination of ways, including lessening materials, transferring away from some heat-trapping anesthetic gases, maximizing instrument reuse or single-use device reprocessing, and lowering down off-hour energy use in the operating room. These methods are able to cut down the carbon footprint of an average surgery by up to 80%.

The results of their study also suggested that common efforts such as recycling have little to no impact in reducing carbon emission in a single procedure. However, it is proven to be useful in terms of reducing hospitals' annual solid waste. To effectively reduce carbon emissions in the operating room, physicians should place their attention on minimising overall material use in each surgery, reusing or reprocessing surgical instruments, and removing heat-trapping desflurane from the formulary as it is confirmed that a combination of approaches helps in yielding significant reductions in GHG emissions per case.

6.4.2 Medical Conferences

Most medical practitioners attend their conferences by plane, and although air travel is not a major contributor to greenhouse gas emissions, it is easily one of the fastest-growing ones. For economy class travel on a commercial airplane, approximately 1 kg of carbon dioxide is emitted per passenger per 10 km (Nathan et al. 2018). Subsequently, a study conducted by Roberts and Godlee (2007) reported that the Intergovernmental Panel on Climate Change (IPCC) estimated that aviation caused 3.5% of human-induced global warming, which could rise to 15% by 2050. A massive change in air travel can easily affect our overall carbon emissions.

The Cochrane Collaboration is an excellent example of an international medical organisation that took part in reducing the carbon footprint of its conferences. With thousands of members in many countries, most of its work is done electronically. Moreover, the organisers attempted to carry out electronic ways of enabling people to "attend" the conference via the internet, and a full session used video conferencing to gather keynote speakers from Papua New Guinea, Tunisia, and Uganda. This indeed is a huge step in the right direction.

Ponette-González and Byrnes (2011) assessed the effectiveness of two additional solutions which included alternating large national meetings that require long-distance air travel with smaller regional meetings that do not incorporate geography into the meeting location selection process. Their results suggested that an alternating timetable of long-distance and short-distance meetings can greatly reduce conference-related carbon dioxide emissions while improving space planning that might result in further reductions.

6.4.3 Hospital Facilities

In the UK, "low carbon travel, transport, and access" is defined as an effort to control the carbon footprint of the National Health Service. A study conducted by Sadler and Guenther (2015) emphasized the important role of modern hospital architectural design and its implications for energy savings. Hospitals should encourage walking or cycling by working together with local authorities on identifying suitable routes and providing secure cycle storage as well as the shower or changing facilities. Staff travel gives a significant impact on the carbon footprint resulting from the healthcare industry as shown in Fig. 6.7 (Tomson 2018). If driving remains necessary for some staffs, reimbursement and parking allocation should promote the use of environmentally-friendly options, e.g. electric or hybrid cars.

Tomson (2018) also stated the most obvious way in which hospitals can contribute to reducing greenhouse gas emissions is by minimising energy usage. For example, actions can be taken on insulation, heating, and lighting, by switching electronics off when not in use, etc. Many estates departments have already attained major financial, as well as carbon savings, in this way. A "Power Down" initiative to turn off all unused OR lights and equipment have amounted to an overall saving of $33,000 and

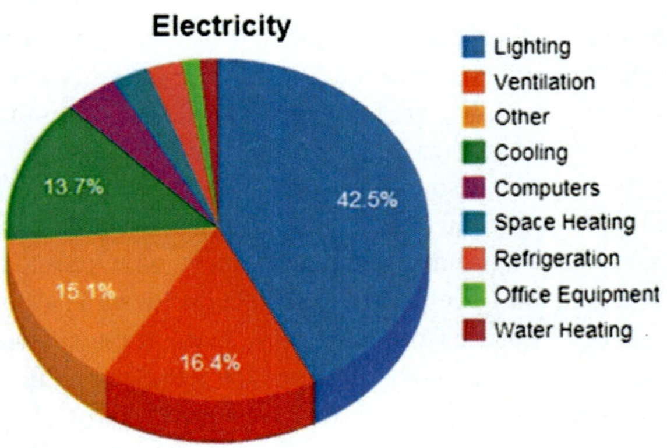

Fig. 6.7 Energy consumption by use in health care organisations (Schroeder et al. 2012)

a reduction of 234.3 metric tons of carbon dioxide emissions per year (Schroeder et al. 2012).

As the carbon footprint of dialysis has been identified completely, the nephrology unit can be taken in as a model to show how both money and carbon could be conserved by carrying out several approaches. These approaches include utilizing heat exchangers on dialysis machines and using a centralised dialysate supply or on-site preparation of dialysate (either of which avoids the use of large quantities of plastic containers to deliver dialysis concentrate to each machine). Not only that, but water can also be saved by reusing the reject water from reverse osmosis plants to prepare ultra-pure water for hemodialysis. If all kidney units adopted all of the sensible practices, about 11,000 tons of CO_2 can be saved up.

6.5 Effects of Carbon Emission to Healthcare Expenditures

First and foremost, Organisation for Economic Co-operation and Development (2018) explained healthcare expenditure in terms of health spending. They said that the end utilization of health expenditure; one of the examples of health care properties and services, also include personal healthcare such as healing care, reintegrate care, long-term care, auxiliary services, medical goods, and collective services like prevention, public health services, and health administration, however, excluding expenditure on investment are measured as health spending. Greenhouse gases are known as gases that capture heat in the atmosphere (EPA 2018a; b).

Figure 6.8 shows some of the examples of greenhouse gas emissions in the U.S. in 2016. Figure 6.9 shows the total U.S. GHG emissions by economic sector in 2016.

Figure 6.10 however, shows the shift in the concentration of GHG from the year 1975–2015. Carbon dioxide (CO_2), methane (CH_4), water vapour, hydrofluorocarbons, nitrous oxide (NO_2), and chlorofluorocarbon (CFC) in the earth's ambient air are the frequent examples of greenhouse gases.

The earth's surface that captured the heat residue was due to the larger concentration of pollutants resulting from the high emission of greenhouse gases. All captured gases could continue to stay trapped in the atmosphere for different periods of time, some can be stay trapped for a few years to thousands of years. The concentrations of the gases throughout the world can be quite alike as the gases are blended well, regardless of their sources (EPA 2018a, b).

6.5.1 Expenditure

Research done by Nicholas et al. (2018) concluded that the emission of carbon dioxide (CO_2) will bring healthcare expenditure to an increasing state. Their findings showed that the U.S. states with more amounts of spending in healthcare expenditures were because they were more affected by the CO_2 emissions from the healthcare. In their

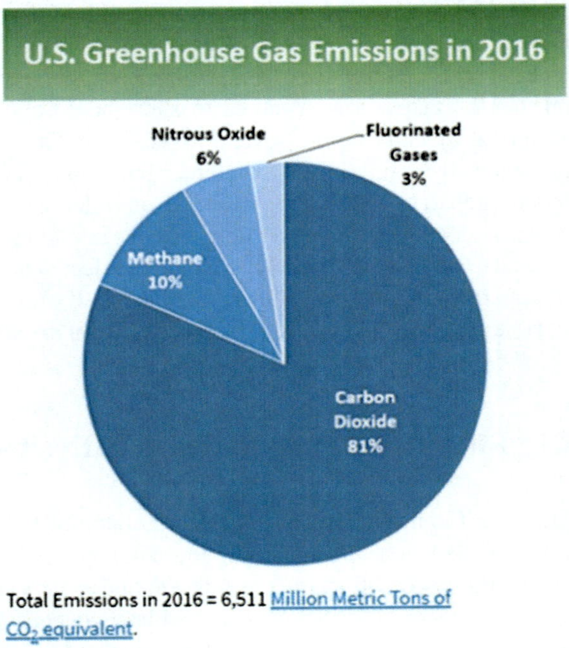

Fig. 6.8 2016 emissions of greenhouse gasses in the U.S. (US Environmental Protection Agency 2020)

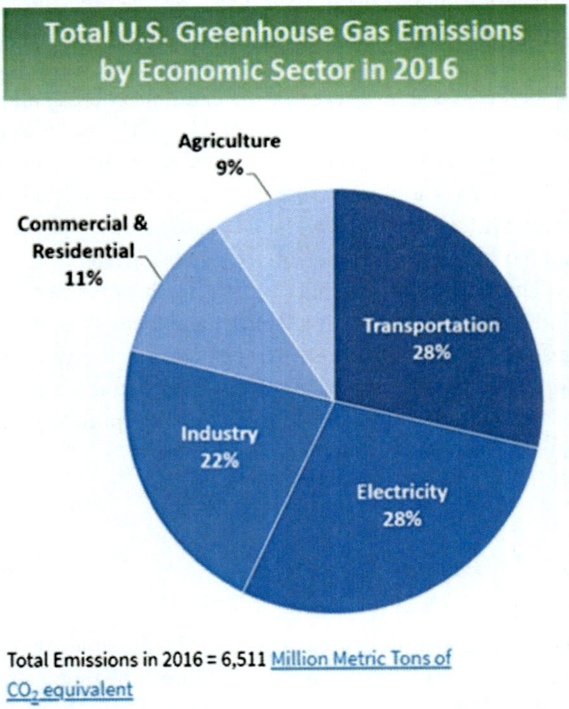

Fig. 6.9 2016 GHG emissions of the U.S. economic sector (US Environmental Protection Agency 2020)

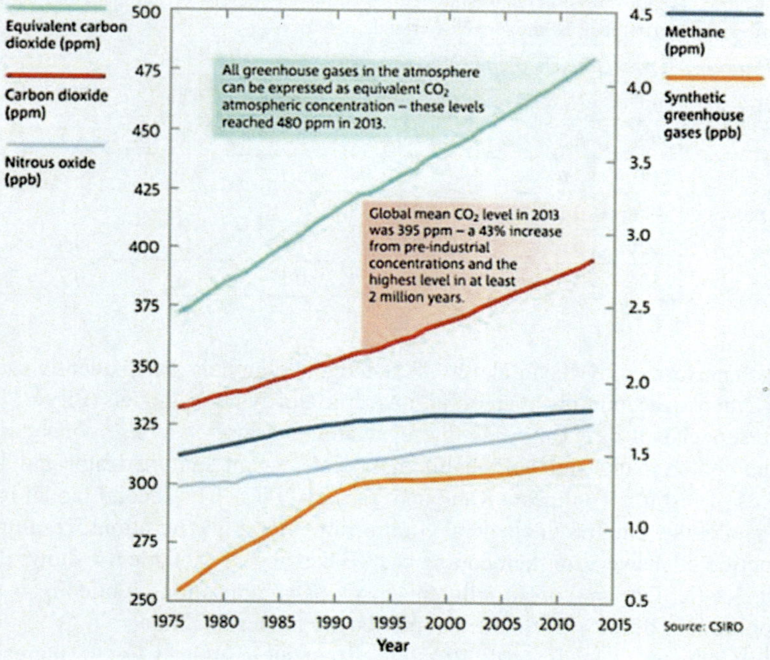

Fig. 6.10 Concentration changes of GHG since 1975 (Bureau of Meteorology and CSIRO 2018)

paper, by using a high-resolution model, they were able to exhibit that chemical and meteorological changes can raise the annual air pollution deaths due to CO_2. At the end of the paper, they concluded that the higher per head healthcare cost was due to the higher amount of per head CO_2 emissions (Nicholas et al. 2018).

Houghton et al. (1996) stated that the use of fossil fuels was initiated by North America and Europe that resulted in severe CO_2 emissions and global warming (Lelieveld et al. 2001). Table 6.1 shows the global anthropogenic CO_2 emission of the India region that includes Bangladesh, Pakistan, Maldives, Nepal, and Sri Lanka, and Myanmar. China region consists of Mongolia, North Korea, Cambodia, Vietnam, and Laos. East Asia covers Japan, the Philippines, Indonesia, Thailand, South Korea, and Malaysia. The changes in CO_2 emissions in Organisation for Economic Cooperation and Development countries and Asia from 2000 to 2020 were increased from 2 to 24% and 41 to 104% respectively, as predicted by the Intergovernmental Panel on Climate Change (Lelieveld et al. 2001). Table 6.6 shows National Institute for Public Health and Environment reports on carbon emission based on nations.

A quantitative study conducted by Thompson et al. (2014) implied that because of air pollution control policies in the states that increase the air quality and lower the health expenses, the cost of the United States carbon policy can be compensated by 26–1050%, showing that those policies are valuable to be implemented (Bai et al. 2018).

Table 6.6 Carbon dioxide emission recorded by global and other countries (National Institute for Public Health and Environment, Bilthoven, Netherlands, 1996)

Source category	Carbon dioxide (Pg of CO_2 per year)					
	Global	North America	Europe	India	China	East Asia
Total	29.8	6.2 (21%)	4.9 (16%)	2.2 (7%)	4.0 (13%)	2.5 (8%)
Fossil fuel use	21.9	5.6	4.5	0.7	2.6	1.7
Industrial process	0.6	0.1	0.2	–	0.1	0.1
Biofuel use	5.5	0.5	0.2	1.4	1.2	0.5
Agriculture	1.8	–	–	0.1	0.1	0.2

Research performed by Bai et al. (2018) stated that 2 models are frequently used for health expenses accounting of air pollution; static and dynamic models (Fig. 6.11). The static model is a direct process that evaluates the dose–response connection between air contaminants and their health final points, e.g. premature deaths and the occurrences of respiratory illness (Kahn and Yardley 2007). The second model is a dynamic model that catches the general equilibrium after a few iterations, resulting in a production of answers for their ending cost (Bai et al. 2018). Table 6.7 shows the price per piece (USD/tonne) of air pollutants by global organisations, while Fig. 6.12 shows welfare loss because of ozone and particulate matter in China.

A study conducted by Jacobson (2008) showed that due to higher CO_2, the increase of water vapour and temperatures increase the ozone. Therefore, in areas that have

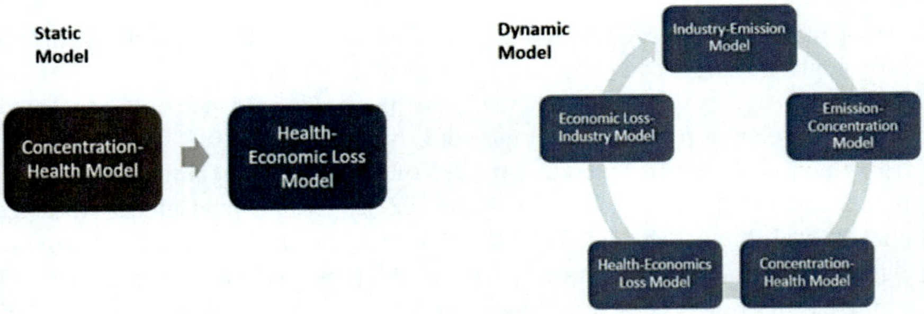

Fig. 6.11 Comparison of the static accounting model versus the dynamic accounting model (Kahn and Yardley 2007)

Table 6.7 Cost per unit in USD/tonne of air pollutants (Wang 1999)

Pollutant	Minimum	Maximum	Mean	Median	Number of organisations
CO_2	2	84	25	20	26
NO_2	42	40,000	8212	4209	36
SO_2	405	21,185	4011	1793	34
TSP	167	8780	3401	2496	20

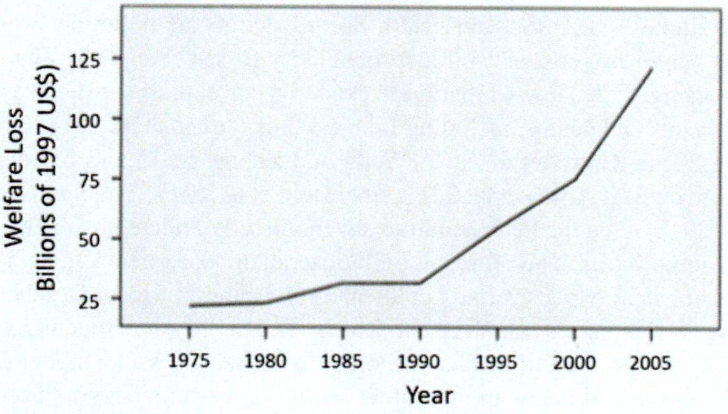

Fig. 6.12 Welfare loss due to ozone and particulate matter in China (Matus et al. 2012)

already been polluted, global warming might get worsen due to ozone. A model known as high-resolution global-regional found that CO_2 might be raised by about 1000 of the U.S. yearly air contamination deaths and by 20–30 pe 1 K rise in CO_2-induced temperature for cancers. From there, additional deaths might rise up to 40% due to ozone, and the rest to particulates, which is high because of CO_2-heighten stability, humidity, and biogenic particle mass (Jacobson 2008).

Some studies about the reactiveness of ozone to temperature were conducted and modeled (Sillman and Samson 1995; Zhang et al. 1998). Other studies performed on the area global effects of climate change from all greenhouse gases on ozone, as well as studies on aerosol particles. The result of water vapour on chemistry had also been highlighted in some studies. Jacobson (2008) also stated that box photochemistry calculation was initially adopted to display the mechanism of the increase of ozone that is influenced by the rise in water vapour and temperature, thus, this explained the relations between CO_2 and health. In his paper, Jacobson summarized how a climate-air pollution model presented by cause and effect that increased the U.S. surface ozone, carcinogen, and particulate matter was due to the high fossil fuel CO_2 emissions, therefore, the number of deaths, asthma, hospitalization, and cancer rates were increasing. He also reported that higher CO_2 from the increased water vapor and temperatures increased the ozone. Lastly, he said that CO_2 increases the stability, relative humidity, and biogenic particle mass, such as PM2.5 (Jacobson 2008).

The increase of 1% annually is subjected to ultra-thin particulate matter (PM2.5) that is in line with the increase of 2.942% of household healthcare expenses. The cause of cardiovascular death and adverse respiratory impacts e.g. reduced lungs function and the development of asthma are said to be caused by continuing contact with PM2.5 (Yang and Zhang 2018). Brauer et al. (2015) and Forouzanfar et al. (2015) conducted a Global Burden of Disease (GBD) study and found that the most recurrent source of environmental-related deaths globally was PM2.5, resulting to a near 2.9 million premature mortality worldwide in 2013 (Yang and Zhang 2018).

Several studies stated that the health hazard due to air pollution brings to an increase in healthcare cost as well as productivity losses (Yang and Zhang 2018). A steadily rising study showed that there existed a correlation between exposure to air contaminants and deaths, including infant deaths (Arceo et al. 2016; Greenstone and Hanna 2014; Knittel et al. 2016; Yang and Zhang 2018) and life expectancy gave a positive result (Chen et al. 2013; Ebenstein et al. 2017; Yin et al. 2017; Yang and Zhang 2018). Healthcare expenditures were directly affected by the biophysical impacts of air pollution. Therefore, air pollution can impose a heavy financial burden in terms of medical expenses like healthcare goods and services (Deschenes et al. 2017; Yang and Zhang 2018). Yang and Zhang (2018) did a heterogenicity analysis that revealed that because of PM2.5 variations, a household with a higher education level paid less on healthcare and the other with a higher number of elderly people paid more on healthcare. They also stated that air pollution had greatly affected the healthcare expenses of the medium-income and low-income households, however, air pollution has no significant effects on expenditures of high-income households (Yang and Zhang 2018).

A study by Ostro and Rothchild (1989) used the Health Interview Surveys and found the correlation between exposure to the fine particulate substance and work loss and bed illness in grownups (Nicholas et al. 2018). Another study by Schwartz et al. (1998) demonstrated that there was an increase in the daily mortality rate when they use the statistics throughout the years 1973–1980 for Philadelphia air contaminants, e.g. total suspended particulate (TSP) and sulphur dioxide (Nicholas et al. 2018). Another study that focused on Canada found that both age and income gave an affirmative reaction on per capita regional healthcare expenses (Matteo and Matteo 1998; Nicholas et al. 2018). Jerrett et al. (2003) found that counties with greater amounts of pollution experienced greater health expenditures, while counties that spend on protection of environmental quality have lesser healthcare expenses (as cited in Nicholas et al. 2018).

6.6 Conclusion

In a conclusion, medical activities contribute to carbon emission that is affecting our ozone layer and contributing to climate change. Different studies conducted in different countries showed that the world healthcare activities and services have dissipated a large amount of greenhouse gasses to the environment every year, which in turn undermines the health of the same population the sector is meant to heal. Many researchers agreed that hospitals are the biggest energy consumer on this planet, leading to the emission of carbon into our environment. Carbon emissions accelerates climate change that gives a significant impact to people and the surrounding. The burning of fossil fuels emits CO_2 and other greenhouse gasses. The release of carbon increases world's temperature because due to the trapping of heat from the sun. These changes water resources and weather patterns, disrupting the production of food and communities who live near the coast with increasing levels of the sea. Although there are many easy ways to reduce healthcare's carbon emission, there are always challenge in executing them. Enforcing modifications might be hard in the beginning, however, in the long run, we will benefit from it. Among ways for the healthcare industry to achieve significant and long-term carbon savings are by managing their operating theatres, altering medical conferences, and properly designing the hospital facilities. Each unit in the health care sector has its own carbon footprint, and the chance to reduce it. We must always remind ourselves that climate change is a risk to public health and that we all are able to significantly reduce our daily carbon footprint. Physicians and other related agencies should work together hand in hand to maintain the health of our planet for generations to come.

Acknowledgements This chapter was written with K. Crystal, K. N. Syazana, M. Y. A. Farah, S. B. N. Hazami, A. S. Melati, J. N. Hidayah.

References

Arceo E, Hanna R, Oliva P (2016) Does the effect of pollution on infant mortality differ between developing and developed countries? Evidence from Mexico City. Econ J 126:257–280

Bai R, Lam CKJ, Li OKV (2018) A review on health cost accounting of air pollution in China. Retrieved from https://www.sciencedirect.com/science/article/pii/S0160412018311176

Brauer M, Freedman G, Frostad J, Van Donkelaar A, Martin RV, Dentener F, Dingenen RV, Estep K, Amini H, Apte JS (2015) Ambient air pollution exposure estimation for the global burden of disease 2013. Environ Sci Technol 50:79–88

Bureau of Meteorology and CSIRO (2018) Climate change in Australia, 2018. GREENHOUSE GASES. Retrieved from https://www.climatechangeinaustralia.gov.au/en/climate-campus/climate-system/greenhouse-gases/

Campbell-Lendrum D, Manga L, Bagayoko M, Sommerfeld J (2015) Climate change and vector-borne diseases: what are the implications for public health research and policy? Philos Trans R Soc Lond B Biol Sci 370:20130552

Chen Y, Ebenstein A, Greenstone M, Li H (2013) Evidence on the impact of sustained exposure to air pollution on life expectancy from China's Huai River policy. Proc Natl Acad Sci 110:12936–12941

Chung JW, Weltzer DO (2009) Estimate of the carbon footprint of the US health care sector. JAMA 302:1970–1972

Costello A, Abbas M, Allen A et al (2009a) Managing the health effects of climate change. Lancet 373:1693–1733

Costello A et al (2009b) Managing the health effects of climate change: Lancet and University College London Institute for Global Health Commission. Lancet 373(9676):1693–1733

Deschênes O, Greenstone M, Shapiro JS (2017) Defensive investments and the demand for air quality: evidence from the NOx budget program. Am Econ Rev 107:2958–2989

Ebenstein A, Fan M, Greenstone M, He G, Zhou M (2017) New evidence on the impact of sustained exposure to air pollution on life expectancy from China's Huai River Policy. Proc Natl Acad Sci 114:10384–10389

Ebi KL, Mills DM, Smith JB, Grambsch A (2006) Climate change and human health impacts in the United States: an update on the results of the US national assessment. Environ Health Perspect 114:1318–1324

EPA (2018a) Overview of greenhouse gases. Retrieved from https://www.epa.gov/ghgemissions/overview-greenhouse-gases

EPA (2018b) Sources of greenhouse gas emissions. https://www.epa.gov/ghgemissions/sources-greenhouse-gas-emissions

Fenner AE, Kibert C, Woo J, Morque S, Razkenari M, Hakim H, Lu X (2018) The carbon footprint of buildings: a review of methodologies and applications. Renew Sustain Energy Rev 94:1142–1152. https://doi.org/10.1016/j.rser.2018.07.012

Forouzanfar MH, Alexander L, Anderson HR, Bachman VF, Biryukov S, Brauer M, Burnett R, Casey D, Coates MM, Cohen A, Delwiche K, Estep K, Frostád JJ, Astha KC, Kyu HH, Moradi-Lakeh M, Ng M, Slepak EL, Thomas BA, … Murray CJ (2015) Global, regional, and national comparative risk assessment of 79 behavioural, environmental and occupational, and metabolic risks or clusters of risks in 188 countries, 1990–2013: a systematic analysis for the Global Burden of Disease Study 2013. Lancet 386(10010):2287–2323. https://doi.org/10.1016/S0140-6736(15)00128-2

Gao T, Liu Q, Wang J (2014) A comparative study of carbon footprint and assessment standards. Int J Low-Carbon Technol 9(3):237–243. https://doi.org/10.1093/ijlct/ctt041

Greenstone M, Hanna R (2014) Environmental regulations, air and water pollution, and infant mortality in India. Am Econ Rev 104:3038–3072

Houghton JT et al (eds) (1996) Climate change 1995: the science of climate change. Cambridge University Press, Cambridge

Jacobson MZ (2008) On the causal link between carbon dioxide and air pollution mortality. Geophys Res Lett 35:L03809. Retrieved from https://agupubs.onlinelibrary.wiley.com/doi/epdf/10.1029/2007GL031101

Jerrett M, Eyles J, Dufournaud C, Birch S (2003) Environmental influences on health care expenditures: an exploratory analysis from Ontario. Can J Epidemiol Comm Health 57:334–338

Junnila SI* (2006) Empirical comparison of process and economic input-output life cycle assessment in service industries. Environ Sci Technol 40(22):7070–7076. https://doi.org/10.1021/es0611902

Kahn J, Yardley J (2007) As China roars, pollution reaches deadly extremes. NY times 26:A1

Knittel CR, Miller DL, Sanders NJ (2016) Caution, drivers! Children present: traffic, pollution, and infant health. Rev Econ Stat 98:350–366

Lelieveld P et al (2001) The Indian ocean experiment: widespread air pollution from south and southeast Asia. Retrieved from https://www.ncbi.nlm.nih.gov/pubmed/11161214

Lenzen M (2000) Errors in conventional and Input–output-based life-cycle inventories. J Ind Ecol 4:127–148

MacNeill AJ, Lillywhite R, Brown CJ (2017) The impact of surgery on global climate: a carbon footprinting study of operating theatres in three health systems. Lancet Planet Health 1(9):e381. https://doi.org/10.1016/S2542-5196(17)30162-6

Malik A, Lenzen M, McAlister S, McGain F (2018) The carbon footprint of Australian health care. Lancet Planet Health 2(1):e27–e35. https://doi.org/10.1016/S2542-5196(17)30180-8

Matteo LD, Matteo RD (1998) Evidence on the determinants of Canadian provincial government health expenditures. J Health Econ 17:211–228

Matus K, Nam K-M, Selin NE, Lamsal LN, Reilly JM, Paltsev S (2012) Health damages from air pollution in China. Glob Environ Chang 22:55–66

Morris DS, Wright T, Somner JEA, Connor A (2013) The carbon footprint of cataract surgery. Eye 27(4):495–501. https://doi.org/10.1038/eye.2013.9

Nathan BJ, Lauvaux T, Turnbull JC, Richardson SJ, Miles NL, Gurney KR (2018) Source sector attribution of CO_2 emissions using an urban CO/CO_2 Bayesian inversion system. J Geophys Res Atmos 123(23). https://doi.org/10.1029/2018JD029231

NHS Sustainable Development Unit (2008, updated 2009, 2016) NHS England carbon emissions carbon footprinting report. http://www.sdu.nhs.uk/documents/publications/1263313924_jgyW_nhs_england_carbon_emissions_carbon_footprinting_r.pdf. Accessed 26 Nov 2018

Nicholas A, Rangan G, Marco LCK, Zinnia M (2018) U.S. state-level carbon dioxide emissions: does it affect health care expenditure? Renew Sustain Energy Rev Elsevier 91(C):521–530

OECD (2018) Health spending. Retrieved from https://data.oecd.org/healthres/health-spending.htm

Ostro B, Rothchild S (1989) Air pollution and acute respirator morbidity: an observational study of multiple pollutants. Environ Res 50:238–47

Pandey D, Agrawal M, Pandey JS (2017) Carbon footprint: current methods of estimation

Ponette-González A, Byrnes J (2011) Sustainable science? Reducing the carbon impact of scientific mega-meetings. Ethnobiol Lett 2:65–71. Retrieved from http://www.jstor.org/stable/26419937

Roberts I, Godlee F (2007) Reducing the carbon footprint of medical conferences. BMJ (Clin Res Ed) 334(7589):324–325

Sadler BL, Guenther R (2015) Ten rules for 21st century healthcare: A US perspective on creating healthy, healing environments. Future Hosp J 2(1):22–27. https://doi.org/10.7861/futurehosp.15.009

Schroeder K, Thompson T, Frith K, Pencheon D (2012) Sustainable healthcare. Wiley, pp 1–280

Scottish Government Climate change (Scotland) act 2009: public bodies duties (2010)

Scottish Government. Climate change (Scotland) act 2009. Edinburgh: The Stationery Office Limited. Retrieved at: http://www.scotland.gov.uk/Topics/Environment/climatechange/scotlands-action/climatechangeact. Accessed 26 Nov 2018

Sherman J, Le C, Lamers V, Eckelman M (2012) Life cycle greenhouse gas emissions of anesthetic drugs. Anesth Analg 114(5):1086–90. pmid:22492186

Sillman S, Samson PJ (1995) Impact of temperature on oxidant photochemistry in urban, polluted rural and remote environments. J Geophys Res 100(D6):11497. https://doi.org/10.1029/94JD02146

Thiel CL, Eckelman M, Guido R, Huddleston M, Landis AE, Sherman J et al (2015) Environmental impacts of surgical procedures: life cycle assessment of hysterectomy in the United States. Environ Sci Technol 49(3):1779–86. pmid:25517602

Thompson TM, Rausch S, Saari RK, Selin NE (2014) A systems approach to evaluating the air quality co-benefits of US carbon policies. Nat Clim Chang 4:917–923

Tomson C (2018) Reducing the carbon footprint of hospital-based care. Retrieved from http://futurehospital.rcpjournal.org/content/2/1/57.full

US Energy Information Administration, Commercial Building Energy Consumption Survey (2007). Retrieved from https://www.eia.gov/consumption/commercial/reports/2007/large-hospital.php

US Environmental Protection Agency, Greenhouse Gas Emissions (2020). Retrieved from http://www.epa.gov/ghgemissions/overview-greenhouse-gases

Wang S (1999) The life cycle ecological evaluation method and case study of complex industrial products in China. Fudan University

Watts N, Adger WN, Agnolucci P, et al (2015) Health and climate change: policy responses to protect public health. Lancet 386:1861–1914

Wiedmann T, Minx J, Barrett J, Wackernagel M (2006) Allocating ecological footprints to final consumption categories with input–output analysis. Ecol Econ 56(1):28–48. https://doi.org/10.1016/j.ecolecon.2005.05.012

Yang J, Zhang B (2018) Air pollution and healthcare expenditure: implication for the benefit of air pollution control in China. Retrieved from https://www.sciencedirect.com/science/article/pii/S0160412018310407

Yin P, He G, Fan M, Chiu KY, Fan M, Liu C, Xue A, Liu T, Pan Y, Mu Q (2017)

Zhang J, Rao ST, Daggupaty SM (1998) Meteorological processes and ozone exceedances in the Northeastern United States during the 12–16 July 1995 Episode. J Appl Meteorol 37(8):776–789. https://doi.org/10.1175/1520-0450(1998)037<0776:MPAOEI>2.0.CO;2

Chapter 7
Safety Issues in Composite Materials

Abstract People who work in the fibre reinforcement industry are susceptible to certain diseases, e.g. respiratory problems, skin irritation, and cancer. Respirable particles of reinforced fibres are the main problems contributing to the health issues. This shows the importance of personal protective equipment (PPE) to minimize health problems among workers. The need to address safety, health, and environmental hazards in the processing and manufacturing of composites and their use are key aspects of continued growth and development within the composites industry. There are four aspects of protective measures and actions, which are engineering controls, administrative control, PPE, and medical surveillance and screening. PPE is a supplemental option in supporting other advanced levels of exposure controls. Several types of PPEs are suitable to be used in composite handlings, such as organic vapor respirator, safety goggles, flock lined latex gloves, and disposable bodysuit. Training is required for each employee at the company. The existing, secure, corporate training system can be modified to guide workers about the exact types of composite products used by the company. This level of specificity is a greater safety benefit than prior general knowledge. In fact, the theory is similar to degree materials but the practice is an opportunity rarely offered or encountered during our working life. As far as reasonably practicable, training programs offer long-life learning that benefits us in many ways.

Keywords Reinforcement fibres · Processing and manufacturing · Composites · Training · Personal protective equipment

7.1 Introduction

A composite is a mixture of two or more materials of different properties that work together to give special properties to the composite. Some various materials and processes can be combined to make the composite more effective and versatile (Composite UK Trade Association 2018). For example, fibre, reinforced plastic, thermoplastic, glass, resin, styrene, and many more. Composites are well known as stronger, lighter, stiffer, non-magnetic, and more resistant to corrosion for their great

S. M. Sapuan et al., *Safety and Health in Composite Industry*,
Composites Science and Technology ,
https://doi.org/10.1007/978-981-16-6136-5_7

properties compared to the materials alone without any mixture (Ambrogio et al. 2018).

Since the manufacturing of composites involves many processes, there are many types of hazards that the handlers or workers might be exposed to. Composite materials are often shaped by different manufacturing processes, e.g. compression moulding, autoclave moulding, filament winding, and pultrusion (Ambrogio et al. 2018). Each of the processes has its own hazards, e.g. high pressure and temperature.

Besides that, there are also hazards from respirable dust, splashes of chemicals used, and accidents involving heat or cut. Respirable dust is one of the most dangerous hazards to all the workers, as they are unseen to the bare eyes. Once the dust is inhaled, it might travel into the deepest area of the respiratory system and flow throughout the body, causing adverse effects to humans' health. Respirable dust is among the deadliest form of air pollutants, because they are able to penetrate the lung deeply, and can transfer through blood, causing heart attacks, premature deaths, and permanent DNA mutations (Hoeflinger and Laminger 2017). Therefore, in processing, fabrication, or handling the composites, safety issues must be identified and tackled to ensure the safety, health, and welfare of the employees or workers.

7.2 Reinforcement Fibres

7.2.1 *Overview*

Reinforcement fibres can be usually found in the aerospace industry because the use of advanced composite materials on aircraft has risen sharply for the past 25 years. Aramid, glass, carbon, and graphite are the types of reinforcement fibres usually found in aerospace applications. Reinforced fibres are also used in many other industries, e.g. automotive, sport, enclosure, ship, and boat. Reinforcement fibres can be fabricated through manual or mechanical layering to obtain properties of high resistance to atmospheric agents and corrosives, low specific weight, and low thermal and electrical conductivities (Abbate et al. 2006).

7.2.2 *Respirable Particles*

After undergoing many complex chains of events, the materials become respirable particles that can be easily inhaled by humans. These airborne particles originating from the materials are a serious concern to human health. This is because, these fibres have a small diameter, long, and thin shape that deeply entrapped in lung tissues. Omran (2018) stated that health issues that are usually observed among those who work in the reinforcement fibres industry are cancers, skin irritation, respiratory problems, and neurobehavioral symptoms.

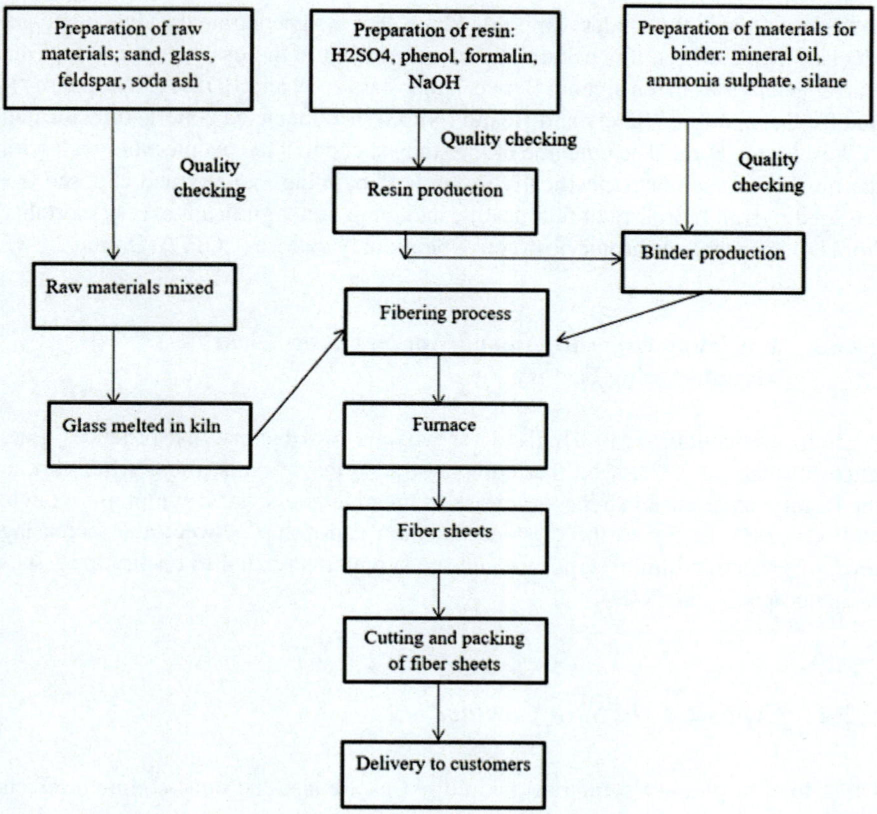

Fig. 7.1 Processes of glass microfibre production (Sripaiboonkij et al. 2009)

People that are usually exposed to the fibres are those who work in microfibre factories, particularly during the process of microfibre production. Workers who are involved with preparing raw materials and chemicals, fibre sheet cutting, packing, quality control, and maintenance (Sripaiboonkij et al. 2009). The production of microfibre involves many steps and techniques, as shown in Fig. 7.1.

7.2.3 Exposure via Inhalation

During all of these processes, workers could be exposed to the materials via two routes; dermal contact and inhalation. The most concerning the route of exposure is via inhalation because it enters the respiratory system. The air inhaled from the mouth and nose transfers to the lungs. The exchange of gases of oxygen and carbon dioxide occurs at the end of bronchial passages in the lungs, where there are tiny sacks named alveoli. The respiratory system supplies oxygen throughout the body. When a person accidentally inhales respirable fibres, they will penetrate the alveoli and get blocked inside. When this occurs, the fibres will injure the tissue and interrupt the exchange of gases in the alveoli. Between the years 1959 and 1978, A National Institute for Occupational Safety and Health (NIOSH) conducted a cohort study located in Washington State. The outcome of the study stated that the people who work with the fibre reinforcement, specifically in the boat-building industry and exposed has reported overall mortality of 598 deaths, including non-significant excess mortality from lung cancer and chronic obstructive pulmonary disease (COPD) (Omran 2018).

7.2.3.1 Inhalation Exposure Among Workers in the Glass Manufacturing Factory

A study was conducted in Thailand for workers of a factory that produces glass microfibre sheets. It showed that compared to office workers, those who work at the factory experienced an increased risk of breathlessness, nasal symptoms, cough, and wheezing. It proved the existence of the relationship between the increasing level of glass microfibres exposure and the symptoms related to respiration issues (Sripaiboonkij et al. 2009).

7.2.4 Exposure via Skin Contact

Other than respiratory symptoms, skin diseases are also the most common effects observed in the fibres reinforcement industry. The workers were exposed via direct contact with the skin. There are few studies carried out to demonstrate that multiple exposures can lead to skin diseases among workers. In Japan, a survey was conducted on the workers (N = 148) of fibre-glass reinforced plastics factories located in Kyushu. There were 87 workers (58.8%) who were reported to have skin problems (mainly itching or dermatitis) since they started to work in FRP manufacturing and 25 of them consulted a physician because of their skin problems; one worker was forced to take sick leave because of his severe dermatitis (Omran 2018).

Therefore, reinforcement fibres are used widely in many industries. So, many people are needed to produce products that involving reinforcement fibres. Every employer needs to comply with exposure limits that are stated in the regulations and decides what types of control need to be done to minimize the exposure among the workers.

7.3 Safety Equipment on Composite

7.3.1 *Overview*

Based on Sumesh et al. (2018), there are four aspects of protective measures and actions which are engineering controls, administrative control, personal protective equipment (PPE), and medical surveillance and screening. PPE acts as a supplemental option in supporting other advanced levels of exposure control (Sumesh et al. 2018). This can be proved by the hierarchy of control, where these protective measures have been bothered, because; the usage of composites and their exposures cause health effects to the workers or any other person exposed to the hazardous substance contained inside the composites. For instance, the use of nanoparticles or nanomaterials in cement production industries that have a larger surface area per unit mass when compared to other larger particles that harm the human body through absorption and inhalation (Sumesh et al. 2018). Thus, evaluation of safety and health risk associated with any potential hazard of the particles, either from the use of topical products or surfaces that might cause direct contact with the body is crucial to reassure which routes of exposure that the hazard came from (Balagna et al. 2018). Figure 7.2 shows the hierarchy of controls.

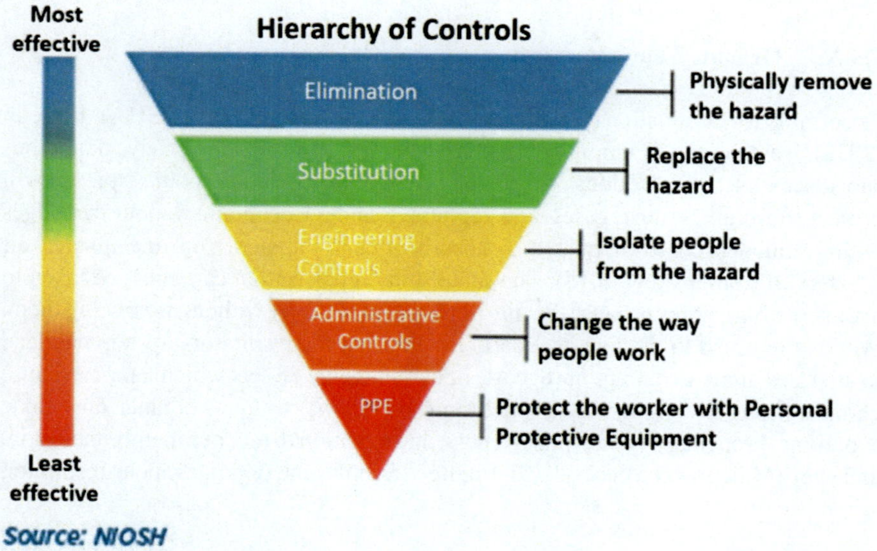

Fig. 7.2 Hierarchy of controls

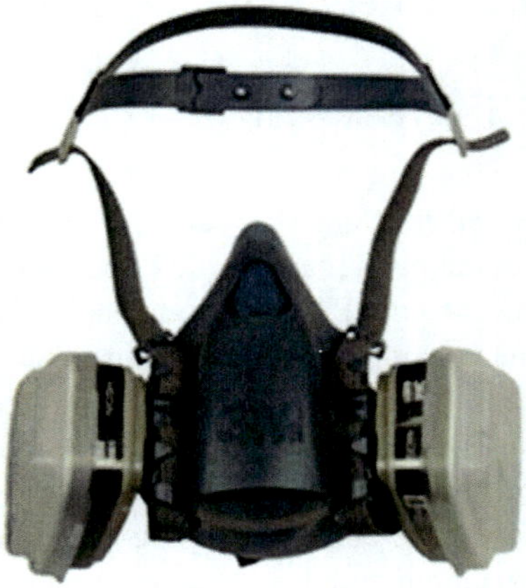

Fig. 7.3 Organic vapour respirator

7.3.2 *Types of Personal Protective Equipment (PPE)*

There are several types of personal protective equipment (PPE) suitable for composites handling that exposed to hazards, e.g. chemical, physical, and health hazards. Examples of the PPE needed are:

7.3.2.1 Organic Vapor Respirator

According to Occupational Safety and Health Administration (OSHA) Bulletin (2018), respirators function to protect workers from inhaling any dangerous substances, such as infectious particles and chemicals. Organic vapour respirator will protect from fumes, dust, gases, and vapours. Besides, it contains vapour cartridges with granulated charcoal treated to absorb organic vapours from the inhaled air ("Safety in Composites" 2018). The same study reported that care must be taken to avoid breathing fibres because the tiny fibres travel in the air without us noticing them. Workers exposed to organic dust particles entering the respiratory system can lead to allergies, toxic irritating, fibrogenic, or carcinogenic effects which later can cause chronic obstructive lung diseases, chronic bronchitis, asthma, organic dust toxic syndrome, bronchial hyperresponsiveness, and irritation to mucous membranes, skin, and eyes (Majchrzycka et al. 2018). Figure 7.3 shows the organic vapour respirator.

7.3.2.2 Safety Goggles

Safety goggles are used to protect workers' eyes from chemicals, vapours, and other substances. These goggles consist of polycarbonate coated lenses that are able to resist any impacts and approximately 100% of harmful UVA/UVB rays ("Safety in Composites" 2018). When workers are not properly protected from composite

Fig. 7.4 Safety goggles

materials such as resin, it can be dangerous as splashes and vapours can lead to unnecessary eye damage and irritation ("Safety in Composites" 2018). For effective protection against all those vapours, chemicals, or other substances, workers need to wear the safety goggles correctly and make sure the safety goggles fit them. Figure 7.4 shows the safety goggles.

7.3.2.3 Flock Lined Latex Gloves

Flock lined latex gloves are suitable to be used for workers who handle composites that contain reinforcements, any resins, or solvents continuously, as this type of gloves has characteristics that will protect workers' hands, e.g. heavy-duty, thicker, and others besides being reusable. According to "Safety in Composites" (2018), different people have different susceptibility in the reaction of certain components in composites, but sensitivity can be developed through time. For instance, there are many cases where workers do not show any particular sensitivity during fibreglass cutting, but later they develop health problems. Skin sensitivity includes rashes, itchy, or both and the intensity varies among individuals. Besides wearing gloves, workers also can wear long sleeves during working as it also prevents or protects workers from contacting hazardous components contained in composites. Figure 7.5 shows the example of glove used, Flock Lined Latex Glove.

Fig. 7.5 Flock lined latex glove

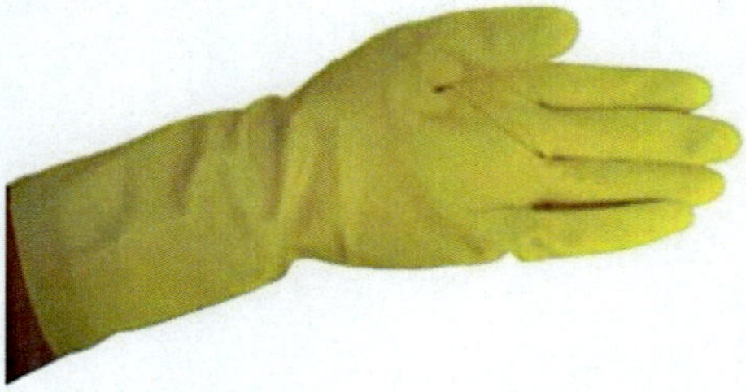

7.3.2.4 Disposable Body Suit

Disposable bodysuits are coverall for the workers that function as double protection to the workers' bodies from the hazardous materials from the composite production. For instance, dirt, aerosols, spills, and splashes of corrosive substances ("Safety in Composites" 2018). There are two types of disposable bodysuit or coverall, which are disposable microporous coveralls and blue polypropylene disposable coveralls (Colony Distribution Inc. 2018). The blue polypropylene disposable coveralls are more breathable than the disposable microporous coveralls. Figure 7.6 shows the disposable microporous coveralls, while Fig. 7.7 shows the blue disposable coveralls.

Therefore, there are several types of personal protective equipment (PPE) for workers handling composites. According to Dong et al. (2018), the misuse of PPE can contribute to hazards, e.g., accidents and occupational diseases like hearing loss, dermatological diseases, head injuries, pneumoconiosis, etc. Therefore, it is vital to ensure the PPEs are properly used as carelessness contributes to hazards to the workers that are everywhere. Hazards cannot be removed, but they can be minimized via the hierarchy of control. So, proper training or program regarding personal protective equipment (PPE) or safety equipment needs to be introduced to the workers for the purpose.

Fig. 7.6 Disposable microporous coverall (Colony Distribution Inc. 2018)

Fig. 7.7 Blue disposable coverall (Colony Distribution Inc. 2018)

7.4 Manufacturing and the Safety Aspect

7.4.1 Overview

In the manufacturing and processing of the composites, safety considerations must be among the priorities to be considered by the users. The workers should possess some knowledge on safety in order to avoid hazards at the workplace. At the basic level, ventilation, storage, product handling, and disposal problems are the main problems that arise from improper building design.

7.4.2 Nano-Titanium

There are several safety considerations for using Nano-Titanium-engineered cementitious composites, although Nano-Titanium (NT) can enhance the durability, mechanical characteristics, and allow the functional properties of cementitious composites. However, NT could cause a toxicological risk to microorganisms which would distrupt the waste degradation and nutrient cycling. They considered that NT would give potential danger to humans, for instance, reactive oxygen species (ROS) that is produced under the UV light could lead to cytotoxicity, DNA damage in mammalian cells, and inflammation. NT is not dispersed into the air in huge quantities during the application as it is combined in the cementitious composites. In the manufacturing of NT-engineered cementitious composites, NT dispersed into the atmosphere resulted in the accumulation of its tiny particles in the manufacturing facility. Wearing masks and gloves is essential to enhance protection against these small particles. However, there is no proof that declares the truth that NT being the

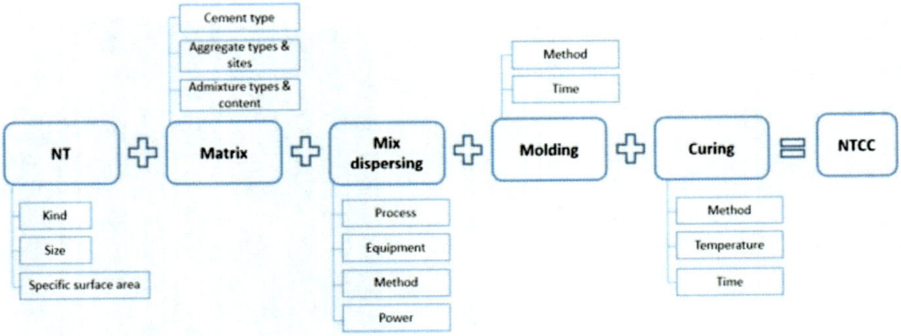

Fig. 7.8 The manufacturing of the NTCC (Li et al. 2018)

instantaneous origin for the damage (Li et al. 2018). Figure 7.8 shows the steps of NTCC manufacturing.

7.4.3 *Additive Manufacturing*

In additive manufacturing of ammonium perchlorate composite for solid rocket propellant application, there is restricted ability to control the local geometry throughout a grain via traditional manufacturing. The design geometry of grain determines the core geometry of a grain that is removed from the mould after the propellant curing. Mcclain et al. (2018) stated that the grain geometry is mostly dependent on the detachment of the mandrel that is made easier through the application of taper and coating. To enhance the capability to fabricate complex parts, additive manufacturing (AM) is utilized. AM is reported to have the effects on the combustion of energetic materials, hence, AM is being explored as a technique to modify the propellants' performances; pyrotechnics, and explosives. High solids and smaller particle size loading have been observed to raise the viscosity of the compound of propellants. To overcome this, significantly high pressure would be needed that dramatically increases safety concerns and costs. Other determinants to be distinguished for quality controls are the start or stop delay times and flow rate consistency (Mcclain et al. 2018). Figure 7.9 shows the adjusted printer depositing propellant into a 7 × 7 × 25 mm strand shape.

7.4.4 *Laminating Object Manufacturing (LOM)*

For laminating object manufacturing (LOM), 3D parts are processed by trimming the 2D cross-sections with a cutter or laser followed by laminating the sheets. Among the materials that can be used in LOM include metals, paper, plastics, synthetic

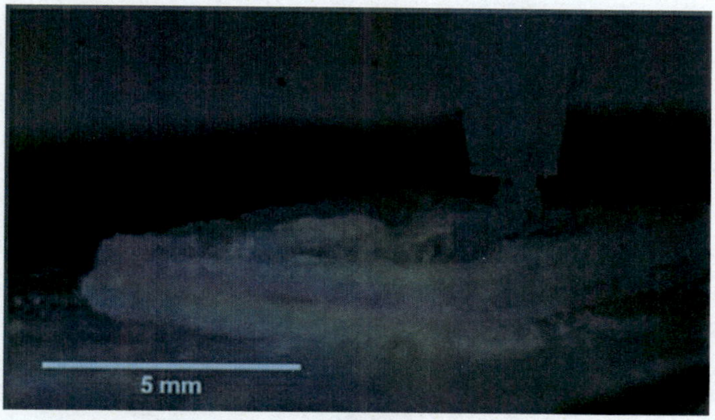

Fig. 7.9 Adjusted printer depositing propellant into a 7 × 7 × 25 mm strand shape (Mcclain et al. 2018)

materials, fabrics, and composites. The inability of the heat roller to initiate parts to full combination and curing was the crucial issue for the LOM operation. Due to small concentrated stress issues, the large roller is more favourable for bonding. Laser-assisted AM, a new method is developed for endless fibre reinforced thermoplastic compound. Prepreg tape was proposed to be used as an alternative or substitute for the pre-cut prepreg sheet. Using a CO_2 laser beam and consolidation roller, the strips were put layer by layer prior to laser cutting of each layer. Superior mechanical properties were exhibited by the fabricated composites from this method due to the high fibre weight ratio, continuous fibre reinforcement, superior interfacial bonding, and minimized void content (Parandoush and Lin 2017). Figure 7.10 shows the schematic of laminated object manufacturing.

7.4.5 Electromagnetic Interference (EMI)

In terms of safety issues on electromagnetic interference (EMI), to have more knowledge on noises generated, avionics and aerospace engineers analysed EMI based problems and established the methods to fabricate current designs for alleviating transmitting noise. In the beginning, most companies favoured swift and cumbersome shielding enclosures. In order to give long-term solutions for the good functioning of electronic tools, the researchers directed their focus towards fabricating, filtering materials, and advanced shielding. Sankaran et al. (2018) claimed that various professional certification standards were organized to ensure compatibility of electrical tools with respect to radiation, emissions, and vulnerability. Owing to the worldwide expansion of the electronic field, there are stern regulations implemented these days that consequently influence the safety of almost all of the avionics, medical, and sensitive reconnaissance apparatus from the strike of catastrophic failure created by noisy

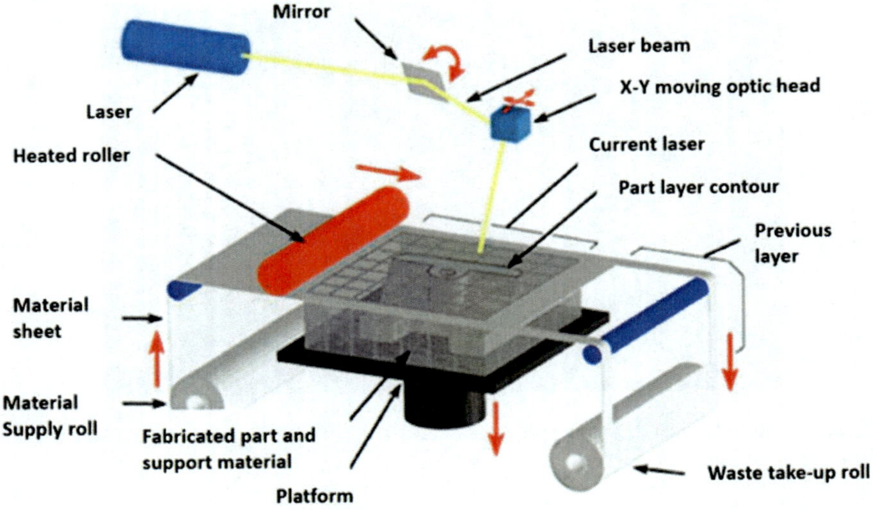

Fig. 7.10 Schematic of laminated object manufacturing (Parandoush and Lin 2017)

EMI. Many electronic systems have been discovered which are simply susceptible to EMI disruption. By using the joining of shielding and filtering matter, such devices can be protected to lower EM fluctuations as well as unpleasant EMI (Sankaran et al. 2018). Figure 7.11 shows the schematic representation of four principal mechanisms of the EMI shielding apparatus, while Fig. 7.12 shows the coefficients of EMI shielding mechanisms.

To summarize, there are many hazards during the production of composites that resulted from a multitude of problems. At the basic level, ventilation, storage, product handling, and disposal issues originating from the improper building design. Wrong placement of equipment and materials could lead to unnecessary hazards even with an adequate building design. In the composites industry, material handling is the primary source of hazards. The proper storage, delivery, transportation distribution, and application of materials are seen as the solution to minimize the hazards. Actual

Fig. 7.11 Easy schematic representation of four principal mechanisms of EMI shielding apparatus (Jagatheesan et al. 2018)

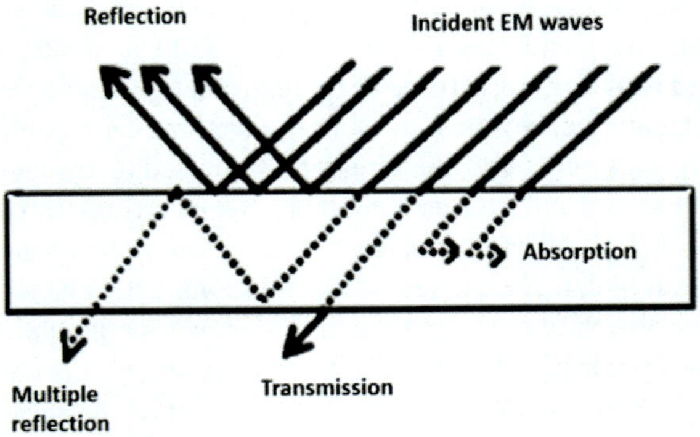

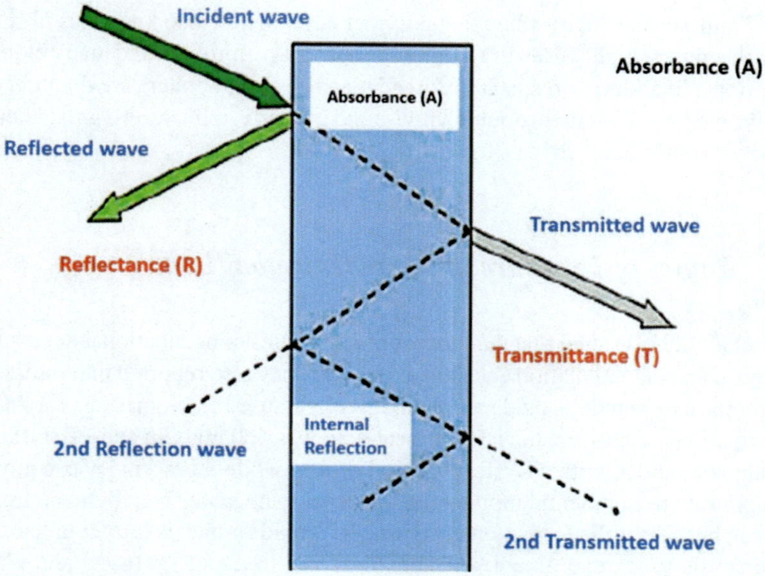

Fig. 7.12 Coefficients of EMI shielding mechanisms (Joshi and Datar 2018)

equipment plays an important aspect in the production hazards control. Many similar hazards are present, although they vary in magnitude, regardless of whether the equipment is a large production autoclave or a small hot press. Many new manufacturing technologies are being introduced because of the increased demand for composite products and this new technology might help to reduce some hazards but consequently, it also introduces several others as well. New hazards appear as a result of mechanical motion, repair, maintenance, and operation of many of these new production media. For completion, time prepregs are often stored earlier. The need to cool or refrigerate the systems to avoid the final curing from taking place also contributes to other issues. Two more hazards that comprise the overall production concerns and storage and delivery are associated with the temperature extremes and storage degradation as well as potential safety hazards once the composite is completed, respectively.

7.5 Styrene Formation

7.5.1 What Is Styrene?

Based on Saamanen (1998), one of the versatile polymers produced is styrene used in the making of various grades of polystyrene and other styrene polymer products, e.g. unsaturated polyester resins that are produced about 0.5 million tonnes (3%) in a year. Tranfo et al. (2012) stated that polymer composites, also known as fibre reinforced plastics are highly used in automobile, aerospace, military, and many industrial applications. In addition, resins, reinforced plastics, and polymers are manufactured from styrene because it has a double vinyl-bond property, indicating a great chemical reactivity (Tranfo et al. 2012).

7.5.2 Routes of Exposure and Concentration Limit

Tranfo et al. (2012) stated that the most common routes of occupational exposure for styrene are through inhalation and skin exposure. They also reported that for the total concentration of mandelic acid and phenylglyoxylic used as biomarkers for ACGIH for occupational exposure, the BEI of creatinine was 400 mg/g in end-of-shift urine. The value coincided with TLV for a period of 8 h, while 85 mg/m^3 was equivalent to 20 ppm for environmental monitoring. Moreover, the major metabolic pathway of styrene in humans is the formation of styrene-7, 8-oxide which is further metabolized by hydrolysis to styrene glycol, and then oxidized to mandelic (MA) and phenylglyoxylic (PGA) acids (Tranfo et al. 2012). In addition, Saamanen (1998) stated the first metabolic oxidation product of styrene are mandelic (MA) and phenylglyoxylic (PGA) acids that are reactive and can form covalent binding to macromolecules which is responsible for the genotoxicity of styrene. The current occupational exposure limits (8 h time-weighted average) vary in different countries. Table 7.1 shows the occupational exposure limits for styrene in different countries.

Table 7.1 Occupational exposure limits for styrene in different countries (Saamanen 1998)

Country/organization	Exposure limit 8 h	Short-term exposure limit
Belgium	50	100
Denmark	25	–
Finland	20	50
France	50	–
Germany	20	–
Norway	25	–
Sweden	20	50
United Kingdom	100	260
USA/ACGIH (TLV)	20	40
USA/NIOSH (REL)	50	100
USA/OSHA (PEL)	100	200

7.5.3 Health Effects of Styrene

Many possible adverse health effects could be experienced among workers in the industry, as documented by various publications and reports. These are sourcing from chronic exposure in particular to low doses that resulted in hearing deficits, neurotoxic effects, mood disorders, nephrotoxic, and hepatotoxic effects (Tranfo et al. 2012). In addition, Tranfo et al. (2012) also stated that evidence did not show that the frequency of sister chromatid exchanges had any association with the exposure to styrene, but it was reported that there was an increase in DNA and hemoglobin adducts and the frequency of chromosomal aberrations among workers who are exposed to styrene.

Saamanen (1998) reported that the most common symptoms of styrene exposure are fatigue, headache, and dizziness. Styrene that also exists in the form of vapour cause irritations to the eyes and mucous membranes at high concentrations. Moreover, as stated from IARC (1994), increased chromosomal damage was also reported in lymphocytes in conjunction with that the critical effect of occupational exposure of styrene, since it could affect the nervous system (cited in Saamanen 1998).

7.5.4 Manufacturing Process of Styrene

According to Tranfo et al. (2012), composites manufacturing processes that are commonly practiced by industries are open and closed mouldings. The most regular operation for the open moulding process is the 'layup' of several layers executed by the workers via hand or mechanical application equipment, such as a spray gun (at high or low pressure). The gel coat and laminate are exposed to the atmosphere during the process; styrene evaporation occurs during the application of gel-coat or resin materials and continues until the gel-coat film and resin laminate are cured during the fabrication (Tranfo et al. 2012).

In addition, Hildeberto Nava et al. (2012) also stated that in open contact moulding, the gelation and curing emit styrene from the wet gel coat or resin. This process is performed in a large area to enable the retainment of the styrene concentration under the current Occupational Safety and Health Administration (OSHA) standard of 100 ppm, making it hard to monitor the styrene emission.

Moreover, as for the closed moulding process, it is believed that this process is successful in controlling workers' exposure to styrene and reducing the emitted styrene concentrations by one order of magnitude as the composite is fabricated in a two-sided mould set, or within a vacuum bag (Tranfo et al. 2012).

7.5.5 *Emission of Styrene from Resin*

The European UP/VE Resin Association stated that unsaturated polyester (UP) resins and epoxy vinyl ester (VE) resins are liquid products containing linear unsaturated polyester polymers, e.g. epoxy vinyl ester polymers (50–70%) and styrene (30–50%). The role of styrene is to dissolve the polymers when adding the reinforcement and to take part in the curing process by cross-linking the linear polymers and creating the final reinforced structure or composite. By choosing different raw materials in building the linear UP/VE polymer, different final properties of the composites can be achieved: high mechanical strength, lightweight, excellent corrosion resistance, electrical properties (isolating), good thermal properties, and flame retardant properties.

Publications have shown that the emission of styrene from resins depends on various factors; application process, ambient conditions, types of product, and types of unsaturated polyester resin (Saamanen 1998). The ambient conditions, e.g. increase in the temperature of the resin or the workroom air increased the emission of styrene (Saamanen 1998). During the application of gel coat and (spraying of the gun set up), the spraying technique contributed to the highest amount of styrene evaporation (Saamanen 1998). In addition, the duration of gelling time during the application of conventional or high solid resins and the types of laminate are crucial factors added to the total emission of styrene in terms of total mass (Saamanen 1998).

7.5.6 *Methods of Control*

7.5.6.1 Engineering Control

General ventilation is normally classified into dilution and local ventilation. Localized well-ventilated zones within the workroom are called zonal ventilation which utilise controlled airflows over the moulds that are useful to control styrene exposure during hand lay-up moulding. The zones are installed with auxiliary fans that have been shown to create a localized airflow and remove styrene vapours from the workers' breathing zone. This kind of ventilation system carried away styrene vapour with the airflow from the wet surface towards the outlet. Systems utilizing vertical local air supply over the lamination zone is called "air shower" and were found to reduce styrene exposure by 73–100% during the lamination of small and medium-size products. Local exhaust ventilation extracts vapours close to the source and operates effectively when the area of the contaminant source is small and the work process is stationary.

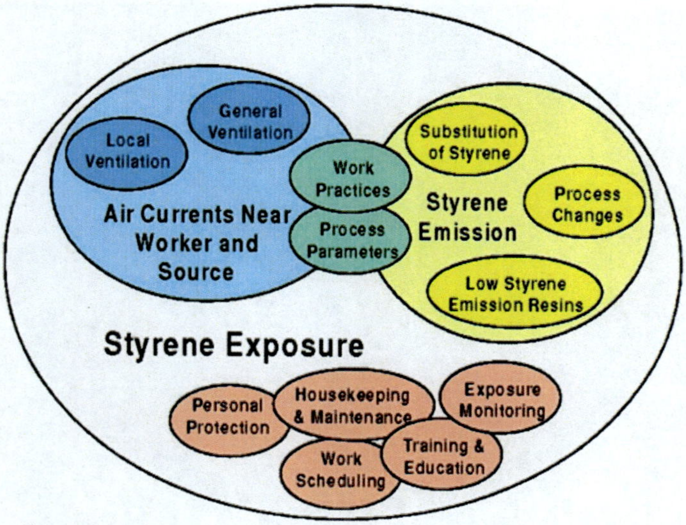

Fig. 7.13 Methods of control of styrene exposure during the moulding of unsaturated polyesters (Saamanen 1998)

7.5.6.2 Training

Training of workers in terms of correct work practices and housekeeping has been shown to reduce styrene exposure markedly (Hopkins et al. 1986). Proper work behaviors include keeping breathing zones away from sources of styrene, working upwind of the sources, as well as enforcing engineering controls. Figure 7.13 shows the methods of control of styrene exposure during the moulding of unsaturated polyesters.

7.5.6.3 Monomer Free Pultrusion Resin System

Nava et al. (2012) described a monomer-free resin system that provides high reactivity and excellent physical properties for pultrusion applications with the approaches to eliminate the problems of monomer emissions during the process and provide similar or better physical properties compared to conventional resins. In addition, this system is designed for pultrusion and especially for close moulding applications that demand enhanced mechanical properties without the compromise of VOCs that also combine features required for processing, including low viscosity, excellent fibre wetting, and high reactivity (Nava et al. 2012). Regulation has been established by the Environmental Protection Agency (EPA) and the Occupational Safety and Health Act (OSHA) for unsaturated polyesters and vinyl esters including unsaturated polyesters, vinyl esters, epoxy, polyurethanes, and phenolic resins (Nava et al. 2012). They stated that the implementation and reduction of volatile organic compounds (VOCs) and

Fig. 7.14 Safe handling guides (The European UP/VE Resin Association 2018)

hazardous air pollutants (HAPS) have been the main focus of the regulation, thus, initiatives and methods have been carried out. These include styrene suppressants, thermal oxidizers, redesigning equipment with improved engineering controls, resins with low styrene content, and replacement of styrene by other reactive diluents with the aim to produce fewer emissions during curing to comply with the EPA and OSHA.

Therefore, under these new legislations, companies need to comply or will be forced to reduce their production or shut their plants down. These have triggered the polymer industry to find other strategies to develop technologies that can provide fewer potential hazards to workers in contact with the thermosetting resins (Hildeberto Nava et al. 2012). Figure 7.14 shows the safe handling guides.

To summarize, statistics have exhibited that styrene has been one of the most used and produced polymers from industries for many uses. Studies have proven that exposure to styrene could lead to many occupational diseases and symptoms, e.g. fatigue, headache, and dizziness. Moreover, exposure to styrene has been observed from the common work practices of industrial workers, which are open and closed mouldings. In addition, efforts to overcome the problems were initiated and established via innovative approaches to develop technologies and enforce the laws to help the workers from exposure to styrene emission during their working hours.

7.6 Safety Training in Handling Composite Products

7.6.1 Overview

Nowadays, composites are widely used in many industries, e.g. aerospace, automotive, and marine industries. However, dealing with composites might give adverse health effects from inappropriate handling of the composites. So, in order to avoid any bad circumstance during the handling of the composites, the employees are required to attend the training depending on the industries they are working at.

7.6.2 Nanosafety Training

In composites used industries, nanosafety training should be made compulsory to the employees. The employees who work as manufacturers, researchers, and developmental employees are advised not to have only skills, but a proper education about composites too. Based on the Industry Code of Practices for Occupational Health and Safety in the Australian Composites Industry, new employees are required to be informed about the health and safety rules at the workplace and the potential hazards they might be facing (Haynes and Asmatulu 2013). In other words, the workers that are exposed to composites must be well equipped with information on the nanotechnology hazards, in both the short-term and long-term. Also, when looking at the packaging of the materials they are working with, the employees should know how to identify hazardous materials so they could be aware of the potential hazards before opening them. Protection methods for handling potentially hazardous engineered nanomaterials in laboratory conditions are as shown in Fig. 7.15.

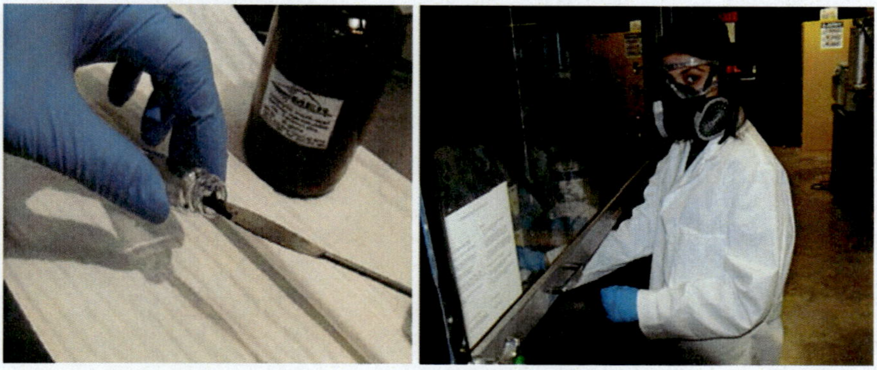

Fig. 7.15 Protection methods for handling potentially hazardous engineered nanomaterials in laboratory conditions (Haynes and Asmatulu 2013)

Companies often give their workers additional training through intracompany programs as they are concern about proprietary data even though the workers already underwent general training on composites at their technical college or other academic and training institutions (Haynes and Asmatulu 2013). Not only in the production sectors, but the employees involved in composite design also require training for them to create parts that require minimal processing and machining. Through this approach, the volume of nanomaterials released and exposure time could be minimized (Haynes and Asmatulu 2013).

7.6.3 Carbon Fibre Training

Not only in nanomaterials, but there is also carbon fibre training that explains the uses of carbon fibre can then be combined with a plastic polymer (resin) to produce very light, strong, durable, and aesthetically pleasing parts (IYRS Composites Technology Training 2018). Carbon fibre is not categorised as hazardous materials, but the secondary component in most carbon composite parts is plastic which usually contains epoxy or resin matrix. Based on Material Safety Data Sheet, the epoxy resin matrix is one of the hazardous materials that pose danger through inhalation of uncured resin (Inc. 2013). So, as precautionary steps, the employees are required to wear well-fitting high-efficiency dust protection. In order to make effective use of personal protection equipment, the employees are required to go for training on correct ways of wearing PPEs in their daily work. This training also benefits in spreading awareness among the workers as they should prioritise their health conditions in both short- and long-terms durations. The training must be conducted by the employers themselves depending on what hazards are associated with their employees' workplace. Table 7.2 shows the industries with composite and carbon fibre training programs.

Table 7.2 Industries with composite and carbon fibre trainings (IYRS Composites Technology Training 2018)

Industries	Specifications
Performance vehicle	Automotive, sports/racing, off-road and energy efficiency
Aerospace and aviation	Aircraft components, satellites, helicopters, drones
Sporting	Bicycles, canoes, paddles, snowboards and skis, surfboards, football helmets, shoes and cleats, hockey sticks
Green/solar energy	Wind turbine and wind turbine blades and support structures
Boating and marine	Boats, keels, rudders, masts and rigging
Instruments	Cellos, violins, guitars
Medical	Operating tables and surgical instruments
Transportation	Car and truck bodies and fairings

Fig. 7.16 Composite repair course (Marsh 2013)

7.6.4 Hands-On Practice

Next, the hands-on practice should be a key element of technician training. When repairing materials, technical skills require a lot of attention, as even the slightest mistake could lead to a huge impact. So, the hands-on practice is helping technician trainees, since they learn under close guidance from expert instructors. Moreover, this practice will soon help the trainees to find new solutions faced with the new generation of composites used in aerostructures. Composites repair courses have a strong focus on practical work, as shown in Fig. 7.16.

The training is exposing the trainees not only to handle epoxy resins, but it also includes many types of other materials, e.g. aramid fabrics and prepregs. For British Airways, the company itself conducts composites training for their employees (Marsh 2013). Generally, the company prepares line and base maintenance technicians to repair not the worst damage but only the slight to moderate damages within their expertise that are included in the scope of the Structural Repair Manual.

Generally, there are many types of training in handling composite products that differ by the industries. Safety must always be the first concern in exploring new technology, and all responsible organizations must carefully consider the necessary precautions for proper nanotechnology use that include training programs. Investment for the future depends on the productivity of their employees, so training by hands-on practice could make them more experienced, creative, and innovative when dealing with problems during handling composite products. Also, carbon fibre training can be used for those working in a variety of composites producing industries.

7.7 Conclusions

In conclusion, numerous types of fibres and resins are used in the industries. The workers who work in reinforcement fibres factories must understand and aware that they are exposed to respirable particles that formed during the process of glass microfibre production for maintaining a safe work environment. Processing of the fibres also has many safety issues that must be concerned by the employer and employees. In the composites industry, material handling is a large source of styrene emission which contributes to hazards. One of the common types of fibres and resins is styrene that causes many implications for employee's health. Nowadays, many technologies are able to reduce the emission of styrene from the workers and the established regulations ensure the employers to have the initiative to reduce the emission of styrene. To maintain the safety and health of workers and the environment during handling the composites, the employers need to be informed about the hazards contained in each work process and hazards management based on the hierarchy of controls. The workers also need to properly use the personal protective equipment (PPE) provided by the employer. For proper and effective handling of the composites by the employees, they need to be adequately trained via the training programs. This training can either be organized by the company itself or by other consultant companies. Therefore, there are many ways to maintain the safety, health, and welfare of the composites handling workers.

References

Abbate C, Giorgianni C, Brecciaroli R, Giacobbe G, Costa C, Cavallari V, Albiero F, Catania S, Tringali MA, Martino LB, Abbate S (2006) Changes Induced by exposure of the human lung to glass fibre—reinforced plastic. Environ Health Perspect 114(11):1725–1729. https://doi.org/10.1289/ehp.8676

Ambrogio G, Conte R, Gagliardi F, De Napoli L, Filice L, Russo P (2018) A new approach for forming polymeric composite structures. Compos Struct 204:445–453. https://doi.org/10.1016/j.compstruct.2018.07.106

Balagna C, Irfan M, Perero S, Miola M, Maina G, Santella D, Simone A (2018) Characterization of antibacterial silver nanocluster/silica composite coating on high performance Kevlar® textile

Colony Distribution Inc. (2018) What are the different types of safety coveralls? Retrieved from: https://colonydistribution.ca/different-types-of-safety-coveralls/

Composite UK Trade Association (2018) Introduction to composite material. Retrieved from: https://compositesuk.co.uk/composite-materials/introduction

Dong S, Li H, Yin Q (2018) Building information modeling in combination with real time location systems and sensors for safety performance enhancement

Haynes H, Asmatulu R (2013) Nanotechnology safety in the aerospace industry. Nanotech Safety, 85–97https://doi.org/10.1016/B978-0-444-59438-9.00007-2

Hoeflinger W, Laminger T (2017) PM2.5 or respirable dust measurement and their use for assessment of dust separators. J Taiwan Inst Chem Eng. https://doi.org/10.1016/j.jtice.2017.07.035

Inc., W. S. (2013). Material Safety Data Sheet Bisphenol A epoxy, pp 1–7

IYRS Composites Technology Training (2018) Carbon fiber training. Retrived from: https://iyrs.edu/composites-technology-program/carbon-fiber-training/

Jagatheesan K, Ramasamy A, Das A, Basu A (2018) Electromagnetic shielding behaviour of conductive filler composites and conductive fabrics—a review. Indian J Fibre Text Res 39(2014):329

Joshi A, Datar S (2018) Carbon nanostructure composite for electromagnetic interference shielding. Pramana 84(6):1099–1116 (2015)

Learning Center—Safety in Composites (2018) Retrieved from: https://www.fibreglast.com/product/safety-in-composites/Learning_Center

Li Z, Ding S, Yu X, Han B, Ou J (2018) Multifunctional cementitious composites modified with nano-titanium dioxide: a review. Compos A. https://doi.org/10.1016/j.compositesa.2018.05.019

Majchrzycka K, Okrasa M, Szulc J, Gutarowska B (2018) The impact of dust in filter materials of respiratory protective devices on the microorganisms viability

Marsh G (2013) Training composite aircraft repairers. Reinf Plast 57(6):30–34. https://doi.org/10.1016/S0034-3617(13)70187-5

Mcclain MS, Gunduz IE, Son SF (2018) Additive manufacturing of ammonium perchlorate composite propellant with high solids loadings. Proc Combust Inst 000:1–8. https://doi.org/10.1016/j.proci.2018.05.052

Nava H, Douglass N, Reichhold YL (2012) Monomer free pultrusion resin system. American Composites Manufacturers Association February 21–23, 2012, Las Vegas, Nevada, USA

Omran FE (2018) Occupational health problems associated with the fibreglass reinforced plastic. Curr Allergy Clin Immun 31(1):32–38

OSHA Bulletin | General Respiratory Protection Guidance for Employers and Workers | Occupational Safety and Health Administration (2018) Retrieved from: https://www.osha.gov/dts/shib/respiratory_protection_bulletin_2011.html

Parandoush P, Lin D (2017) A review on additive manufacturing of polymer-fiber composites. Compos Struct 182:36–53. https://doi.org/10.1016/j.compstruct.2017.08.088

Saamanen A (1998) Methods to control styrene exposure in the reinforced plastic industry. Technical Research Centre of Finland. VTT Publications 354, 83 p +app. 63 p

Sankaran S, Deshmukh K, Ahamed MB, Pasha SKK (2018) Recent advances in electromagnetic interference shielding properties of metal and carbon filler reinforced flexible polymer composites : a review. Compos A. https://doi.org/10.1016/j.compositesa.2018.08.006

Sripaiboonkij P, Sripaiboonkij N, Phanprasit W, Jaakkola MS (2009) Respiratory and skin health among glass microfiber production workers : a cross-sectional study. Environ Health 10:1–10. https://doi.org/10.1186/1476-069X-8-36

Sumesh M, Alengaram U, Jumaat M, Mo K, Alnahhal M (2018) Incorporation of nano-materials in cement composite and geopolymer based paste and mortar—a review

Tranfo G, Gherardi M, Paci E, Gatto M, Gordiani A, Caporossi L, Capanna S, Sisto R, Papaleo B, Fiumalbi C, Garofani P (2012) Occupational exposure to styrene in the fibreglass reinforced plastic industry: comparison between two different manufacturing processes

Upresins.org (2018) Safe handling guides—the European UP/VE Resin Association. [online] Available at: http://www.upresins.org/safe-handling-guides/. Accessed 4 Dec 2018

Chapter 8
Fire Safety in Polymers Composites

Abstract The applications of polymeric materials have been widely used to meet the usage and requirements of conventional fire retardants. However, the presence of halogen-based and phosphorus compounds generates corrosive and toxic combustion products. The compounds, which are organic pollutants, have been persistent global environmental concerns as they are harmful to humans, living things, and the environment. The window of options to combat the issue has become narrower as the incorporation of non-toxic nanofillers shows the positive potential towards flame retardants. Next, this review paper consists of qualitative methods and results but, arguments about short-term and long-term effects still exist. Thus, it is crucial to fundamentally understand the fire responses which comprise the experimental results with theoretical modeling simulation. The eco-friendly agents used in conventional fire retardants will be highlighted as they cause so much impact on the ecological system. The use of various types of nanofillers was also explored and their performance was simultaneously compared with traditional systems, which creates an insight into different testing standards and combustion mechanisms.

Keywords Flame retardancy · Nanofillers · Pyrolysis · Decomposition temperature

8.1 Introduction

The interaction between humans and the environment could come in many ways because they are multi-faceted. According to research, this can cause direct impacts such as air pollution, poor ambient air, and insufficient sanitation. WHO reported 13 million deaths every year from environmental pollutions. The advancement of technology has added some seriousness to this matter with many direct or indirect effects. This review aims to show the global widespread usage of petrochemical-based polymeric materials. The two main objectives of this research are to emphasize the significance of switching to eco-friendly materials and to know the real properties or characteristics of the nanofillers. It is essential to understand the mechanism of fire retardants to adopt green and environmentally friendly materials.

S. M. Sapuan et al., *Safety and Health in Composite Industry*,
Composites Science and Technology ,
https://doi.org/10.1007/978-981-16-6136-5_8

8.2 Principles of Flame Resistance in Polymers: Conventional Flame Retardants (FR) and Their Mechanisms

The incineration of polymeric resources can be a complex technique concerning combinations of synchronous transmission/diffusion of heat and mass, fluid dynamics, and deprivation chemistry. In common, four core steps distress the incineration of compounds: ignition, combustion, pyrolysis, and feedback. In the existence of an acceptable amount of heat (to induce the splitting of the bonds), the polymers can decay or "pyrolyze" the combustible fuels (Lewin and Weil 2001). Figure 8.1 is exothermic reaction of pyrolysis process.

Ignition occurs by flash or automatic ignition that relies on the material flammability limit. After ignition, if the fire releases heat, an independent burning phase will be formed. It is adequate to preserve the polymer disintegration rate advanced than that compulsory to maintain the concentration of unstable fuel compounds (Price et al. 2001). Nevertheless, several difficulties are carried on in every mentioned appliance that creates, ample protection besides the hearth is not probable, though, the utmost mutual method is to extend the flame retardant (FR) in demand to increase the flame resistance's performance of a polymer. They act with chemicals and/or physically inside different matters and hinder burning in completely different phases (ignition, heating, decomposition, or flame development) having the type/nature of the FR. This can be reflected in the fire behavior of the fabric assessed in terms of resistance to ignition, flame propagation, time to escape, and containment of the flame and fuel degradation product, etc. as mentioned in the past study Zaikov and Lomakin (2002) and, therefore, to avoid duplication.

8.2.1 *Halogen-Based Agents*

In common, halogenated FR, Cl, or Br are usually not strong and capable of changing or destruction when subjected to heat, carbon-bromine/chlorine as shown in Fig. 8.2.

$$H^* + O_2 \rightarrow OH^* + O^*$$

$$O^* + H_2 \rightarrow OH^* + H^*$$

The main exothermic reaction that provides most of the energy to maintain the flame is:

$$OH^* + CO \rightarrow CO_2 + H^*$$

Fig. 8.1 Exothermic reaction (Dasari et al. 2013)

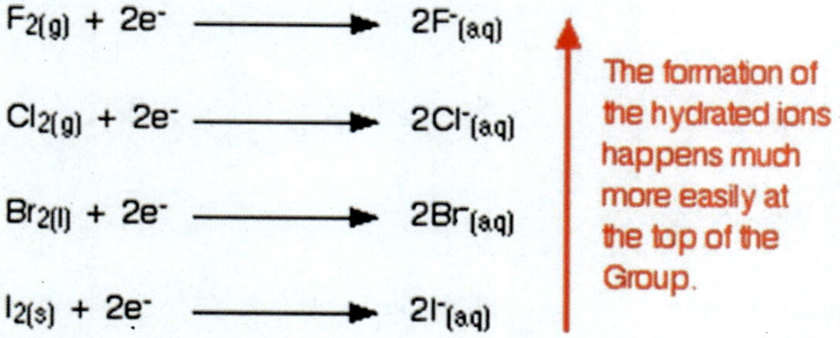

Fig. 8.2 Halogen-based (Bocio et al. 2003)

Though, tremendously stable (heat) radical fluorine-based compounds release at normally performed polymer decomposition temperatures; on the opposing, the process temperatures are unsteady in iodine-based compounds as in most polymers. Specifically, deca-BDE, HBCD, octa-BDE, Penta-BDE, and TBBPA are the five reinforced compounds of FR (or FB) that are widely used for their outstanding performance. Approximately 97% of decabromodiphenyl ether make up Deca BDE; octa-BDE contains from 10 to 12% of hexabromodifeniléteres, 31 to 35% octabromodiphenyl ethers, and 43 and 44% heptabromodiphenyl ether; and Penta-BDE, ~ 50–62% of pentabromodiphenyl ether and 24–38% of tetra-ethers bromodiphenyl (Dasari et al. 2013). Nonetheless, the use of some FBs is restricted due to their essential toxicity (e.g. octa-BDE and Penta-BDE), and the potential to produce polybrominated dioxins and furans (e.g. Bromo-dibenzodioxins, chlorine, and dibenzofurans) has severe affect to the ecosystem. They are capable of releasing a huge sum of eroding gases (hydrogen halide), corrosive metal components, and destruction of sensitive electronic mechanisms, and more terrible impacts in narrow spaces, e.g. a hulled marine hull or a usual aircraft fuselage.

8.2.2 *Phosphorous-Based Compounds*

The most marketed and accessible non-halogenated substitutes are composed of reinforced elements. These compounds function to neutralize the condensed pyrolytic pathway of the polymeric material and decrease the number of gaseous fuels (dehydration and carbon formation are the main modes of action). Mutual examples of this category include FR red elementary phosphorus, phosphine oxides, phosphine, phosphonium compounds, phosphates, phosphinates, phosphites, and phosphonates. They alter the state of the dehydration reaction of compound end chains (and likely reactive groups) and stimulate the formation of slag. It could lead to bioaccumulation and response with other gases, creating an extra type of gas that is dangerous to the environment. The discharged water dilutes the oxidizing gas phase. In other cases, the FR-based phosphorus volatilized in the gaseous phase creates active radicals ($pO^{2\bullet}$, $^{\bullet}PO$ and $^{\bullet}HPO$), and act as a scavenger of radicals $H^{\bullet}$ and $OH^{\bullet}$ (Levchik and

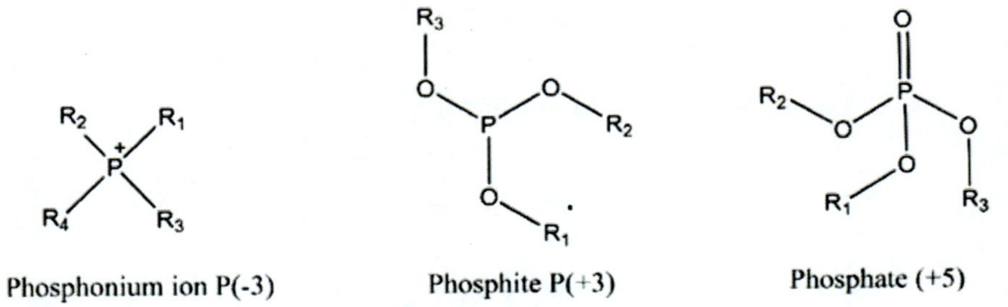

Fig. 8.3 Type of phosphate (Gilman 1999)

Weil 2006). The early researchers on the phosphorus toxicity compounds have recognized these neurotoxicants as FR as they decompose in the environment. Still, after volatilization and condensation, stress corrosion phenomena of various components will occur. Figure 8.3 shows types of phosphate.

8.2.3 Metal Hydroxides

Mutual samples of this FR category are MH and ATH. The pyrolysis zone can be cooled when these fillers disintegrate endothermically and discharge free H_2O. Besides, they indicate a dilution result (to some extent) within the gas part, and a protective coating was formed (consisting of Al_2O_3 or MgO) on the incineration surface of the compound once the action mode has been drained of water release. Still, when betting on their disintegration temperatures for heat absorption, composite systems with appropriate decomposition temperatures should be selected (Zaikov and Lomakin 1998). For these systems, MH with a heat absorption decomposition temperature > 300 °C is the most suitable. Furthermore, the boehmite chemistry supported with a newly introduced flame agent compound (aluminum monohydrate, AlOOH) is also appropriate when the process temperature exceeds 200 °C since its heat absorption decomposition temperature is 340–350 °C (Laachachi et al. 2009). However, a known major disadvantage is their high demand for load levels of (generally > 40% by weight) for adequate flame resistance, which often causes process difficulties and deterioration of the flame characteristics, alternative crucial compounds.

8.2.4 Boron-Based Flame Retardants

The Zn borates (hydrates) are widely used. Boric acid and a chemical compound (B_2O_3) are released when heat absorption decomposition (~ 503 kJ/kg) is in the temperature range of 290–450 °C. The establishment of a protective glass layer is a result of B_2O_3 softening and flowing at 350 °C and more than 500 °C, respectively.

There are many variations of the Zn salt which differ in the ratio between zinc/boron magnitude and water content. Synergists are mainly used for polymers of halogenated polymers by the interaction of Zn^{2+} ions with halogen radicals. For instance, the acid generated by the PVC reacts with the salt to produce Zn volatile chloride ($ZnCl_2$) and oxychloride, plus boron in boric acid chloride and volatile as shown in Figs. 8.4 and 8.5. The first carbon ion is decomposed into alkene, acid, and zinc chloride. Next, the formulated carbon ions react with the unsaturated PVC equipment to produce coal through a series of crosslinking reactions and ultimately reduce the formation of smoke. In a previous study, borate zinc weighing 0.5–7 by parts ($4ZnO \cdot 6B_2O_3 \cdot 7H_2O$) improved the flame resistance and oxidation of Pu (Yildiz et al. 2009). An ultraviolet absorber had clearly helped PU to increase its oxidative stability in the presence of zinc oxide. In addition, the PU combustion time was significantly improved, even with 0.5% by weight of borate zinc, from 15 s in pure PU to 39 s. The combustion time was increased further to 84 s with 7% zinc borate.

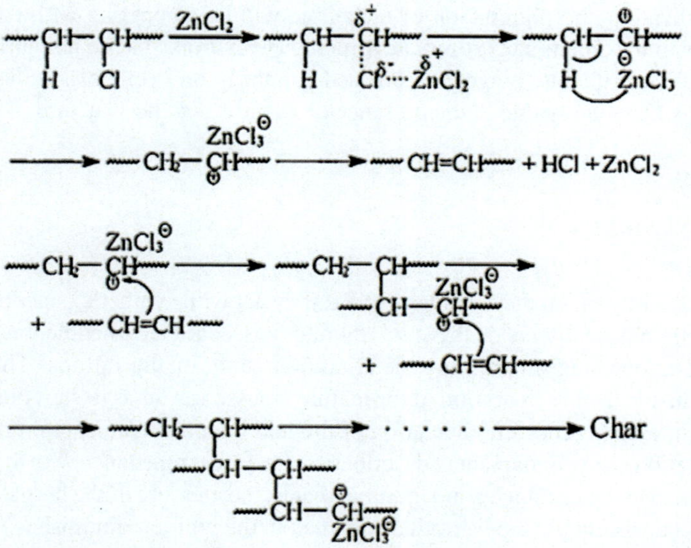

Fig. 8.4 Cross-linking of PVC during its decomposition, leading to the formation of char (Dasari et al. 2013)

$$2ZnO{\cdot}3B_2O_3 + 12HCl \rightarrow Zn(OH)Cl + ZnCl_2 + BCl_3 + 3HBO_2 + 4H_2O$$

Fig. 8.5 The carbonium ions reaction with unsaturated teams in PVC to manufacture char by a series of cross-linking reactions (Bourbigot and Duquesne 2007)

8.2.5 Nitrogen-Based Flame Retardants

The nitrogen-based flame retardants are pretty less poisonous, discharge a small sum of smoke through the fire, and are environmentally friendly despite their low flame retardant power compared to the halogen-based FR. Among the compounds containing nitrogen, the usage of melamine (or its salts, e.g. melamine polyphosphate, cyanurate, melamine phosphate, melamine pyrophosphate, and melamine pyrophosphate) predominates (Horacek and Grabner 1996). Melamine can be an even crystal-like product and comprises 67% by weight of nitrogen atoms. Melamine decomposes at higher temperatures with the removal of ammonia and the dilution of combustible oxygen/gases and thermally stable condensates, melon, melem, and melam are formed. Energy is absorbed in a huge amount during sublimation that takes place at ~ 350 °C, decreasing the temperature. In many polymers, triazines (and their derivatives) are used alone or together with APP as processing or carbonizing agents with positive results (Li et al. 2008). If the volatilization of melamine is hindered, the volatilization and domination of melamine will be competing with those reactions. Melamine and its salts stimulate dripping (with drops that do not burn), unlike standard FRs. Alternative agents are based primarily on cyclohexane that is used together as FR, mainly due to the presence of N and its cyclic structure.

8.2.6 Summary

In summary, relying on the sort/idea of FRs, they act with synthetic concoctions and additionally physically inside the solid, fluid, or gas stages and meddle with burning at totally unique stages (warming, deterioration, start, or fire unfurl). This can be reflected inside the fire execution of the texture assessed as far as to start obstruction, fire spread, escape time, and regulation of fire and flammable/corruption item. This demonstration is, by fundamental, describes as substance impedance with the extreme chain system in the gas stage amid ignition; that is, the measure of the burnable matter that stays consistent yet the warmth discharged in the ignition diminishes. The most broadly promoted and realistic non-halogenated choices are bolstered component mixes. Basic examples of this class of FRs are basic red phosphorus, phosphines, phosphine oxides, phosphonium mixes, phosphonates, phosphites, phosphinates, and phosphates. The carbonium particles molded in PVC respond with unsaturated groups to facilitate the progression of burning of cross-connecting responses and, at last, diminishing smoke arrangement. Despite their lower fire retardant strength than halogen-based FRs, these mixes are relatively less toxic, release low measures of smoke all through the flame, and are ecologically amicable. Thusly, gas specialists like cyanuramide, triazine, carbamide (phosphate), and guanidine-based for the most part mixes are getting continuously standard.

8.3 Environmental Impact of Conventional Flame Retardants

The world demand for traditional halogen and phosphorus-based total flame retardants used to be at 2.2 million metric lots in 2011, but there was a 4.7% rise in the amount from the year before. Unfortunately, the compounds are the contributors to explosive reaction in the environment that create destructive and harmful combustion by-products, e.g. dioxins and furans. The compounds' structures control the poisonous and steadiness qualities despite there are more than two hundred halogenated mixes that exist in the world (D'Silva et al. 2004). They have been seriously contemplated and were classified as human carcinogenic compounds by IARC in 1997 due to the fact that they have the capability to disturb numerous pathways of the endocrine system (Grassman 1997).

8.3.1 *Major Source and Accumulation*

The improper ways of burning or incinerating halogenated-based wastes, e.g. hospital waste and sewage release halogen compounds. They exist as the by-products of some chemical processes that can accumulate in soils, waste disposal sites, or organic matter due to their hydrophobicity. Human beings and other living things might be exposed to this through breathing, eating, or dermal absorption. Since they are lipophilic, the probability of getting these compounds accumulates and becoming resistant to metabolism is higher in invertebrate species. In human beings, they can be observed in blood, serum, adipose tissue, placental tissue, milk, and in the brain.

8.3.2 *Unlawful Disposal of Consumer Products*

In the present day world, textiles, furniture, packaging, and electrical or electronic units are petrochemical-based polymeric substances. With the advanced developments in science and technology, the lifespans of electronic products are getting shorter and significant amounts are being discarded internationally every year. According to the Environmental Protection Agency (EPA), an estimated amount of over 315 million computers were get rid of between the years 1997–2004 in the USA alone (National Safety Council). 140 million mobile phones were identified to be made in 2007 and only 10% of them were recycled. Figure 8.6 shows the E-waste generation and recycling from 2000 to 2013.

This electronic waste dumping contributes to soil contamination and broadly affecting the rivers and farms since the components are manufactured from flame retardants materials; the outer sheaths of cable, connector, and printed circuit

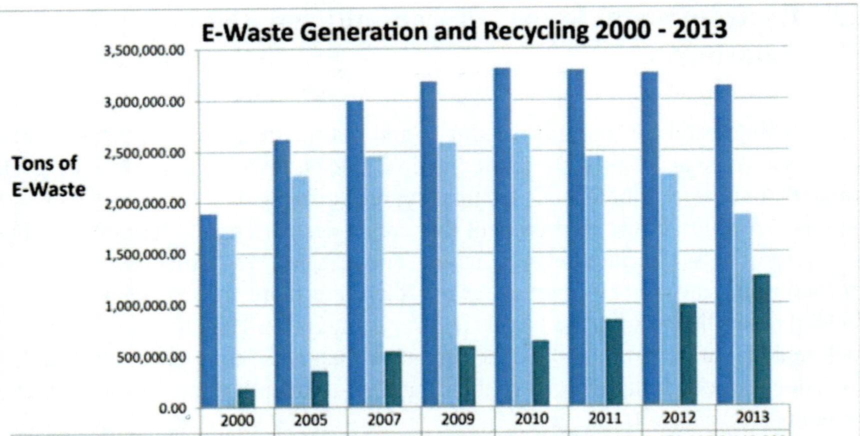

	2000	2005	2007	2009	2010	2011	2012	2013
Total e-waste generated	1,900,000	2,630,000	3,010,000	3,190,000	3,320,000	3,300,000	3,270,000	3,140,000
E-waste trashed	1,710,000	2,270,000	2,460,000	2,590,000	2,670,000	2,450,000	2,270,000	1,870,000
E-waste recycled	190,000	360,000	550,000	600,000	650,000	850,000	1,000,000	1,270,000
Percent Recycled	10.0%	13.7%	18.3%	18.8%	19.6%	25.8%	30.60%	40.40%

Fig. 8.6 E-wastes in landfills generation and recycling (U.SEPA 2015)

boards 7576 (Gregory 2009). Most recycling centres in some countries might be lacking sufficient services in terms of electronic wastes recycling and use crude techniques that harm the environment and human well being.

8.3.3 *Control and Monitoring Efforts*

Most countries have banned or restricted the use of organochlorine or bromine compounds, together with PCBs/BDEs and some chlorinated pesticides. As for Europe, they limit all usage of penta-BDE and octa-BDE. Basel Convention on the Control of Transboundary Movements of Hazardous Wastes and Their Disposal was held to lessen the mobility of hazardous waste into and out of the nations, and specifically to mitigate the switch of hazardous waste from modern to much less modern nations in every other effective pace.

Despite many toxic flame retardants are still extensively used at present, these control measures of examining and inspecting the use of halogenated compounds are some steps in phasing out toxic flame retardants.

8.3.4 *Summary*

Hence, exposure to people or other living creatures might happen by means of a few ways; inward breath, dermal retention, and ingestion (nutrition). Notwithstanding the recyclability of polymers, the present foundation is lacking to oblige the generation transfer rates of different sorts of blended polymer squanders. Electronic wastes such as FRs have created global concern as they are used in essential electronics parts, e.g. printed circuit sheets, as connectors, in external sheaths, and links. Researches have demonstrated that their ill-advised dumping/burning resulted in contamination of soils and the environment. Moreover, the disposal fee of e-waste to various nations, especially, East and South Asia alongside Africa has prompted the improvement of new sources of ecological polybrominated diphenyl ether. For instance, 60,000 tons of second-hand e-wastes were yearly imported by Nigeria, which consist of 18,000 tons of plastics. These preparatory methods for checking and controlling the creation/utilization of halogenated mixes are certainly a stage towards eliminating dangerous fire retardants. This electronic waste dumping contributes to soil and environmental pollutions, e.g. in rivers and farms since the components are formed from flame-retardant materials, for instance, in the outer sheaths of cable, connector, and printed circuit boards 7576 (Gregory 2009). Most of the recycling centres in some countries are lacking in the appropriate approaches for recycling-waste.

8.4 Nanofillers as Fire Retardants in Polymers

In order to increase fire safety, the utilization of high non-combustibility polymer materials in construction, cars, aircraft, ships, and electrical devices are being used. Nanofillers are the fire retardant additives in polymers that are used to reduce heat release and mass loss rates. Utilizing nanofiller added substances is a probable method for improving fire retardancy and diminishing the feast of fire of polymeric constituents amid flame. The same study used a cone calorimeter to identify the virtual execution of fire-resistant added substances and the flame response properties of the polymer.

8.4.1 Layered Silicates

The study of flame retardancy was conducted via a cone calorimeter experiment to give various sorts of important data and demonstrates fire retardancy. The sample in cone calorimeter basically utilizes the heat to ignite the sample at 35 or 50 kW/m^2. Cone calorimeter experiments yield data on numerous properties including total heat released, mass loss rate, time to ignition, heat release rate, and peak of heat release. The heat release rate and mass loss rate are principally utilized as the most significant indicators of fire retardancy. A study on layered silicates as nanofillers in the polymer was conducted by Dasari (2013) who reported a major problem to the formation of a thick and continuous barrier and the soil platelets when the surface was burnt. Aside from this, the heat release rate values were being significantly reduced. After the structural collapse, different opacity of the samples were producing the intermittent and

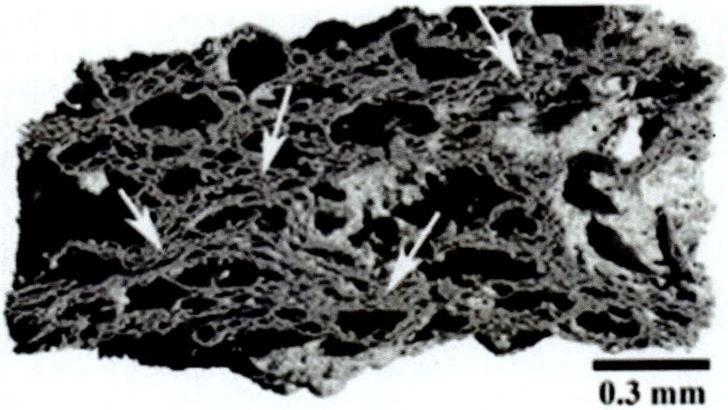

Fig. 8.7 PLA/silicate nanocomposite residue under the X-ray tomogram (Gonzalez et al. 2012)

thick inorganic barriers. This was observed under the volume imaging approach as shown in Fig. 8.7 (Gonzalez et al. 2012). Due to the intermittent carbonaceous-silicate barriers, the nanocomposites flame retardancy was increased (retard in burning and the HRRs/MLRs of nanocomposites were decreased by more than 50% compared to neat polymer). Hence, it constrained the hot temperature, volatile, or oxygen in navigating the convoluted way in the polymer grid encompassing hindrances. This had significantly expanded the constructive way length for dissemination.

Figures 8.8, 8.9, 8.10 and 8.11 show the result from cone calorimeter experiments. In the test, the volume of the organoclay did not appear to assume the noteworthy part

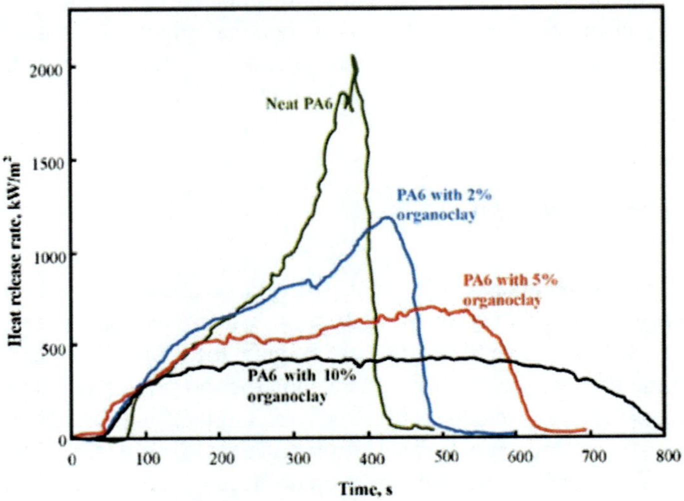

Fig. 8.8 Heat release rate curves showing the effect Polyamide, PA 6 nanocomposites with 2, 5, and 10% organoclay according to the organoclay loading rate at 50 kW/m^2 (Dasari et al. 2007)

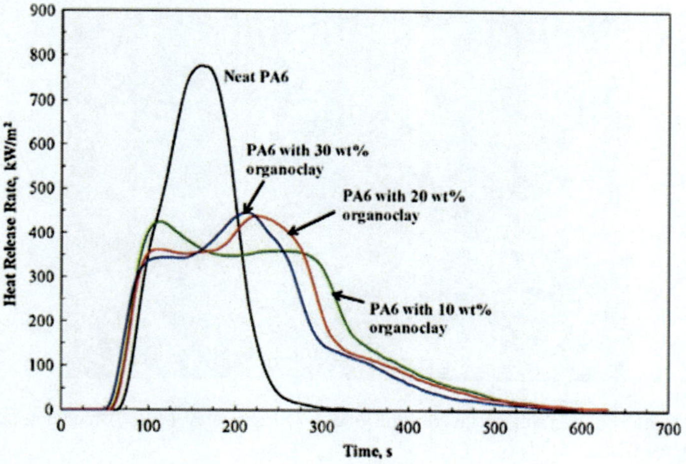

Fig. 8.9 Heat release rate curves showing the effect Polyamide, PA 6 nanocomposites with 10, 20, and 30% of organoclay according to the organoclay loading rate at 50 kW/m^2 (Bourbigot et al. 2006)

Fig. 8.10 Digital photograph of the residue of PA 6/organoclay (90/10) nanocomposites after combustion test (Dasari et al. 2007)

whenever the heat release rates were decreasing, as displayed in Figs. 8.8 and 8.9. This was regardless of the residue's surface, numerous discrete island-like structures were perceived at 10 wt%, as exhibited in Fig. 8.10. For bigger volumes, the residues in all accounts were seemed to become ceaseless and solid with the least fine cracks identified in Fig. 8.11. Physically, the results recommended the development and arrangement of adequate intermittent structures as stated above, to be conceivable as one of the causes. Those factors are, for example, the burning was mixed up and inconsistent, and the surface had huge cracks that were moderately insignificant in lessening the heat release rates. In any case, the estimated relative rates were decreased for about 40, 65, and 77.5% for 2, 5, and 10 wt% of organoclay in Fig. 8.8

Fig. 8.11 Digital photograph of the residue of PA 6/organoclay (70/30) nanocomposites after combustion test (Dasari et al. 2007)

and 45% of each amount of 10, 20, and 30 wt% organoclay in Fig. 8.9. It is an obvious sign that the decreasing heat release rate relied upon different causes, for instance, the matrix viscosity. Polyamide 6 utilized in Fig. 8.8 possessed a low consistency review, meanwhile, in Fig. 8.2, a high consistency review was observed. Hence, the physical crumple and development of discontinuous hindrances happen quicker in low consistency review contrary with high consistency review material offering bigger changes (Dasari et al. 2013).

8.4.2 *Needle-Like Sepiolite*

One of the nanofillers fire retardants utilized is needle-like sepiolite with open nanotunnels. Sepiolite is a hydrated magnesium silicate that is characterised by a needle-like shape (Grim 1962). Usually, the platelet-like filler has more effective diffusion than fibrous fillers (Billoti et al. 2009). Despite the fact that molecules conglomeration relies upon various reasons; processing conditions, compound structure, size, and shape, the platelet-like fillers have a bigger exterior extent than fibre-like fillers. A study on sepiolite (Billoti et al. 2009) stated that some health problems had arisen associated with the nanoparticles usage. Hence, it is essential to study how sepiolite harms humans health. Nevertheless, the in vitro, in vivo, and epidemiological studies related to sepiolite reported that no health problems were found linked to sepiolite. In case of health issues from sepiolite, they might be caused by inhalation. Studies demonstrated that unbent fibres with a length of more than 8 μm and thickness of less than 0.25 μm are possible to cause health problems, by bio-persisting in biological tissues. Figure 8.12 shows the crystal structure of sepiolite.

The HRRs values become significant when sepiolite consolidates with magnesium hydroxide. The sepiolite is not creating a significant drop in heat release rate values. Despite that, Table 8.1 has shown that a significant decrease of time-to-ignition with

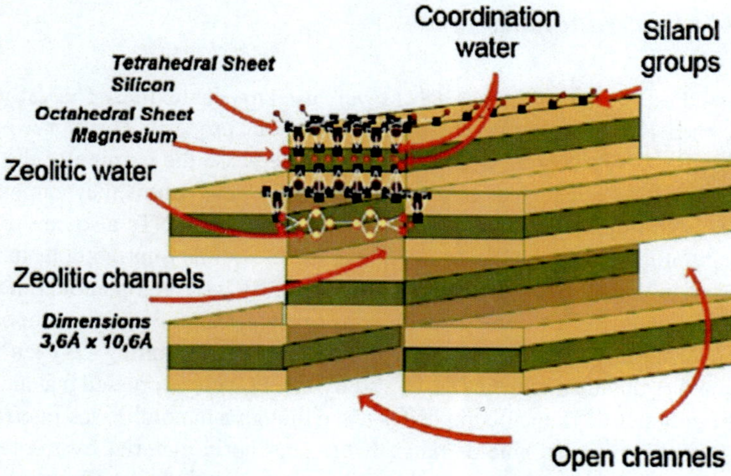

Fig. 8.12 Crystal structure of sepiolite

Table 8.1 Collected data of the cone calorimeter of polypropylene and the mixtures (Marosfoi et al. 2008)

Material	Time-to ignition, s	Peak HRR, kW/m^2	Time to reach peak HRR, s	Total heat released, MJ/m^2
Neat PP	37	584	167	75.6
PP/Mg(OH)2 (90/10)	33	471	158	65.9
PP/sepiolite (95/5)	24	533	111	68.1
PP/organo-sepiolite (95/5)	23	515	124	66.1
PP/organo-sepiolite/Mg(OH)2 (85/5/10)	29	328	202	62.1
PP/organo-sepiolite/organoclay (95/2.5/2.5)	62	417	213	63.7
PP/Mg(OH)2/organo-sepiolite/organoclay (85/10/2.5/2.5)	54	246	294	56.3

sepiolite (Marosfoi et al. 2008). This happened due to the inter-channel networks that hold the hot temperature and utilize it as a source to accelerate the decay from the outside heat. When mixing with organoclay, the pinnacle heat-releasing rate value was further diminished as shown in Table 8.1. This was due to the structure of bonds between nano and micro-scale particles in sepiolite (Dasari et al. 2013).

8.4.3 Carbon Nanotubes

In recent studies, carbon nanotubes, CNTs were used to coat textiles. This effort was to create a barrier layer on the composite for fire retardant property. This can be proved by studying the CNTs coating which was able to decrease the burning of the foam by 35%. Among the advantages of using CNTs as fire retardant is they can be used in lower concentrations compared to the other nanofillers. CNTs also can provide fire-resistant properties to the material while in the meantime improving its material quality or giving an electrostatic release impact. CNTs have comparable outcomes of diminished heat release rates and delayed blazing, relying on the proportions on fillers, yet negative Underwriters Laboratories 94 and Limiting Oxygen Index information was obtained with CNTs. In a study, CNTs were expressed that amid the ignition of polymer/CNT nanocomposites, even though a nanotube was interrelated to linkage coating created shields beneath the polymeric material by the heating transition. For several circumstances, the polymer was totally burnt, deserting just a layer of CNTs. Moreover, the excess amount was almost as the initial amount of the CNTs in the original nanocomposites. This situation showed that the development of char was not affected by the network layer (Dasari et al. 2013). Figure 8.13 shows CNT-based char layer structure amid ignition of a CNT-altered coating.

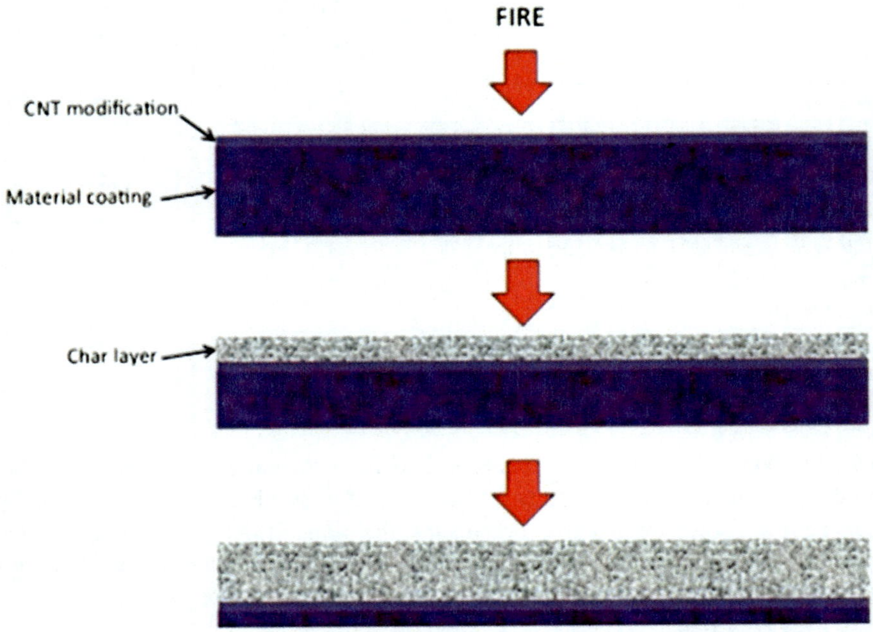

Fig. 8.13 CNT-based char layer structure amid ignition of a CNT-altered coating (Kashiwagi et al. 2005)

8.4.4 Summary

There are various types of nanofillers studied by researchers. These included layered silicates, needle-like sepiolite, and carbon nanotubes (CNTs). Nanofillers are additives added into composites that function as fire retardants. Based on the literature, nanofillers had been proven to improve mechanical, thermal, and electrical properties. Nonetheless, nanofillers themselves were unable to demonstrate excellent fire retardancy such as self-extinguish properties. Nanofillers need to be mixed with other substances to minimize the heat release rate in some amount of loadings, as explained above. Hence, only some of the nanofillers mixture possessed positive effects as fire retardants. In some combinations, negative effects were obtained.

8.5 Issues with Fire Performance of Polymer Nanocomposites

8.5.1 Thermal Stability

Improving the thermal stability of polymer nanocomposites does not come with the assurance that common fire reduces the speed of execution. This is to improve the starting decay temperature that can affect the start by changing the mud surface with the normal surface. This is done by replacing it using hydrophobic cations to upgrade its comparability with polymers. In the perspective of the cation-exchange limit of mud, the alkylammonium surfactant content is normally more than 30 wt.% (Fina and Camino 2012). Nevertheless, these blends are thermally inconsistent and rot, generally from 180 °C. Most parts of fabricating polymers require a temperature of significantly higher than 180 °C. This leads to a warm breaking down of the normal salts that are unavoidable and ominously impacts the warm quality of the nanocomposites.

There are many ways to create elective routes to transfer nanocomposites without using ordinary alkyl ammonium surfactants. The methods include:

(I) in part traded frameworks that diminish the measure of surfactant required (Auad et al. 2007)

(II) sol–gel innovation comprising of direct crystallization of naturally changed layered silicates by aqueous treatment with a gel containing organics and organometallics (Olewnik et al. 2012)

(III) utilization of thermally stable mixes dependent on imidazolium, phosphonium, pyridinium, and iminium to expand the underlying deterioration temperature (Olewnik et al. 2012)

(IV) water-helped approach (Gilman et al. 2000)

(V) sans surfactant strategy (Gilman et al. 2000).

8.5.2 *Time-to-Ignition and Thickness of Samples*

For polymers that are non-charring, in radiative warming, the time-to-ignition (TTI) was based on the imposed heat flux. A few examples corresponding to TTI and warmth motion have likewise been expected to accepting no adjustments in the physical of the material preceding start. Because of the absence of contemplations, for example, heating of polymer can decompose the polymer and others that incorporate surface changes and absorb/emit properties influencing heat flux that are reliably vulnerable to the anticipation of TTI. Its validity for polymer nanocomposites, which is the time to ignite, is completely unpredictable and has variable patterns from the observed neat polymers. In any event, a subjective start was observed for nanocomposites in the vast majority of tests on polymer/clay nanocomposites compared to their neat polymers.

The thermal instability of nanoparticles is one of the imaginable aims on early starting. In addition, it was assumed in a study that clay nanoparticles assemble the oxidation of gasses caused by the dissolution of the polymer at an early start (Fina and Camino 2012). The surface temperature was estimated by contacting the upper surface of the example throughout the burning test. After the process of melting was done, an increment in temperature was distinguished to another level at around 400 °C.

8.5.3 *Migration of Nanoparticles*

Regardless of the idea of nano restriction, the basic crumple and movement of nanoparticles to the consuming surface is another problem affecting the consistency and strength of the shaped nanoparticles/carbon structure of the system. There are many key thrusts for nanoparticles, such as temperature-inclined convection power, liquefied consistency slopes (from the consuming surface to the base), and lower surface-free mud layer vitality than carbon-based polymers. Besides that, a rising air pockets hypothesis was studied that occurred when the surface temperature of the polymer turned out to be sufficiently high and the degradation created pyrolysis elements, which were monomers, direct, cyclic oligomers, and small amounts of gas volatiles (contingent on the polymer). These items were overheated, nucleated, and become softer because the polymer's ignition temperature was much higher than the monomer's bubbling temperature. The air pockets were easily rising and exploded into the vaporous stage as fuel water vapor on the warm surface. In order to improve fire execution, a basic arrangement was to stifle the percolation rate and delay the supply of fuel. During the movement, however, the air pockets also forced the nanoparticles to form a defensive warmth shield (Gilman 1999).

Moreover, regardless of the main thrusts for movement, there were different variables that impact this procedure, for example, interface communication between polymer and clay layers and the temperature of ignition of natural surfactants with regard to the polymer. It should also be noted that only a few of the nanoparticles were moving to the best surface, hence, evidence of the proximity of different mud layers stacked together in the middle and bottom parts of the building was uncovered.

In other words, if the clay structure falls completely before its movement, it is conceivable that these clusters/stacks could not be pushed by rising bubbles and move according to any of the other main thrusts. As mentioned in the first, these irregular and thick inorganic obstructions at various thicknesses help to reduce the pinnacle estimation of HRRs and have a strong impact on the tank.

8.5.4 *Homogeneity/Uniformity of the Barrier*

For the best fire retardancy execution, its defensive boundary has to cover the entire surface of the example consistently to completely shield/secure the polymer evaporation, adequate physical quality (packaging density) should not be broken or exacerbated by bulbing and stay in place throughout the entire burning period.

The defensive boundaries, whether they appear smooth without a major crack, or is constantly absorbent in mass/heat energy are probably due to the proximity of fine cracks. Along these lines, the hazard elements develop to the gas stage through such cracks. The use of a gasification contraction showed that, especially at the beginning of the process, the physical quality of the structure of the dirt molecule organization was insufficient to maintain the development of cracks in the surface layer (Date 2009).

8.6 Conclusion

The use of conventional FRs is very important as they ultimately produce persistent organic pollutants that can be harmful to the environment. Facets regarding these issues are focused on both materials and fire exposure techniques. The investigation of fire retardancy and the limited studies are discussed in this literature. Hence, it is necessary to understand the replacement of conventional polymeric materials. The involvement of biodegradable polymers as alternatives to conventional materials is vital and incorporated in this study. However, lack of infrastructure, limited knowledge and awareness among consumers, and poor performance of the existing biodegradable polymers are among the main challenges faced by scientists. They are now still proliferating their efforts to find the most suitable preferences on achieving the fire safety standards regarding the polymers and keeping the environment safe from pollutants.

Acknowledgements This chapter was written with Eizzat Hilmi, Akhmal Nawwar, Azizi Hilmi, Faiz Afzal, Nufail Naim.

References

Auad ML, Nutt SR, Pettarin V, Frontini PM (2007) Synthesis and properties of epoxy-phenolic clay nanocomposites. Express Polym Lett 1(9):629–639. https://doi.org/10.3144/expresspolymlett.2007.86

Bocio A, Llobet JM, Domingo JL, Corbella J, Teixidó A, Casas C (2003) Polybrominated diphenyl ethers (PBDEs) in foodstuffs: Human exposure through the diet. J Agric Food Chem 51(10):3191–3195. https://doi.org/10.1021/jf0340916

Bourbigot S, Duquesne S (2007) Fire retardant polymers: recent developments and opportunities. J Mater Chem 17(22):2283–2300. https://doi.org/10.1039/b702511d

Bilotti E, Zhang R, Deng H, Quero F, Fischer HR, Peijs T (2009) Sepiolite needle like clay for PA6 nanocomposites: an alternative to layered silicates ? Compos Sci Technol 69(15–16):2587–2595. https://doi.org/10.1016/j.compscitech.2009.07.016

Bourbigot S, Duquesne S, Jama C (2006) Polymer nanocomposites: how to reach low flammability? Macromol Symp 233:180–190

Dasari A, Yu ZZ, Cai GP, Mai YW (2013) Recent developments in the fire retardancy of polymeric materials. Prog Polym Sci 38(9):1357–1387. https://doi.org/10.1016/j.progpolymsci.2013.06.006

Dasari A, Yu ZZ, Mai YW, Liu S (2007) Flame retardancy of highly filled polyamide 6/clay nanocomposites. Nanotechnol 18:445602/1–10

Dasari A, Yu Z-Z, Mai Y-W (n.d.) Polymer nanocomposites: towards multi-functionality. Google Books. Retrieved from https://books.google.com.my/books?isbn=1447168097

D'Silva K, Fernandes A, Rose M (2004) Brominated organic micropollutants—igniting the flame retardant issue. Crit Rev Environ Sci Technol. https://doi.org/10.1080/10643380490430672

Fina A, Cuttica F, Camino G, Time I (2012) Flame Ignition mechanisms in polymer nanocomposites: experimental evidences and interpretation, June, 1–4

Gilman JW (1999) Flammability and thermal stability studies of polymer layered-silicate (clay) nanocomposites. Appl Clay Sci 15(1–2):31–49. https://doi.org/10.1016/j.molstruc.2004.10.036

Gilman JW, Jackson CL, Morgan B, Harris R, Giannelis EP, Wuthenow (2000) Flammability properties of polymer—layered-silicate. Chem Mater 12(5):1866–1873. https://doi.org/10.1021/cm0001760

González A, Dasari A, Herrero B, Plancher E, Santarén J, Esteban A, Lim SH (2012) Fire retardancy behavior of PLA based nanocomposites. Polym Degrad Stab 97:248–256

Grim RE (1962) Applied clay mineralogy. McGraw-Hill, New York

Grassman J (1997) Acquired risk factors and susceptibility to environmental toxicants. Environ Toxicol Pharmacol. https://doi.org/10.1016/S1382-6689(97)10013-8

Gregory MR (2009) Environmental implications of plastic debris in marine settings—entanglement, ingestion, smothering, hangers-on, hitch-hiking and alien invasions. Philos Trans Royal Soc B Biol Sci. https://doi.org/10.1098/rstb.2008.0265

Horacek H, Grabner R (1996) Advantages of flame retardants based on nitrogen compounds. Polym Degrad Stab 54(2–3):205–215. https://doi.org/10.1016/S0141-3910(96)00045-6

Hornsby PR (n.d.) Fire 203:203–214

Kashiwagi T, Du F, Winey KI, Groth KM, Shields JR, Bellayer SP, Kim H (2005) Douglas JF Flammability properties of polymer nanocomposites with single-walled carbon nanotubes: effects of nanotube dispersion and concentration. Polymer 46(2):471–481. https://doi.org/10.1016/j.polymer.2004.10.087

Laachachi A, Ferriol M, Cochez M, Lopez Cuesta JM, Ruch D (2009) A comparison of the role of boehmite (AlOOH) and alumina (Al_2O_3) in the thermal stability and flammability of poly(methyl methacrylate). Polym Degrad Stab 94(9):1373–1378. https://doi.org/10.1016/j.polymdegradstab.2009.05.014

Levchik SV, Weil ED (2006) A review of recent progress in phosphorus-based flame retardants. J Fire Sci 24(5):345–364. https://doi.org/10.1177/0734904106068426

Lewin M, Weil ED (2001) Mechanisms and modes of action in flame retardancy of polymers. Fire Retardant Materials. Woodhead Publishing Ltd. https://doi.org/10.1533/9781855737464.31

Li Y, Li B, Dai J, Jia H, Gao S (2008) Synergistic effects of lanthanum oxide on a novel intumescent flame retardant polypropylene system. Polym Degrad Stab 93(1):9–16. https://doi.org/10.1016/j.polymdegradstab.2007.11.002

Marosfoi BB, Garas S, Bodzay B, Zubonyai F, Marosi G (2008) Flame retardancy study on magnesium hydroxide associated with clays of different morphology in polypropylene matrix. Polym Adv Technol 19:693–700

Olewnik E, Garman K, Piechota G, Czerwiński W (2012) Thermal properties of nanocomposites based on polyethylene and n-heptaquinolinum modified montmorillonite. J Therm Anal Calorim 110(1):479–484. https://doi.org/10.1007/s10973-012-2380-9

Price D, Anthony G, Carty P (2001) Introduction: polymer combustion, condensed phase pyrolysis and smoke formation. Fire Retardant Materials. Woodhead Publishing Ltd. https://doi.org/10.1533/9781855737464.1

USEPA (2015) National overview: facts and figures on materials, wastes and recycling. Retrieved from https://www.epa.gov/facts-and-figures-about-materials-waste-and-recycling/national-overview-facts-and-figures-materials

Yildiz B, Seydibeyoğlu MÖ, Güner FS (2009) Polyurethane-zinc borate composites with high oxidative stability and flame retardancy. Polym Degrad Stab 94(7):1072–1075. https://doi.org/10.1016/j.polymdegradstab.2009.04.006

Zaikov GE, Lomakin SM (1998) Ecological aspects of polymer flame retardation. Oxid Commun 21(2):149–161. https://doi.org/10.1002/vnl.10301

Zaikov GE, Lomakin SM (2002) Ecological issue of polymer flame retardancy. J Appl Polym Sci 86(10):2449–2462. https://doi.org/10.1002/app.10946

Chapter 9
Health Hazard from Composites

Abstract There are increasing concerns regarding the health hazards from composites these days. There are a few types of composites discussed in this study, e.g. carbon fibre, dental composite resin, nanocomposite of Low-density Polyethylene (LDPE), Medium Density Fibreboard (MDF), and concrete compound. Health hazards from carbon fibre include the fibrous dust and airborne particles released during burning that can be inhaled and deposited in the deep lung parts. Next, health hazards in dental composite resin arise due to degradation of the composite that releases (Hydroxyethyl)methacrylate (HEMA) and Triethylene glycol dimethacrylate (TEGDMA) molecules. The third issue is health hazards in composites used for the food packaging industry such as low-density polyethylene (LDPE) nanocomposites. LDPE can leach estrogenic chemicals when in contact with high temperature. According to the International Agency for Research on Cancer (IARC), formaldehyde and wood dusts released from MDF has been classified as Group 1 human carcinogen alongside the non-carcinogenic effects. Last but not least, concrete compounds have been found to cause silicosis in humans through inhalation of the particles.

Keywords Health hazards · Carbon fibres · Dental resin · LDPE · Nanocomposite · MDF · Silicosis

9.1 Introduction

Composite materials are engineering materials having unique properties. Composites are defined as materials consisting of at least two distinct materials combined and results in a new material with different performance as compared to the original constituent materials. These constituent elements are usually reinforcement matrices and fibres which maintain individual properties in the composites.

Nowadays, these composite materials can be been found in use in various applications and products (Sapuan 2017). Examples of industries that use these composites are the healthcare and manufacturing industries.

S. M. Sapuan et al., *Safety and Health in Composite Industry*,
Composites Science and Technology ,
https://doi.org/10.1007/978-981-16-6136-5_9

The main contributing factor in the usage of composites in many different industries is because of its incomparable lightweight performance add-on to the exceptional stiffness and also strength. Glass, aramid fibres, and glass are the major reinforcing fibres that are used in composites. As for matrices, thermoplastics and thermosetting are the ones that hold the fibres together. However, the usage of metal and ceramics as matrices in advanced composites has been dramatically increasing in the past years. High-performance composites, for instance, nanocomposites are the latest advancement in composite technology. Examples of nanocomposites are graphene-based composites and also composite nanotubes. These latest advancements of composites display extraordinary service performance, for instance, graphene-based composites can withstand the weight of an elephant (Sapuan 2017).

In this book chapter, we want to identify the hazards that possibly come from the usage of composites such as carbon fibre, dental resin composites, LDPE nanocomposites, MDF, and concrete compounds which come from different settings such as manufacturing and food packaging.

9.2 Health Hazards of Combustion Products from Aircraft Composite Materials (Carbon Fibre)

9.2.1 Introduction to the Hazard of Combustion Products from Aircraft

Aircraft mishaps involving the burning and explosion of fibre-reinforced composite materials can cause unusual environmental, safety, and health hazards that are different from normal accidents. Fibre-reinforced polymer composites in an air crash situation such as the sharp splinters from exposed materials, the distribution of fibrous dust by the blast and fire and the production of toxic gases during the polymer matrix resin burning are the health hazards that should be given prioritization. The rise in usage of fibre composites in aircraft which involve primary and secondary structural components has become a concern, especially the health hazards because of the exposure to carbon fibres with airborne characteristics that were discharged during a post-crash fire (Gandhi et al. 1999).

Response staffs are likely to be exposed to hazards caused by the fibrous particulates following an accident and the fire from the accident due to spillage of fuel. Personnel that participating in post-crash events such as emergency crew have the possibility of health effects in a long run from exposure to carbon and are required to be given special assistance in extinguishing and handling combustion of fibre composites (Gandhi et al. 1999).

The possible health risks related to fibre-reinforced polymer composites during fire comprise of releasing of fibrous dust and airborne particles that can easily enter humans bodies through inhalation and deposited in the deep lung regions, while the sharp fibre pieces could penetration the skin. Accidents involving fire and impact or explosion will increase the concentration of respirable airborne microfibres.

Another possible risk after an air crash is the occurrence of short-circuiting of electrical and electronic equipment because of the electrical conductivity characteristic of carbon fibres could affect nearby transformers, power distribution lines, and other electrical equipment and cause electrical hazards (Gandhi et al. 1999).

9.2.2 *Background of the Carbon Fibre Usage*

Carbon fibre composites have lightweight functional characteristics that have the potential to substitute metals in aircraft industries with a top-notch strength-to-weight ratio and corrosive resistance. The applications of composites in primary and secondary structural parts escalate due to its weight savings and the efficiency of fuel utilization over the lifespan of the aircraft (Soutis 2005). The steady increase in reinforcing the components with carbon fibre composites from 1981 involves two of the major suppliers for commercial aircraft. The operating empty weight of the aircraft affects the percent structural weight (Gandhi et al. 1999).

9.2.3 *Composition of Polymer Composites in Carbon Fibre*

Polymer structural composites in carbon fibre consist of constant, high-strength fibres permeated with polymer matrix producing reinforced layer (ply), which afterwards bind together with other layers to form an orthotropic laminate with the help of heat and pressure. The orientation of the fibres with respect to the loading direction and their volume fraction in the composites determine the stiffness and strength of the laminate. Fibres consist of around 55–60% in volume of the laminate, while the rest is polymer resin in a typical polymer matrix composite (Gandhi et al. 1999).

Reinforcement using continuous carbon fibres is more desirable due to their high strength-to-weight ratio that can provide improvement for better performance. Using fibres alone or as a mixture with other fibres (hybrids) in different forms such as continuous fibre fabrics, tapes, and discontinuous chopped strands. Carbon and graphite fibres are synthesized in a continuous process from polyacrylonitrile (PAN). This process involves controlling pyrolysis at around 1000 to 3000 °C of the precursor so that carbon and graphite fibres formed contain around 93–95 and 99% carbon atoms, respectively. Carbon fibres with a diameter around 6–8 μm are commonly used in composite manufacturing (Gandhi et al. 1999).

9.2.4 Carbon Fibre Toxicology

Fibre can enter our body through two main routes, which are by skin and inhalation. A response to sharp, fragmented fibres of diameter greater than 4–5 μm can cause skin and eyes irritation, which are common effects when exposed through dermal. The degree of severity of exposure relies on the size of fibres and their stiffness (Gandhi et al. 1999).

The exposure through inhalation has the greatest possibility for adverse effects on humans and depends on the physical dimensions of the fibres and their total dosage. Particles that are smaller tend to be deposited in the lower regions of the lungs and also related to chronic toxicity. The analysis showed that when the length of carbon fibres was less than 100 μm, the likelihood of oxidation-induced thinning of the fibres from composites burning was increased and deposited within the lung regions (Gandhi et al. 1999).

The deposition of fibres in bronchial airways are divided into 5 modes. The airway areas in the human's lungs composed of bronchi and bronchioles, a series of branching airways that becomes smaller as they progress deeper. The multiple division of the bronchi significantly boosts the total cross-sectional area of the airways available for fibre deposition. The mechanism of the deposition in the airways is determined by its aerodynamic length. Fibres with higher terminal velocity are deposited against the inside walls of airways in larger airways by a process called impaction. This process takes place when the fibres are unable to adjust or change the angle and velocity of the airflow. As airways become smaller, the airflow velocity also reduces, thus, the sedimentation process takes place. There are other mechanisms of fibre loading in lungs which are interception, electrostatic deposition, and diffusion (Lippmann 1990).

A study showed that fibres can be eliminated from the pulmonary region through dissolution in lung fluids or phagocytosis, a defensive mechanism that is self-regulated and carried out by macrophages (Hesterberg and Hart 2001). Macrophages are cells that are usually found close to the thin walls of the alveolus that help to get rid of dead cells, bacteria, foreign particles, and fibres by ingestion (Warheit et al. 1995). Macrophages assist remove the particulates in the lung walls through the peripheral mucociliary drainage pathways. Its diameter of 10–15 μm can affect the removal efficiency. Fibres with lengths less than 10 μm can be cleared easily by the macrophage. Nonetheless, the fibres that are longer (20–30 μm) cannot be phagocytized completely by the macrophage cells. The uncompleted ingestion of long fibres caused the alveolar macrophages to be overloaded and it takes a longer time than it supposed to for the whole fibres clearing process. Partial ingestion also might result in the death of the macrophages (Singh et al. 2017).

The inhalation of fibres such as asbestos can cause health effects that show that there is an association between fibres with its specific critical length and diameters. Table 9.1 displays the critical fibre dimensions for each of the asbestos-related diseases; asbestosis, lung cancer, and mesothelioma. Asbestosis is related to surface areas of the inhaled fibres with a length around 2–5 μm and a maximum of 10 μm

Table 9.1 Critical fibres dimensions for asbestos-induced diseases (Gandhi et al. 1999)

Diseases	Diameter (μm)	Length (μm)
Asbestosis	0.2–2.0	2.0–5.0
Lung cancer	0.3	> 10.0
Mesothelioma	< 0.1	> 5.0

when the retention is the highest while the diameter is around 0.2–2 μm. The upper limit of the diameter of the fibres is 3 μm. The chances of getting lung cancer rise when a considerable amount of long fibres present (> 10 μm). The critical limits of the diameter of fibre are around 0.3–0.8 μm. It is expected that fibres with longer diameters (> 10 μm) have more influences on the lung malignancy because of the longer retention time. Shorter fibres, for instance, are more easily cleared from the lungs as macrophages carried out ingestion and dissolution process in lung fluids.

Table 9.2 shows a sum-up of studies on hazards due to inhalation of carbon fibres. These studies were planned to measure whether the workers are exposed to chronic

Table 9.2 Summary of carbon fibre toxicity from different studies (Gandhi et al. 1999)

Species	Fibre parameter			Exposure conditions				Results
	D (μm)	L (μm)	Conc. (f/cm^3)	Hour/day	Day/ week	Weeks	Post-exposure recovery	
Rat	7	20–60	40	6	5	16	32 weeks	No adverse effect on lung function No fibrosis
	3.5	3500	40–80	1	5	2	1, 14 days	No carbon fibres in any tissue No abnormality of pulmonary function
	3	10–60	40	6	5	16	35, 80 weeks	Some non-fibrous particles in lung tissue No abnormal pulmonary functions
	1–4	–	50–90	6	5	–	4, 12 weeks	Temporary lung inflammation Reversible after 10 days No fibrosis of the lung tissue

risks in an occupational environment. Waritz et al. (1998) exposed rats to 7 μm PAN-based carbon fibres at aerosol concentrations of 40 fibres per cubic centimetre (f/cm^3) for 6 h per day, 5 days per week for 4 months. The animals were observed based on the physiological responses for 32 weeks. The results showed there were no pulmonary inflammation, fibrosis effects, and lung function impairment detected. Another study carried out tests using carbon fibres with 3.5 μm in diameter and 3.5 mm long by exposing the rats to different concentrations of fibre which were 40, 60, and 80 f/cml for 1 h a day that lasted for 14 days. From the observation, there were no changes in physiological response after post-exposure periods between 1 and 14 day. Then, they carried out another experiment using carbon fibres with a diameter of more than 3 μm and length of more than 80 μm. Based on the density and the physical dimensions of the carbon fibres, an estimation was made that D was more than 10 μm from the studies. Fibres (Da > 10 μm) are usually viewed as being non-respirable (Thomson et al. 1990).

9.2.5 *Toxicity from Combustion of Carbon Composites*

The carbon fibre composites when burned will produce heat and combustion products that comprise a mixture of gases and visible products from incomplete combustion or smoke. Smoke compositions differ drastically with fire development stages as there are changes in burning conditions and growth rate of fire. During any stages, the smoke stream consists of a mixture of evolved gases, vapours, and some solid particles. Smaller particles tend to stay in the air longer because they have a greater surface area which is more probably to adsorb chemical vapours present in the smoke. The physiological effects of the exposed human rely on the distribution size and solubility characteristics of the aerosols that will determine the degree of the lung penetration and the absorption rate in the body (Gandhi et al. 1999).

A study discovered that heating the surface of a composite material (carbon fibre) resulted in thermal decomposition and surface thinning because of oxidation when the temperature reached around 500–550 °C. The decomposition rate was rapidly increased while there was a rapid loss in the thickness of fibre when the increasing temperature of higher than 550 °C. Surface decomposed by oxidation removed the modulus contribution of the highly-oriented layer planes, thus causing the fibre to become softer. The strength of the fibre became poor because of thermally-activated damage, assuming that the sub-micron flaws at the surface were developed. The reduction in the strength of fibre was affected by temperature which explained that the application of tensile stress on hot, decomposing composite materials do not surpass the fibre strength then failure will not happen (Inthavong et al. 2013).

9.2.6 *Risk Mitigation for Carbon Fibre*

The applications for fibre-reinforced polymer composites in aircraft industries have risen up drastically as the time pass by and will continuously grow in the future. The main causes of fire hazards in interior and secondary composites used in the aircraft cabin and fuselage components are the heat release rate and the toxicity because of the gaseous that comes from combustion products during the burning of the polymer matrix. The aircraft cabin crews and passengers are likely to be exposed to these hazards during an impact-survivable accident.

Some safety measures can be taken is providing the best equipment during emergency or that involved combustion of aircraft. All departments working in close proximity to a crash site are compulsory to wear self-contained breathing apparatus, leather gloves, chemical specialized protective clothing, and neoprene coveralls to decrease the exposure to all airborne species. Guidelines for handling the mishap of aircraft must be established to ensure the safety of the personnel involved and prevent any problems from arising. For example, The US Air Force Advanced Composites Program has established a guideline that maintains minimum safety and health protection requirements for firefighters, clean-up crew, and investigators that involved in accidents of aircraft with advanced composite materials.

9.2.7 *Summary of Health Hazards of Carbon Fibre*

From the aforementioned studies, we can conclude that the thinning due to oxidation of carbon fibres because of fire boosted the probability of fibre to be deposited within the lung regions when their length is less than 100 μm. Longer carbon fibres are usually trapped within the nose region and therefore unlikely to cause severe health problems. Modelling also showed that glass fragments which are wider and heavier than carbon fibres, are mainly trapped in the nasal cavity due to the airway that exhibits curvatures, and accelerated flow areas play the role as a natural filter to high inertial fibre particles. The outcome from this study indicates that there is a need to concern for the release of fibre fragments (especially carbon filaments) into the smoke plume of burning composites that usually occurs in any post-crash aircraft accident.

9.3 Health Hazards of Dental Resin Composites in Medical Industry

9.3.1 Background of Dental Resin Composites

The filling is a technique used to restore damaged teeth by decay or other reasons to its normal shape and function. Materials that have been used for fillings include composite resin and amalgam. Dental composites resin has been used since the 1960s as an alternative to silver amalgam for filling cavities and fissures that form in the teeth. Dental composites are made of polymerizable resin matrix, silane coupling agent, and reinforcing glass particle fillers (Drummond 2008). Quality, aesthetic properties, and durability are the reasons for the wide utilization of these glass particle/resin matrix composites as materials for the restoration of anterior teeth. The polymerizable resin matrix usually consists of one or more monomers, such as triethylene glycoldimethacrylate (TEGDMA), urethane dimethacrylate (UDMA), and bis-phenol-A-diglycidyl dimethacrylate (Bis-GMA). Polymerization of the resin matrix can be chemically started in light-activated, "self-cure" composites or both combinations. Several inorganic materials, like glass filler, are being used as reinforcing components and utilized as fine or micro-fine particles. These fillers made up the majority of the composites and they differ in size (Drummond 2008).

Numerous clinicians usually use at least two types of composites; microfill and hybrid. Microfills are mainly used when clinical situations need stain resistance and optimal aesthetics, while the hybrid is used to tackle fracture resistance and mechanical strength issues. Hybrids are highly filled, whereas microfills are lesser filled. Table 9.3 shows that actual filler particles divided into several groups by size, e.g. microfiller, minifiller, midfiller, and macrofiller (Bayne et al. 1994).

The likely routes of systemic intake of chemical substances released from resin-based composites can be through (i) oral mucosa directly, (ii) absorption of volatile components in lungs, (iii) ingestion of released components in the gastrointestinal tract, and (iv)diffusion to a pulp via dentinal tubules (Saxena et al. 2012).

Table 9.3 Classification of dental composites by filler particle sizes (Bayne et al. 1994)

Category	Particle size	Homogenous	Hybrid	Heterogeneous
Megafill	0.5–2 mm		(XXX)	
Macrofill	10–100 μm	XXX		
Midfill	1–10 μm	XXX	XXX	XXX
Minifill	0.1–10 μm	XXX	XXX	XXX
Microfill	0.01–0.1 μm	XXX		XXX
Nanofill	0.005–0.01 μm	(XXX)	(XXX)	

9.3.2 *Cause of Degradation of Dental Resin Composites*

All materials can be corroded and degraded by time. In the case of dental composites, this can be contributed by a few factors such as saliva components, mastication forces, temperature changes, changes of chemical dietary, and oral microbes. The first factor is the saliva components, where water is the main component of saliva. Water particles will simply enter into the polymer network permitting the dispersion of boundless or uncured monomers and additives from the material network since resin composites could be a polar material (Saxena et al. 2012). The second factor is the chewing forces. Fatigue contributed by moderately weak repetitive loads such as normal chewing force can cause degradation throughout exposure to the oral environment. Continuous degradation, crack beginning, and growth can happen due to a nonstop application of environmental and mechanical loads causing in terrible degradation of the resins. This process is then aided by recurrent voids introduced during the residual stresses and processing of the material (Drummond 2008).

The third cause for degradation of dental resin composites is temperature changes. Repetitive drinking and eating might encourage alterations in intraoral temperature. Extreme environment for the materials might be produced when the temperature changes. Fluctuations of temperature experienced in vivo can cause an impact of stress on the surface due to the increasing temperature gradients near the surface. These actions can lead to the degradation of these composites (Bettencourt et al. 2010). The next factor is the changes in chemical dietary. Everyday intake of drinks and foods might affect dental composites by their direct effect or their capability to alter the intraoral pH values.

Lastly, oral microbes also play a role in the degradation of dental resin composite. There is a lack of current information available regarding the interaction that occurs between the polymer network and oral microbes, as shown in Fig. 9.1. An in vitro study has proven that bacteria can inhabit on the surfaces of composite resin. The

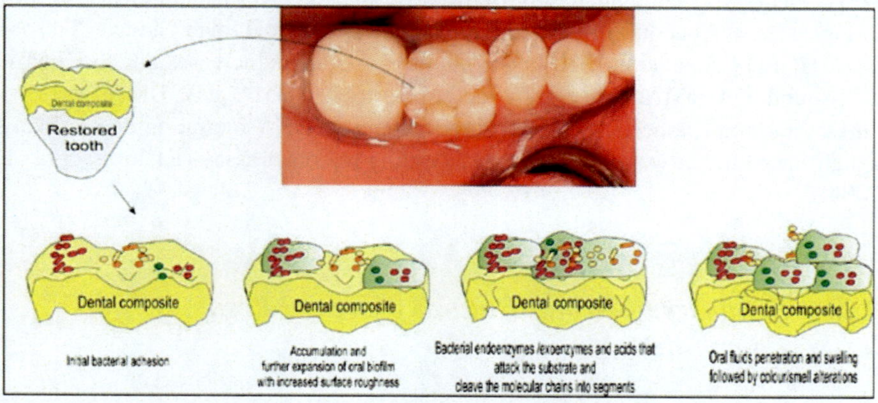

Fig. 9.1 Schematic illustration of complex interactions between dental composite and oral microbes (Zhang et al. 2017)

roughness of the material's surface increased after growth of bacteria; suggesting some surface degradation (Whillerhausen et al. 1999). The bacteria that are known to release acids that are likely the cause of composites surface degradation. In general, all these factors are accountable for the degradation of polymer together. For instance, the effect of temperature changes might be compounded with saliva and oral microbes (Saxena et al. 2012).

9.3.3 General and Local Toxicity of Dental Resin Components

MTT test is a broadly used approach for assessing cytotoxicity of composite resin, despite the fact that alkaline phosphatase and succinic dehydrogenase (SDI) responses have also been practised. Resinous monomers such as HEMA and TEGDMA can interrupt the odontogenic differentiation of SCAP cells, physiological repair and/or development of human permanent teeth (Bakopoulou et al. 2012). HEMA also can induce cell death by damaging the mitochondrial activity in human gingival fibroblasts (Spagnuolo et al. 2006). TEGDMA can react with intracellular molecules and infiltrate membranes. This hydrophilic compound can interfere with oral tissues. Exactly, a mechanism of reducing cellular detoxifying potential are formed from glutathione-TEGDMA adducts. A contemporary study showed that TEGDMA encourages important intracellular glutathione level (GSH) reduction and results in excessive cytotoxicity in cultures of human periodontal ligaments fibroblasts (HPLF). There is no protective consequence against TEGDMA-induced cytotoxicity even in high concentrations of glutathione. Hence, the investigated hypothesis that the proposed exogenous GSH can reduce and prevent TEGDMA-associated cytotoxicity in HPLF needs to be excluded (Martins et al. 2012).

Assessment of the mutagenicity has proven that TEGDMA can result in the large DNA deletions in mammalian cells (Schweikl et al. 2006). Resin monomers as shown in Fig. 9.2 could encourage DNA double-strand breaks in Human Gingival Fibroblasts (HGFs) (Urcan et al. 2010). The number of mutants increases due to HEMA, GMA, and TEGDMA induce, by a factor of 2 to 8. UDMA, TEGDMA, and Bis-GMA can tempt important but less development of DNA migration and is likely a sign for restricted genotoxic effects according to comet assay (Kleinsasser et al. 2004).

9.3.4 Summary of Health Hazards of Dental Resin Composites

In summary, concerns have been increasing due to the safe clinical practises of these substances and to their biodegradation beneath the oral environment. The complex

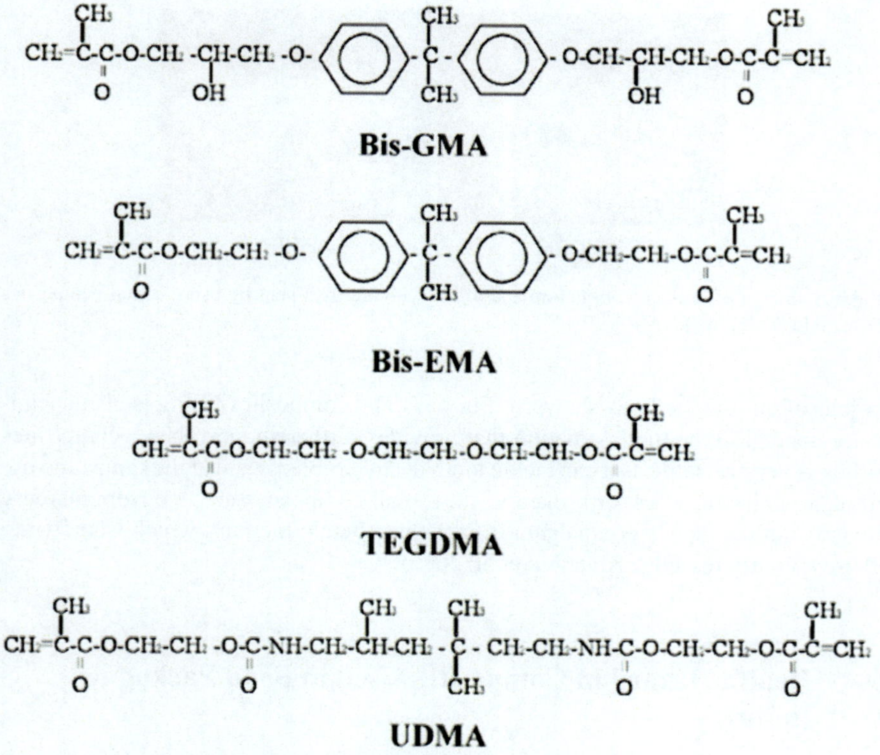

Fig. 9.2 Mutual monomers used to form cross-linked dental resins (Ferracane 2006)

interplay of interactions is being identified from the huge quantity and variety of processes by which composite resins might be degraded in the oral cavity. Causes and reasons for biodegradation include a few factors such as saliva components, oral microbes, chemical dietary changes, mastication forces, and temperature changes (Saxena et al. 2012).

After considering all of the searched statistics concerning the toxicity of resin composites, one could come to a conclusion that the recent literature confirmed that the materials released to be carcinogenic, mutagenic, and toxic but the latest facts showed the development. On the scientific and clinical point of view, this review strongly recommends the following of the manufacturer's instructions and technical consideration as regards to the polymerization of resin materials. It can be light-curing time, light intensity, compatibility between light and brand of composites, space between the surface of composites, and source of light and shelf life of the material. No direct skin contact should be allowed with the materials and cavity coating must be completed in regions with deep dentin (Saxena et al. 2012).

Although there have been some health issues derived from using dental composites, it still serves some advantages to the patients, such as aesthetics properties because the colour or shade of the composites filling can be closely matched to the

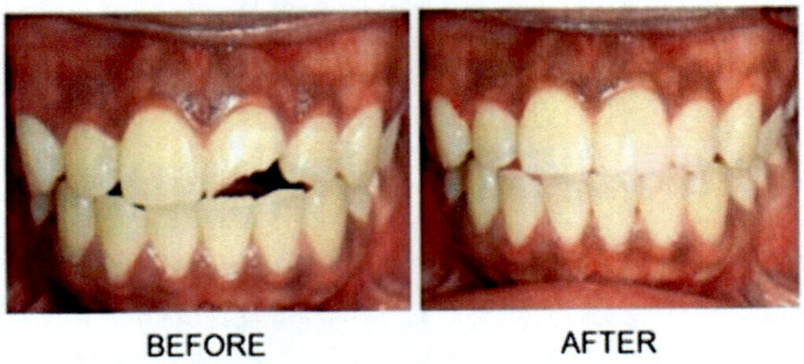

Fig. 9.3 Image of broken teeth before and after receiving treatment by using dental composites (Khurshid et al. 2015)

colour of current teeth, as shown in Fig. 9.3. The composites filling is also chemically bonded to the tooth structure that provides further support. Next, composites filling is very versatile as it can tackle tooth decay problems and at the same time use to repair chipped, worn, or broken teeth. Dental composites are free from mercury toxicity, unlike the silver amalgam fillings that contain mercury which is known to cause mercury toxicity (Khurshid et al. 2015).

9.4 Health Hazard in Composites Used in Food Packaging Industry

9.4.1 Background on Composite Food Packaging

A common composite used for food packaging is low-density polyethylene (LDPE). But this compound alone does not give enough protection for food, since food, whether ready to eat or raw is often susceptible to microorganism infection and growth, especially bacteria. Besides, the surrounding exposure such as moisture/humid, sunlight and heat might react with the food, as certain packaging are translucent or transparent. Furthermore, the addition of nanoparticles can also increase the mechanical strength of the food packaging (Ekielski 2018; Han et al. 2018).

LDPE is often combined with other nanoparticles to produce nanocomposites. This is to improve food protection as packaging, for example, Ag/LDPE, TiO_2/LDPE, and clay-LDPE (Ekielski 2018). As can be seen in Fig. 9.4, these types of plastics are highly available in the online market and commonly used by consumers. These plastic composites are often used in the food industry as food packaging. A study that Meal Ready to Eat (MRE) are packed using LDPE-clay (Han et al. 2018). But the concern is on the food which might be hot, acidic, or stored in the package for long durations, as it might cause the packaging materials to react differently. Not only that, environmental factors for example weathering, might also cause the packaging materials to react (Han et al. 2018).

Fig. 9.4 LDPE plastic roll and plastic sheet

9.4.2 Health Hazard Concerns from Food Packaging of LDPE Nanocomposite Material

9.4.2.1 LDPE-Ag

Ag addition in food packaging material allows the packaging to inhibit the growth of bacteria. However, nano-silver can migrate to food from the packaging, and this can be more significant with the increase in temperature and storage time. In a study, it was noted that high temperature and acidity of the liquid in contact with the nanocomposites containing silver might release silver ions (Jokar and Abdul Rahman 2014). A separate study showed the influenced of storage temperature and storage time towards silver migration from packaging to the surface of the food (Cushen et al. 2013). A 10-day examination for silver migration rates of 4 commercial food packaging on food stimulants showed that the exposure to 3% acetic acid released 3.1 nanograms of silver per centimetre square; thus, acidic food contact to packaging containing silver has a higher potential for silver to migrate into food (Mackevica et al. 2016).

9.4.2.2 LDPE-Clay

On top of that, the addition of clay as a nano-filler to LDPE improves its resistance to fire, mechanical strength, transparency, and also reduces its gas permeability. The clay used in food packaging has the potential to leach out inorganic biocides. Since clay is a mineral mined from the earth, it is bound to have impurities; impure mined materials are also very difficult or near impossible to achieve 100% purity. Mineral clays are carriers for biocides, for example, silver, zinc, copper, and magnesium. In addition, Farhoodi et al. (2017) observed that clay in a nano-composite can migrate out and this process depends on the temperature and the duration of storage. Farhoodi et al. (2017) also reported that at 45 °C, the migration of substance from the tested nanocomposite was 23% more. In a separate study, it was found that the analysed LDPE had nano-clay that leached out and through observation, large particles were detected, that seemed to have the same features as TiO_2 compounds; thus, it was suspected that TiO_2 might have served as an additive in the preparation of the food packaging material (Han et al. 2018). The large particles that resemble TiO_2 compounds is a concern because it is a health hazard (Castro et al. 2012). This matter is further explained in the next paragraph pertaining to TiO_2.

9.4.2.3 LDPE-TiO_2

TiO_2 has bactericidal properties, therefore, it has been used in some food packaging industries. The migration of nano-TiO_2 due to temperature elevation was investigated and it was found that the higher concentration of the nanoparticle additive in the packaging material increased the migration of the added nanoparticles from the cut edges of the film and the surface of the film into the food in liquid form (Lin et al. 2014). In a study of transported TiO_2 in M cells, it was found that the TiO_2 transported was higher for food-grade TiO_2 as compared to the general grade of TiO_2; but, most of the TiO_2 can be eliminated through defecation (Jo et al. 2016). At the same time, it was also found that the chemical reactions that occur on the surface of the material affect the dispersion of the nanoparticles (Jo et al. 2016). But the TiO_2 still pose a risk, as in a study conducted by Castro et al. (2012) on nano-TiO_2. It was signified that this compound can induce toxicity in human beings; exposure to pollution from the atmospheres such as ozone will cause reactions to the compound even at room temperature resulting to high amounts of radicals being produced in a short duration of time.

9.4.2.4 LDPE-CuO and LDPE-ZnO

Furthermore, the CuO addition gives the food packaging antibacterial properties. It works by releasing ions; but, this cause redox reaction that can lead to radical formations which is a concern for health hazards. Therefore, the matter concerning the radical formation of CuO added to food packaging material on food needs to be researched on. On the same topic of health hazards of LDPE-CuO, a study reported that Cu particles were found to have migrated in significant amounts from the polyethylene food packaging (Ekielski 2018). In relation to the findings of this study, copper poisoning can happen when high amounts of Cu are ingested due to the Cu migration to foodstuff originating from the food packaging.

The addition of zinc also gives the food packaging material antibacterial properties and UV absorbing ability. In a study conducted by Wang et al. (2014), the reaction of cells to the introduction of ZnO nanoparticles and 300 mg/L of vitamin C was increased cell death; which showed that the combination of ZnO and vitamin C produced a synergistic effect that is significantly cytotoxic. Thus, ZnO is a health hazard if leached out form the food packaging to food containing vitamin C.

9.4.2.5 LDPE- SiO_2 and LDPE-Spherical Carbon Nanoparticles

Moving on, the addition of nanoparticle of SiO_2 reinforces the packaging material. Nevertheless, the nanoparticles of SiO_2 are still hazardous to health through the 3 routes of entry to the human body, which are the dermal, inhalation, and ingestion. However, there is still not enough strong evidences to prove whether there are interactions between this nanoparticles and food (Ekielski 2018).

Spherical carbon nanoparticles are also used to reinforce food packaging. The carbon nanotubes are able to increase tensile strength and also improve certain chemical bonds of the food packaging. Moreover, the carbon properties also increase thermal stability. In a study on carbon black nanoparticles that was an additive to plastic food packaging, it was found to not migrate into food (Bott et al. 2014).

9.4.3 Summary on LDPE Nanocomposites

As a sum up, the LDPE nanocomposites that are used in the food industry with those spherical carbon nanoparticles addition are stable and safe to use as food packaging, but they might still lacking in terms the antimicrobial properties. Thus, there is a need for more researches to be performed on antimicrobial properties of LDPE and spherical carbon nanoparticles. The other types of LDPE nanocomposites besides the composite LDPE with carbon might pose long-term or chronic risks if used as food packaging, as leaching of the additives were found. However, more extensive studies is essential to confirm the health hazards.

9.5 Health Hazards of Medium Density Fibreboard (MDF) in the Manufacturing Industry

9.5.1 General Information on the Composition of MDF and Its Production Process

Medium Density Fibreboard or MDF is defined as a wood type composite material. It is primarily composed of hardwood and softwood combined or bonded with formaldehyde-based resin, usually urea–formaldehyde resin (Burton et al. 2011; Selikoff 2010). The production of the composites requires wood material; consisting of fibre, which is treated with adhesives and compressed under the heat of up to 220 °C. The wood fibres undergo structural modification during the process of hot-pressing. Varying raw materials and pressing temperature produces various characteristics of MDF. The properties and structure of the wood composites are dependent on these characteristics (Walther et al. 2018). The process for making MDF furniture, however, includes processes such as sawing, drilling, sanding, polishing, and

Fig. 9.5 MDF wood panel (Araújo et al. 2019)

glueing. Workers handling the manufacturing processes and workers working in the same workspace are at risk of inhaling formaldehyde and MDF dust released from the work tasks (Thetkathuek et al. 2016). Figure 9.5 shows an example of an MDF wood panel.

9.5.2 Emission of MDF Dust and Formaldehyde

The formaldehyde released from MDF wood is emitted from the usage of, for instance, the urea–formaldehyde resin. Song et al. (2015) studied the emission of formaldehyde from MDF wood, which the Chinese National Centre for Quality Supervision supplied. One of the test methods conducted in the study was the method of perforator extraction, which is a practically conventional test method used by European countries; thus, having the standard range to be compared to. According to the European EN 120, an E1 grade ($\leq$ 8 mg/100 g) and an E2 grade ($\leq$ 30 mg/100 g) were assigned to the perforator test. Products with the grade of E1 is safe for use and are suitable to be used in an enclosed or indoor environment. Figure 9.6 shows the formaldehyde emission by the perforator extraction method. Song et al. (2015) recorded a value of greater than E1, thus, the MDF studied was given the E2 grade. This shows that the MDF wood studied was unsafe for usage according to the European standard.

The sanding process on MDF releases more dust compared to natural woods, however, sawing releases the same amount (Chung et al. 2000). Another study was concerned with occupationally related concentrations from MDF dusts from work processes such as sawing and sanding of MDF. The results showed a significant level of dust is emitted (Hursthouse et al. 2004). Both the softwood and hardwood dust emitted from the MDF were above the concentration of the eight-hour time-weighted average set by Malaysia's Use and Standards of Exposure of Chemicals Hazardous to health (USECHH) Regulations 2000.

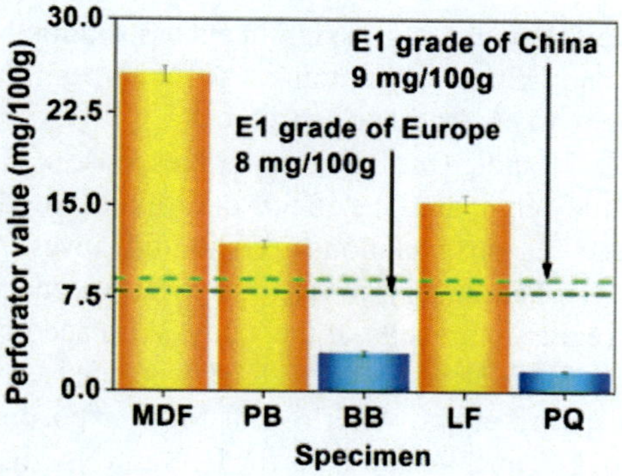

Fig. 9.6 Formaldehyde level of emission of varying wood materials categorised in accordance with the process of perforator extraction (Song et al. 2015)

9.5.3 Health Hazards of Emissions from the MDF Work Process

According to the IARC, wood dust classified as Group 1 is carcinogenic to humans. Bruschweiler et al. (2016) conducted a comet assay study to detect the DNA damage and assess cancer risk in workers exposed to MDF dust. The study compared the score of the comet assay between groups exposed to composite wood products, natural woods, and controls. The visual characterisation of cells quantified the degree of DNA damage. There were four categories of visual characterisation (0–3). No DNA damage was represented with the tail size of 0, tail size of 1 signified the lower level of DNA damage, 2 denoted the medium or moderate level of DNA damage, and 3 indicated the higher level of DNA damage. A higher comet assay score represented a higher degree of DNA damage. The result of the study showed a higher level of DNA damage occurred among workers who were exposed to composite wood dust compared to natural wood dust ($P < 0.001$). However, the study did not determine the formaldehyde concentrations, which was a limitation of the study.

Another comet assay test showed sufficient evidence that wood dust causes the nasal cavity, paranasal sinuses, and nasopharynx cancers ("IARC monographs on the evaluation of carcinogenic risks to humans" 2012). Formaldehyde is identified as Group 1 carcinogenic to humans by the IARC (Marsh et al. 2014).

Besides carcinogenic effects, MDF dust and formaldehyde also have noncarcinogenic effects, including dermatitis and also respiratory illnesses, e.g. asthma and bronchitis (Bell and King 2002; Hulin et al. 2012). There was a positive association of exposure to high levels of MDF dust (>5 mg/m^3) with allergic symptoms and respiratory irritation symptoms. High concentrations of formaldehyde were also associated with the symptoms. The risk of allergic symptoms and irritation in the respiratory system rose if the concentration of MDF dust was > 5 mg/m^3 (Thetkathuek et al. 2016). Other factors contribute or amplify to respiratory irritation when exposed to MDF dust such as atopic history. People with atopic history also had an increased risk of respiratory allergic symptoms. A significant association between the concentration of MDF dust and irritation symptoms was found from the same study when excluding the allergic workers from the analysis.

Bornholdt et al. (2007) studied the inflammatory response of different wood dust in epithelial cells. The study compared the effects of different wood dust by calculating the regression of a dose–response relationship. The study investigated the expression of mRNA for the pro-inflammatory interleukin 6 (IL-6) and 8 (IL-8) cytokines. Table 9.4 shows the regression results of the study. With accordance to the cellular expression of IL-6 and IL-8, after 3 h and 6 h of exposure to wood dust, there was a significant dose–response effect. After 6 h of MDF exposure, the R^2 value was increased to 0.81 for IL-6 mRNA and 0.86 for IL-8 mRNA. IL-6 and IL-8 mRNA had similar responses, both had higher responses after 6 h, as compared to after 3 h. The first line of defence against pathogens is the lung epithelium, which produces cytokine. The increase in IL-6 and IL-8 showed an inflammatory response upon exposure to MDF dust.

Exposure of formaldehyde towards rats resulted in bronchial constrictions in the rats (Qiao et al. 2009). From the exposure, it is possible to say that after inhaling, formaldehyde containing dust has the ability to enter lower respiratory tract. This might cause respiratory system disorders and reduction in lung function. A study by Burton et al. (2011) found that among a group exposed to MDF dust, lung function tests showed a significant reduction in forced expiratory volume in 1 s (FEV_1) and also forced vital capacity (FVC) when the predicted value was used as a comparison. After a few months of redeployment of the workers to an area with lower MDF levels, the symptoms were improved progressively. This showed that exposure to MDF dust might also have long-term effects on lung function.

9.5.4 *Recommendations to Reduce Exposure to Formaldehyde and MDF Dust*

The primary control measures for controlling the exposure of employees to formaldehyde are through elimination and substitution. Only MDF with low formaldehyde emission should be used to ensure the formaldehyde emission rating of the wood is rated as E0 or E1. For further security, a safety data sheet should be requested before purchasing any MDF ("Formaldehyde in pressed wood products—NICNAS" 2016). A proper ventilation system at the work area will reduce the concentration of formaldehyde and MDF dust (Chung et al. 2000). Usage of tools with build-in vacuums which extracts dust and formaldehyde could also reduce the concentration significantly ("Formaldehyde in pressed wood products—NICNAS" 2016). Training should be given to employees exposed to MDF dust and formaldehyde to detect symptoms of poisoning and upon the presence of symptoms of poisoning by dust

Table 9.4 Linear regression curves of IL-6 and IL-8 cellular content for different wood dust (Bornholdt et al. 2007)

Wood species	Exposure time	IL-6 mRNA regression line				IL-6 mRNA regression line			
		p-value	Slope (mg/ml)-1	Intercept	R^2	*p*-value	Slope (mg/ml)-1	Intercept	R^2
Teak	3	0.004	69.2	1.32	0.58	0.002	25.3	1.47	0.65
	6	< 0.001	231.9	0.89	0.83	< 0.001	374.8	0.39	0.81
Pine	3	0.005	12.4	0.91	0.56	0.001	27.1	1.02	0.67
	6	0.006	116.5	1.31	0.55	< 0.001	104.2	1.04	0.82
Spruce	3	0.087	25.2	1.43	0.26	0.02	29.3	1.09	0.43
	6	< 0.001	45.6	0.10	0.85	< 0.001	64.7	1.16	0.82
MDF	3	0.015	40.1	1.56	0.46	0.07	52.2	1.74	0.28
	6	< 0.001	161.2	-0.01	0.81	< 0.001	180.6	0.13	0.86

and formaldehyde, employees should cease activity and move to a fresher air environment. Respirators with suitable ratings could be given if all of the prior mentioned methods are insufficient.

9.5.5 Summary of Findings on MDF

Overall, MDF causes various health hazards ranging from the carcinogenic health effects to the non-carcinogenic ones. The non-carcinogenic health effects range from respiratory symptoms such as asthma, bronchitis, and irritation. MDF dust and formaldehyde emitted from the MDF wood composites cause these health effects. Different activities such as sawing and sanding emit different levels of MDF dust and formaldehyde. However, these emissions can be controlled in many ways, ranging from purchasing MDF with E0 or E1 rating, installation of the LEV system, and the usage of PPE. There are still rooms for improvement for current research and studies on health hazards of MDF to humans health. Most studies on wood dust are about natural woods, and not about composite woods like MDF. As seen in the comet assay and regression of IL-6 and 8, the degree of health effects varies between natural woods and composite woods such as MDF. Further researches need to be conducted on composite woods.

9.6 Health Hazards of Concrete Compounds in Construction Industry

9.6.1 Background of Concrete Compounds

It has been 2000 years since the first-ever recorded history regarding the making of concrete. A short history with respect to concrete, the Romans during third century B.C. created a cement mixture with lime, which was used afterwards in the future to construct clay bricks. Afterwards, innovation of silica and alumina materials that made a more grounded bond came in. During the downfall of the Roman Empire, the utilization of cement were overlooked and forgotten. Up to the era of Renaissance in 1824, a well-known brick artisan (Joseph Aspdin) who was located in Leeds, England, planned another bond with the shade of the stone at Portland, an island located near the English Channel. The concrete was named as Portland cement, refering its origin. Majority of its compound is calcium silicates. This specialised concrete was made primarily of limestone and mud that were blazed at around 2,700 ± °F, then a little bit of gypsum was added later on. Detailing or added substances utilized might fluctuate contingent upon the accessibility of raw materials required by Portland cement. The mixture of concrete was made by blending cement with stone, silica sand, and water. Aggregates cement consist of 75% while sand makes up to 33% to the total mixture of the compound. Cement can be enhanced for specific purposes by including different mixtures. In order to consume less water, hydroxylated carboxylic corrosive was used which is a water-reducing chemical, subsequently producing more grounded, progressively useful cement. Calcium chloride is utilized as a quickening agent to lessen the production time of cement (Linch 2002).

Another compound that might possess high chances of hazards is the concrete dust. To be more specific, concrete dust is one of the major sources of health hazards to human and environment. Concrete is a very fundamental compound in the construction industry and nowadays, almost every construction requires the use of concrete cement for establishing infrastructure. Without the application of concrete, there is almost no better way to construct a building. However, every concrete utilization can release vast quantities of respirable dust from various processes, including removing, repairing, or altering existing concrete structures. Moreover, the use of concrete is not permanent, since after certain years, the concrete buildings start to corrode and become fragile. At this particular moment, lots of silica dust are released from the concrete and if inhaled by any human being, many diseases can struck such as silicosis. The aerodynamic diameter of concrete particles are ranging from 0.05 to 5.0 μm. These particles can easily penetrate the lungs. Particles with that size can travel to the section in the lung known as tracheobronchial respiratory zone. The main route of entry of cement particles in the body is via the respiratory or the gastrointestinal tract, or both by inward breathing and even swallowing. The two paths, particularly the respiratory tract are presented to various conceivably hurtful substances in the cement mill surroundings. The significant physical properties include molecule size and thickness, shape and vulnerability, surface area, electrostatic charge, and hygroscopicity (Meo 2004).

9.6.2 Health Hazards from the Usage of Concrete Compounds

Silicosis is a disease associated with the lung caused by inhalation of small particles of silica, a type of mineral that usually can be found in stone and rock, sand, mineral ores, and also compound like concrete. This disease mostly affects occupational workers exposed to silica dust which include fields like glass or concrete manufacturing, mining, or other industrial backgrounds. Long term effects of exposure of silica particulates might cause scarring in the lungs wall and at the end might harm our ability to breath properly. Silicosis is divided into three types; acute, chronic, and accelerated. The most unfortunate event regarding silicosis is that once you have been diagnosed with silicosis, there will be no cure, hence, prevention is better than cure in the case of cilicosis. Silica presence is associated to illnesses among the development and construction workers. However, more silicosis deaths are related to development and construction than some other industry. Furthermore, fundamentally elevated

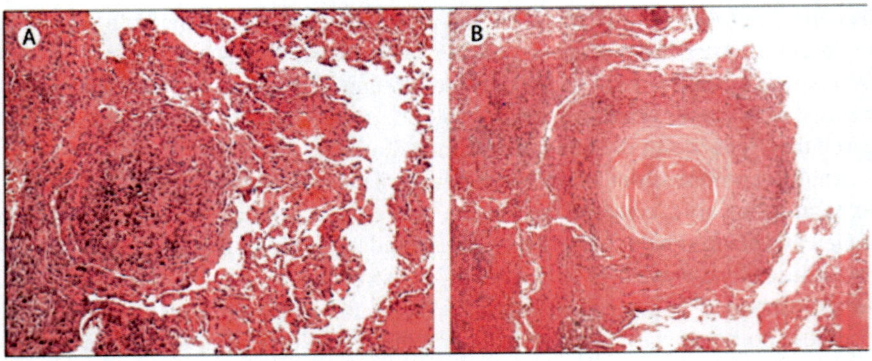

Fig. 9.7 Early silicotic symptoms in the lung cells, B; X40 (Leung et al. 2012)

mortality hazard from silicosis has been observed from development workers. Aspiratory tuberculosis, known to be increasingly pervasive among silicosis patient, was coming from development and construction labourers and a general development populace (Flanagan et al. 2003).

In developed nations, silicosis is an occupational health concern. From 1990 to 1993, it was reported that 600,000 labourers in the United Kingdom and over 3 million specialists in Europe were exposed to crystalline silica. In 1996 and 2009, for the most part, less than 100 cases were accounted for consistently in the UK (Leung et al. 2012). Figure 9.7 shows the condition of the early silicotic symptoms occurring in the cells of the lungs.

9.6.3 *Effects of Silicosis to Human Body*

Firstly, the silica dust might enter the lung through the airways and at the same time it causes damages to the lining of the lung air sacs. The process leads to scarring and it might get worse, causing massive fibrosis. But this only happened when the conditions are met which severe scarring and stiffening of the lung are causing breathing difficulty. Unintentionally, this might reduce work performance of workers in the workplace. Besides, deaths can be resulted if the cause is serious enough. Labourers associated with concrete grinding are possibly exposed to large amounts of crystalline silica dust that occasionally surpass in excess of multiple times the criterion, setting these labourers in danger for an assortment of respiratory illnesses, e.g. silicosis, lung cancer, rheumatoid arthritis, scleroderma, Sjogren's disorder, lupus, and renal illness. High concentration or prolonged inhalation, or both of cement dust in concrete industry labourers can incite clinical side effects and incendiary reactions that might result in utilitarian and basic anomalies. The most often revealed clinical highlights in concrete factory specialists are perpetual hack and mucus creation,

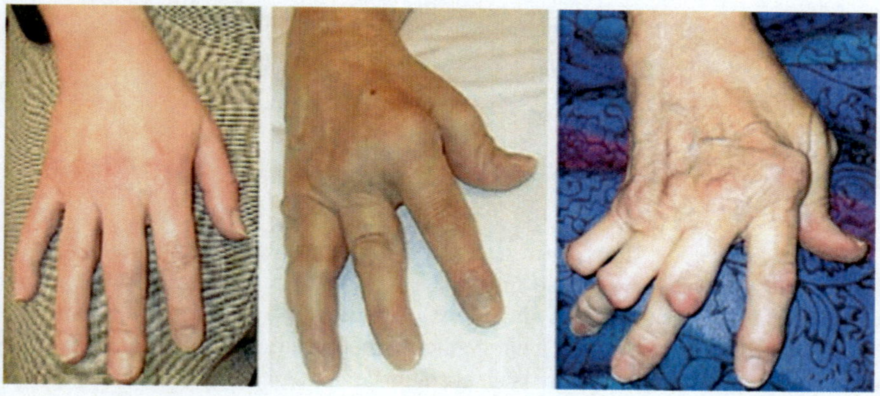

Fig. 9.8 Rheumatoid arthritis (Sparks 2019)

the impedance of lung work, chest snugness, obstructive and prohibitive lung sickness, skin aggravation, conjunctivitis, stomachache, cerebral pain, weakness, weariness and carcinoma of lung, stomach, and colon (Meo 2004). Figures 9.8 show the conditions experienced by rheumatoid arthritis patient.

9.6.4 Prevention of Contacting Silicosis from the Concrete Compound

NIOSH has proposed a guideline or steps need to be taken in order to prevent silicosis in workers in the workplace, especially in the concrete industry. Several steps can be taken such as avoid working in or near dust at work and avoid the possibilities of breathing in crystalline silica dust. These steps are crucial for any workers in the industry, as it is very important to have knowledge regarding various hazards. Then, always remember that even the dust are invisible, they still might be present around us and cause risk. Proper ventilation for confined space and installing water spray systems should be noted by every manager and responsible person in the workplace. Unless the water spray systems installed and ventilation are inadequate, the manager or top board should prepare appropriate and certified PPE (Personal Protective Equipment) for their workers as a final means of prevention. The PPE should provide the workers with protection from any crystalline silica dust from the concrete in this case. The employer also should conduct health and lung screenings to the employee. It is necessary and advantage to early detect any health problems, especially when it comes to workplace health effects. As for personal initiative, every employee must wash their hands before eating or drinking, especially after working in a dusty area. This is to avoid any unwanted or excessive dust presenting in the food or drinks. Last but not least, every employee is advised to shower and always change into new clean clothes before leaving from work. Through this way, we can avoid contamination of external assets such as car and home.

9.6.5 *Summary of Health Hazards of Concrete*

To sum up, silica dust from concrete compound has been a true hazard and a silent killer to those who work in the industry. Every worker should always be aware of the danger possessed by the silica dust from the concrete. Workers in concrete background gain added risk in having silicosis because they are highly exposed to the emission of silica dust. There is no absolute cure after having silicosis making in more dangerous and helpless once being struck by the disease. It is crucial for each and every workers in the related industry to always take the prevention method.

9.7 Conclusion

To conclude, composite materials have different properties from their individual compositions. These differences of characteristics pose different levels of health hazards, hence, further study should be conducted to fully understand the differences.

- Fire causes the thinning of carbon fibres induced by oxidation. Carbon fibres with the length of below100 μm have an increased likelihood of being deposited into the lungs. The concerns arising from the release of the carbon filaments from burning composites are justifiable due to the health hazards.
- Dental composite resin is the best material that been used for fillings compared to amalgam. Although there have been some health issues derived from using dental composites, it still serves the advantages to the patients such as the fillings is invisible, provide easy adhesion and preserves the maximum amount of tooth.
- Among the review LDPE nanocomposites; those with carbon-nanotubes and TiO_2 addition are stable and safe to use as food packaging. The other types of LDPE nanocomposites besides carbon-nanotubes and TiO_2 might pose long-term (chronic) risks as leaching of the additive are found, but more researches are needed to confirm the health hazards.
- MDF emits dust and formaldehyde upon reconstructing the composite wood, either through sawing or sanding. Different processes emit different levels of dust and formaldehyde. From this review, it can be seen that MDF causes DNA damage and might lead to cancer. MDF also causes respiratory illnesses and symptoms.
- Every worker should always be aware of the danger possessed by the silica dust from the concrete. Workers in concrete background gain extra risk in having silicosis because direct exposure to the silica dust emission. Controlling these health hazards should be started by having awareness or the health hazards these

composite materials could impose on humans health and then taking precautionary measures to reduce the risk of exposure.

Acknowledgements This chapter was written with A. B. M. Amirul Hakim, N. M. Syafiq, A. Brendon, I. M. Najmi, N. A. M. Zikri Aizat.

References

Bakopoulou A, Leyhausen G, Volk J, Koidis P, Geurtsen W (2012) Effects of resinous monomers on the odontogenic differentiation and mineralization potential of highly proliferative and clonogenic cultured apical papilla stem cells. Dent Mater 28(3):327–339. https://doi.org/10.1016/j.dental.2012.01.002

Bayne SC, Heymann HO, Swift EJ (1994) Update on dental composite restorations. J Am Dental Assoc (1939) 125(6):687–701. https://doi.org/10.14219/jada.archive.1994.0113

Bell HK, King CM (2002) Allergic contact dermatitis from urea-formaldehyde resin in medium-density fibreboard (MDF). Contact Dermatitis 46(4):247. https://doi.org/10.1034/j.1600-0536.2002.460417.x

Bettencourt AF, Neves CB, de Almeida MS, Pinheiro LM, Oliveira SAe, Lopes LP, Castro MF (2010) Biodegradation of acrylic based resins: a review. Dental Mater. **26**(5), 171–180.https://doi.org/10.1016/j.dental.2010.01.006

Bornholdt J, Saber AT, Sharma AK, Savolainen K, Vogel U, Wallin H (2007) Inflammatory response and genotoxicity of seven wood dusts in the human epithelial cell line A549. Mutation Res—Genetic Toxicol Environ Mutagenesis 632(1–2):78–88. https://doi.org/10.1016/j.mrgentox.2007.04.016

Bott J, Störmer A, Franz R (2014) A comprehensive study into the migration potential of nano silver particles from food contact polyolefins. ACS Symp Ser 1159:51–70. https://doi.org/10.1021/bk-2014-1159.ch005

Bruschweiler ED, Wild P, Huynh CK, Savova-Bianchi D, Danuser B, Hopf NB (2016) DNA damage among wood workers assessed with the comet assay. Environ Health Insights 10:EHI.S38344. https://doi.org/10.4137/EHI.S38344

Burton C, Bradshaw L, Agius R, Burge S, Huggins V, Fishwick D (2011) Medium-density fibreboard and occupational asthma. A case series. Occup Med (Oxford, England) 61(5):357–363. https://doi.org/10.1093/occmed/kqr090

Castro CA, Osorio P, Sienkiewicz A, Pulgarin C, Centeno A, Giraldo SA (2012) Photocatalytic production of 1O2 and OH mediated by silver oxidation during the photoinactivation of Escherichia coli with TiO_2. J Hazard Mater 211–212:172–181. https://doi.org/10.1016/j.jhazmat.2011.08.076

Chung KYK, Cuthbert RJ, Revell GS, Wassel SG, Summers N (2000) A study on dust emission, particle size distribution and formaldehyde concentration during machining of medium density fibreboard. Ann Occup Hygiene 44(6):455–466. https://doi.org/10.1016/S0003-4878(00)00005-3

Cushen M, Kerry J, Morris M, Cruz-Romero M, Cummins E (2013) Migration and exposure assessment of silver from a PVC nanocomposite. Food Chem 139(1–4):389–397. https://doi.org/10.1016/j.foodchem.2013.01.045

de C Araújo CK, Salvador R, Piekarski CM, Sokulski CC, de Francisco AC, Camargo SK de CA (2019) Circular economy practices on wood panels: a bibliographic analysis. Sustainability (Switzerland) 11(4). https://doi.org/10.3390/su11041057

Drummond JL (2008) Degradation, fatigue, and failure of resin dental composite materials. J Dent Res 87(8):710–719. https://doi.org/10.1177/154405910808700802

Ekielski A (2018) Interactions between food ingredients and nanocomponents used for composite packaging. Reference module in food science. Elsevier. https://doi.org/10.1016/B978-0-08-100596-5.21850-7

Farhoodi BM, Mousavi SM, Sotudeh-gharebagh R, Emam-djomeh Z, Oromiehie A (2017) Migration of aluminum and silicon from PET/clay nanocomposite bottles into acidic food simulant. Packag Technol Sci (November 2013):161–168. https://doi.org/10.1002/pts

Ferracane JL (2006) Hygroscopic and hydrolytic effects in dental polymer networks. Dent Mater 22(3):211–222. https://doi.org/10.1016/j.dental.2005.05.005

Flanagan M, Seixas N, Majar M, Camp J, Morgan M (2003) Silica dust exposures during selected construction activities. Am Ind Hyg Assoc J 64(3):319–328. https://doi.org/10.1080/15428110308984823

Formaldehyde in pressed wood products–NICNAS (2016)

Gandhi S, Lyon R, Speitel L (1999) Potential health hazards from burning aircraft composites. J Fire Sci 17(1):20–41. https://doi.org/10.1177/073490419901700102

Han C, Zhao A, Varughese E, Sahle-Demessie E (2018) Evaluating weathering of food packaging polyethylene-nano-clay composites: release of nanoparticles and their impacts. NanoImpact 9(November 2016):61–71. https://doi.org/10.1016/j.impact.2017.10.005

Hesterberg TW, Hart GA (2001) Synthetic vitreous fibers: a review of toxicology research and its impact on hazard classification. Crit Rev Toxicol 31(1):1–53. https://doi.org/10.1080/20014091111668

Hulin M, Simoni M, Viegi G, Annesi-Maesano I (2012) Respiratory health and indoor air pollutants based on quantitative exposure assessments. Eur Respir J 40(4):1033–1045. https://doi.org/10.1183/09031936.00159011

Hursthouse A, Allan F, Rowley L, Smith F (2004) A pilot study of personal exposure to respirable and inhalable dust during the sanding and sawing of medium density fibreboard (MDF) and soft wood. Int J Environ Health Res 14(4):323–326. https://doi.org/10.1080/09603120410001725667

IARC monographs on the evaluation of carcinogenic risks to humans (2012) International Agency for Research on Cancer, 100(Arsenic, metals, fibres, and dusts), 407–443

Inthavong K, Mouritz AP, Dong J, Tu JY (2013) Inhalation and deposition of carbon and glass composite fibre in the respiratory airway. J Aerosol Sci 65:58–68. https://doi.org/10.1016/j.jaerosci.2013.07.003

Jo M-R, Yu J, Kim H-J, Song J, Kim K-M, Oh J-M, Choi S-J (2016) Titanium dioxide nanoparticle-biomolecule interactions influence oral absorption. Nanomaterials 6(12):225. https://doi.org/10.3390/nano6120225

Jokar M, Abdul Rahman R (2014) Study of silver ion migration from melt-blended and layered-deposited silver polyethylene nanocomposite into food simulants and apple juice. Food Addit Contaminants A Chem Anal Control Exposure Risk Assess 31(4):734–742. https://doi.org/10.1080/19440049.2013.878812

Khurshid Z, Zafar M, Qasim S, Shahab S, Naseem M, AbuReqaiba A (2015) Advances in nanotechnology for restorative dentistry. Materials 8(2):717–731. https://doi.org/10.3390/ma8020717

Kleinsasser NH, Wallner BC, Harréus UA, Kleinjung T, Folwaczny M, Hickel R, Kehe K, Reichl FX (2004) Genotoxicity and cytotoxicity of dental materials in human lymphocytes as assessed by the single cell microgel electrophoresis (comet) assay. J Dentistry 32(3):229–234. https://doi.org/10.1016/j.jdent.2003.11.002

Leung CC, Yu ITS, Chen W (2012) Silicosis. The Lancet 379(9830):2008–2018. https://doi.org/10.1016/S0140-6736(12)60235-9

Lin QB, Li H, Zhong HN, Zhao Q, Xiao DH, Wang ZW (2014) Migration of Ti from nano-TiO_2-polyethylene composite packaging into food simulants. Food Addit Contaminants A Chem Anal Control Exposure Risk Assess 31(7):1284–1290. https://doi.org/10.1080/19440049.2014.907505

Linch KD (2002) Respirable concrete dust—Silicosis hazard in the construction industry. Appl Occup Environ Hyg 17(3):209–221. https://doi.org/10.1080/104732202753438298

Lippmann M (1990) Effects of fiber characteristics on lung deposition, retention, and disease. Environ Health Perspect 88:311–317. https://doi.org/10.1289/ehp.9088311

Mackevica A, Olsson ME, Hansen SF (2016) Silver nanoparticle release from commercially available plastic food containers into food simulants. J Nanopart Res 18(1):1–11. https://doi.org/10.1007/s11051-015-3313-x

Marsh GM, Morfeld P, Collins JJ, Symons JM (2014) Issues of methods and interpretation in the National Cancer Institute formaldehyde cohort study. J Occup Med Toxicol 9(1):1–9. https://doi.org/10.1186/1745-6673-9-22

Martins CA, Leyhausen G, Geurtsen W, Volk J (2012) Intracellular glutathione: A main factor in TEGDMA-induced cytotoxicity? Dent Mater 28(4):442–448. https://doi.org/10.1016/j.dental.2011.11.022

Meo SA (2004) Health hazards of cement dust. Saudi Med J 25(9):1153–1159. https://doi.org/10.15537/4918

Thomson SA, Hilaski RJ, Wright R, Mattie D (1990) Nonrespirability of carbon fibers in rats from repeated inhalation exposure. Chemical Research Development and Engineering Center Aberdeen Proving Groundmd

Qiao Y, Li B, Yang G, Yao H, Yang J, Liu D, Yan Y, Sigsgaard T, Yang X (2009) Irritant and adjuvant effects of gaseous formaldehyde on the ovalbumin-induced hyperresponsiveness and inflammation in a rat model. Inhalation Toxicol 21(14):1200–1207. https://doi.org/10.3109/08958370902806159

Sapuan SM (2017) Introduction. Compos Mater Concurr Eng Approach 22:1–27. https://doi.org/10.1016/B978-0-12-802507-9/00001-5

Saxena P, Pant A, Gupta S, Pant V (2012) Release and toxicity of dental resin composite. Toxicol Int 19(3):225. https://doi.org/10.4103/0971-6580.103652

Schweikl H, Spagnuolo G, Schmalz G (2006) Genetic and cellular toxicology of dental resin monomers. J Dent Res 85(10):870–877. https://doi.org/10.1177/154405910608501001

Selikoff IJ (2010) Mount Sinai-Irving J. Selikoff Center for Occupational & Environmental Medicine MDF Safety for Carpenters.

Singh S, Khanna VK, Pant AB (2017) Development of In Vitro Toxicology: A Historic Story. In Vitro Toxicology. Elsevier Inc.https://doi.org/10.1016/B978-0-12-804667-8.00001-8

Song W, Cao Y, Wang D, Hou G, Shen Z, Zhang S (2015) An investigation on formaldehyde emission characteristics of wood building materials in Chinese standard tests: product emission levels, measurement uncertainties, and data correlations between various tests. PLoS ONE. https://doi.org/10.1371/journal.pone.0144374

Soutis C (2005) Carbon fibre reinforced plastics in aircraft construction. Mater Sci Eng A 412:171–176. https://doi.org/10.1016/j.msea.2005.08.064

Spagnuolo G, D'Antò V, Cosentino C, Schmalz G, Schweikl H, Rengo S (2006) Effect of N-acetyl-L-cysteine on ROS production and cell death caused by HEMA in human primary gingival fibroblasts. Biomaterials 27(9):1803–1809. https://doi.org/10.1016/j.biomaterials.2005.10.022

Sparks JA (2019) Rheumatoid arthritis. Ann Internal Med 170(1):ITC1. https://doi.org/10.7326/AITC201901010

Thetkathuek A, Yingratanasuk T, Ekburanawat W (2016) Respiratory symptoms due to occupational exposure to formaldehyde and MDF Dust in a MDF furniture factory in Eastern Thailand. Adv Prev Med 2016:1–11. https://doi.org/10.1155/2016/3705824

Urcan E, Scherthan H, Styllou M, Haertel U, Hickel R, Reichl FX (2010) Induction of DNA double-strand breaks in primary gingival fibroblasts by exposure to dental resin composites. Biomaterials 31(8):2010–2014. https://doi.org/10.1016/j.biomaterials.2009.11.065

Walther T, Thömen H, Donath T, Beckmann F (2018) Microstructure of medium density fiberboard (MDF), d, 1–2

Wang Y, Yuan L, Yao C, Ding L, Li C, Fang J, Sui K, Liu Y, Wu M (2014) A combined toxicity study of zinc oxide nanoparticles and vitamin C in food additives. Nanoscale 6(24):15333–15342. https://doi.org/10.1039/c4nr05480f

Warheit DB, Driscoll KE, Oberdoerster G, Walker C, Kuschner M, Hesterberg TW (1995) Contemporary issues in fiber toxicology. Toxicol Sci 25(2):171–183. https://doi.org/10.1093/toxsci/25.2.171

Waritz RS, Ballantyne B, Clary JJ (1998) Subchronic inhalation toxicity of 3.5-μm diameter carbon fibers m rats. J Appl Toxicol 18(3):215–223. https://doi.org/10.1002/(SICI)1099-1263(199805/06)18:3<215::AID-JAT499>3.0.CO;2-W

Whillerhausen B, Callaway A, Ernst CP, Stender E (1999) The influence of oral bacteria on the surfaces of resin-based dental restorative materials—an in vitro study. Int Dental J, 231–239. https://doi.org/10.1111/j.1875-595X.1999.tb00527.x

Zhang N, Ma Y, Weir MD, Xu HHK, Bai Y, Melo MAS (2017) Current insights into the modulation of oral bacterial degradation of dental polymeric restorative materials. Materials 10(5):1–13. https://doi.org/10.3390/ma10050507

Chapter 10
Safety and Health Issues Associated with Fibre Reinforced Polymer Composites in Various Industrial Sectors

Abstract For more than 3000 years, natural fibres are used to reinforce materials generally, or biopolymer composites specifically. Water hyacinth, coir, sisal, oil palm empty fruit bunch, ramie, kenaf, grass reeds, sugarcane (sugar and bamboo), oats, rye, barley, wheat, rice husks, wood fibre, straw, jute are the examples of natural fibres. Hemp, flax, pennywort, kapok, paper-mulberry, Raphia, banana, pineapple leaf, and papyrus are among the types of natural fibres that have been investigated for use in plastics. The natural fibres are advantageous since they have marketing appeal and originate from renewable resources. For example, jute is a common reinforcement in India and have been commercially used in Asian markets for many years. It is increasingly used in many industries such as automotive, packaging materials, textile, agriculture, and also marine activities. The agricultural waste provides the biggest source of natural fibre reinforced biopolymer composites for commercial use. These are associated with the easy availability at a low cost of the natural fibres. Despite all the advantageous stated, this article reviews the health and safety concerns of natural fibre reinforced biopolymer composites.

Keywords Biopolymer composites · Natural fibre · Agriculture · Automotive · Medical and dental practices · Environment · Ethical issues

10.1 Introduction

Malaysia is one of 192 countries that ratified the Kyoto Protocol that was announced in Kyoto, Japan together with the Philippines, Myanmar, Indonesia, India, Brunei, etc. According to Kyoto Protocol, the countries need to reduce emissions of sulphur hexafluoride (SF_6), hydrofluorocarbons (HFCs), methane (CH_4), carbon dioxide (CO_2), perfluorocarbons (PCFs), and nitrous oxide (N_2O) which are the six main greenhouse gases (United Nation 2018). Due to that, Malaysia needs to reduce the use of neither of the gases. Many alternatives had been employed in many sectors to substitute the materials or the processes of the work to make sure that the emissions of those gases can be reduced. Among the alternatives was the use of natural fibres (NFs) that were believed to be as CO_2 neutral resource that can lead to greener planet

S. M. Sapuan et al., *Safety and Health in Composite Industry*,
Composites Science and Technology ,
https://doi.org/10.1007/978-981-16-6136-5_10

upon their usage (Wen and Ma 2012). In addition, NFs were first used in composites due to their advantages they offer that include low cost, environmental benefits, sustainability, and easy accessibility. Apart from NFs being one of the renewable energy sources, NFs can replace the man-made fibres that have many risks to human health. Even though there are many benefits of the NFs, so little NFs were produced because of the escalation of modern agriculture activities (Dungani et al. 2016).

Other than that, cellulosic fibre reinforced polymer composites have been used for many applications such as automotive components, aerospace parts, sports goods, and building industry. This is due to the advantageous properties of these materials, such as low self-weight, high strength, free formability, and substantial resistance to corrosion and fatigue (Jawaid et al. 2011).

Natural fibres are really useful in various fields as its usage now has gone worldwide. Numerous advantages can be stated out for this fibres as they are biodegradable and have minimal health hazards. As in the engineering field, the high tensile strength and percentage of elongation breaking compared to the pure biopolymer composites are desirable. In the aspect of environmental concern, NFs utilisation is useful in controlling the waste accumulation since it is biodegradable, thus, it will not harm the flora or fauna. However, despite the benefits, there are also disadvantages of this natural fibres. For instance, they are unable to withstand extreme temperature and start to degrade at high-temperature environment over time.

Fibre reinforced biopolymer composites with high strength of filling materials are desirable for various applications, e.g. in medicine and dentistry. Furthermore, in the few past decades, many researchers found the potential of the fibre reinforced biopolymer composites in a variety of clinical products. This is because, they have a good biocompatible, translucent, bondable, easy, and with aesthetic value. According to their unique characteristics and properties, fibres became one of the potential materials to be used in dentistry (Mahaveer et al. 2014).

Food packaging is used as a protection to maintain the quality and safety of food products during transport and storage to extend the shelf life and preserve the nutritional value. With this goal, the use of bio-nanocomposites appears as a promising solution to improve the properties of the existing biopolymer-based packaging materials, and for development of novel functionalities, while permitting replacement of non-biodegradable plastic packaging materials (Ramos et al. 2018).

10.2 Agriculture

10.2.1 Overview

Malaysia is among the 192 countries that signed the Kyoto Protocol together with the Philippines, Myanmar, Indonesia, India, Brunei and other Asian countries. Kyoto Protocol introduced the need to reduce emissions of greenhouse gases (GHG) (United Nation 2018). Malaysia has come out with many alternatives in many sectors towards the substitution of the materials or the processes that produce GHG with the aim to minimise GHG emissions. Natural fibres utilisation is seen as one of the efforts that were believed to be a CO_2 neutral resource (Wen and Ma 2012). In addition, the great benefits of NFs have motivated the idea of them replacing the man-made fibres that have many risks to human health. Despite the benefits of the NFs, the production rate of NFs are declining due to escalation of modern agriculture activities (Dungani et al. 2016).

The fibres are classified according to the length, plant type, natural origin, and physical and chemical structures. NFs, also known as the vegetable fibres are extracted from plants, especially the plants that contain high lignin, cellulose, hemicellulose, pectin, and waxy substances (Marques 2010). In general, natural fibres are extracted from primary and secondary plants (Rijswijk 2018). Primary plants are the plants that are grown for their fibre contents like hemp, jute, sisal, and cotton. Furthermore, secondary plants produces fibre as a by-product, e.g. stalk, agave, pineapple, and palm oil. The plant fibre are categorized based on their extraction method from plants. Plant fibre can be extracted from plant stems (bast fibre), fruits, leafs and seeds depending on the potential ability and function of fibres.

10.2.2 Fruit Fibre

As mentioned above, fruit fibres are the fibres extracted from the fruit of the plant. Cotton is an example of a fruit fibre. Based on cotton properties, it is a cellulose type fibre that cannot be formed at temperature of higher than 70 °F or 21 °C. The cotton fibre are extracted from seeds inside its fruit boll (Textile School 2018). Moreover, cellulose fibre is the main fibre wastes that come from agriculture activities. The dominant natural fibre in cotton is the seed hair of this plant, or Gossypium, that is present in the pure form of cellulose that exist in nature. The multi-layered structure contains primary wall, secondary wall, and lumen (Kozłowski 2012).

Cotton is being used widely in the world as it is the most significant fibre used in the textile industry, however, the process of harvesting of the cotton is still using the traditional method. There are machines for harvesting the cotton called strippers and pickers but to avoid damage to the quality of the cotton caused by the wind or rain, it needs to be picked quickly when the bolls open (Weigmann 2018). This does not happen simultaneously that has made handpicking more effective and preferred. Usually, the plants of cotton were in large scale area and it became high labour-intensive for cotton harvesting. This process in large scale area can create many risks in terms of safety and health concerns, e.g. ergonomic factor. Workers need to maintain the correct posture while performing the work. The effects that workers might get in the case of wrong posture is back pain due to bending of the body or standing for a long time. Sometimes, hand-harvesting also can cause heavy skin damage or cut. The worst case when dealing with cotton is the risk of exposure to cotton dust. The disease that related to cotton dust exposure is byssinosis caused by the exposure of cotton dust through inhalation as the route of exposure (Bakirci et al. 2004). Commonly, the person with byssinosis will show symptoms such as shortness of breath, coughing, wheezing, and chest tightness (Cooke 1979). All of the symptoms result from inflammation inside the lungs when the cotton dust were inhaled. It is crucial to make sure that the workers wear proper personal protection equipment (PPE), e.g. face mask to minimise the exposure risk.

The usage of the picker machine make the work easier and increase the production of the harvesting. In addition, human exposure can also be reduced if the machine does the harvesting work.

Another fruit fibre is coconut fibre (coir) obtained from the husk of the coconut. The fibres are light, durable to heat and saltwater, and also strong. Coir-based industries have been developed in many countries, especially coconut producing ones. According to statistics, Indonesian is the number one coconut producer in the world with 183,000,000 tones followed by the Philippines and India (Burton 2018). There was a study conducted to produce lightweight concrete using coconut fibres and they conclude that coconut fibre has high potential for the purpose that also lead to low-cost structures in the construction sector (Rahman and Sharma 2018). The production of coir involved many steps and every step have its own concerns in terms of safety and health. The first step is to pick the coconut from the coconut trees. The coconut trees are very high and can reach up to 25 m (80 ft) (Augustyn et al. 2018). Usually, monkeys are trained to climb the tree and pick the coconut, but sometimes the job is done by skilled people. There are risk for the workers to fall from the trees and get injured. When extracting the fibres, a machine was used to separate the fibre from the pith. Another high-risk step in producing the coir is the weaving process. Coir yarn is treated in dilute sulphuric acid solution (Ferdus Alam 2018). Sulphuric acid is a strong acid that is corrosive to the hand and skin upon contact. Sulphuric acid can cause serious burns with improper handling. Furthermore, direct contact of sulphuric acid also can cause permanent blindness and accidental ingestion resulted in irreversible organ damage, internal burns, and the worst case is death. Even the mist containing sulphuric acid also can affect the respiratory system, where the mist will accumulate at the area of the vocal cord and caused inflammation and lead to laryngeal cancer. Specific protection needs to be worn by the workers that perform this job to prevent any injuries or health effects.

10.2.3 Stem Fibres

Jute is the short length and weakest stem fibres. It is obtained from the stem that was cut from the ground using special equipment, sickle knife. Jute is grown every year mainly in Bangladesh and India and it is categorised under lime tree family. It sizes around 5 mm and usually used as a bag to pack things, ropes, wall decoration, carpet backing, and yarns. Jute being called "The Golden Fibre" because of its yellowish-brown and shining colour (Athalye 2018). Jute has low extensibility, high tensile strength, and can make good breathability of fabrics. However, apart from all of the jute benefits, it can cause health effects to those that handling it as it involved chemical like alkaline, acid, and bleaches such as sodium chlorite and hypochlorite that usually used as bleaches for jute. Both sodium chlorite and hypochlorite eliminate lignin from jute to make it whiter by changing the colour from natural yellow to reddish colour (Roy and Lutfar 2012). Ingestion of sodium chlorite in a high dose can lead to fatigue, headache and nausea, diarrhoea, insomnia, and also lowered blood pressure. There are also serious health problems related to this chemical, e.g. nosebleed, bronchitis, shortness of breath, skin burns, and coughing (Snyder 2018). Workers practising that kind of work will be exposed to the ergonomic hazard. In long term, workers can get serious back pain and maybe other ergonomic-related health effects.

Next is a flax fibre from the plants that are pulled from the ground when it is ripe. The plants are not being cut because the fibre loss will result from the stubble left in the field. Flax fibre is sturdy and will become stronger by 20% in wet conditions. Flax fibre can also captivate 20% humid without having a wet feeling (van Rijswijk et al. 2018). In the past few years, the function of flax fibres as strengthening agent in composite is rising alongside the demand for emerging sustainable materials. Moreover, flax fibres also offer specific mechanical properties and cost-effective. Flax fibres can reinforced the glass fibre and have high potential for future raw materials in structural request for automotive, infrastructure, and household products. Flax composites used in construction building materials will be lighter in weight other than lower in cost (Yan et al. 2014). As in construction, the materials used are usually heavy, hence, the flax composites can help making construction easier and reducing the workers' exposure to risk in the construction site. Heavy materials can increase the risk of fall or trip of workers during working and the materials from above.

The usage of agriculture fibres really give benefits to many industries and also can reduce the usage of biopolymer composites that have many negatives effects to human and also environment. They are among the great initiatives to reinforce biopolymer free from safety and health concern.

10.3 Automotive Industry

10.3.1 Background

Extended load from ecological activists, protection of essential resources, and stringency of laws continued by developing countries are prompted to the development and enhancement of regular materials with consideration on maintainable raw materials (Anandjiwala and Blouw 2007; Wittig 1994). Therefore, composites manufacturers need to search for plant-based natural fibres for reinforcement purposes, e.g. flax, hemp, kenaf, etc. Natural fibre reinforced polymer composites have been utilised for some applications, e.g. car segments, aviation parts, wielding products, and construction. This is due to the beneficial properties of these materials, for example, lightweight, high quality, free formability, and significant protection from erosion and weariness (Jawaid et al. 2011).

10.3.2 DPF in Automotive Safety and Health

Moreover, enforcing date palm fibre-normal fibres composite DPF-NFC in car manufacturing sector will have a huge environmental impacts through accomplishing an effective sustainable waste administration practice. The usage of date palm fibres (DPF) in the automotive industry will be an additional esteem stop towards upgrading the sustainability and production of such industry and also clearing up and dealing with an ecologically waste problem issue (AL-Oqla and Sapuan 2014). Other than that, depending on the way that date palm strand has a moderate estimation of prolongation to break property, it can be viewed as appropriate for an extensive variety of industrial applications from the mechanical execution point of view including the car industry (AL-Oqla and Sapuan 2014). Therefore, date palm fibres can be considered as free from safety and health issue of natural fibre reinforced polymer in automotive industries.

10.3.3 Lignocellulosic Fibres in Automotive

Additionally, natural fibres were able to get a wonderful weight decrease of 20% of the door panel to further enhance travellers' assurance in the event of a mishap (Kalia et al. 2011). Besides, based on Marsh (2003), by compression or infusion moulding of natural fibre reinforced polymers, it is potentially lower cost, including acoustic protection, safety management, and easier fabrication. Furthermore, according to Monteiro et al. (2009), polymer matrix composites (PMC) are used in vehicle parts, and the generally more prominent sturdiness of fibre PMC segment can retain the high energy impact during a car accident. Polymer-matrix composite (PMC) manufacturing from natural fibres, especially lignocellulose fibres extracted from plants has drawn a substantial amount of attention because of the natural fibres' advantages over synthetic fibres. Furthermore, natural materials are viewed as ecologically well-disposed as well as sustainable and biodegradable making them advantageous in terms of CO_2 productions (Monteiro et al. 2009). According to Jawaid et al. (2011), lignocellulose filaments were likewise anticipated to cause lesser medical issues for the workers delivering the composites compared to glass strand-based composites as they were not causing skin irritations and not associated with lung malignant growth. There are several other advantages and disadvantages of lignocellulose fibres as shown in Table 10.1 (Sreekumar 2008). The use of natural fibres mechanical application gives difficulties to the analysts to develop reasonable strategies to acquire great quality aspects for use as support for polymer composites (Jawaid et al. 2011). The utilisation of natural fibres reinforced biopolymer composite in the automotive industry is able to avoid health and safety issues for the users and employees in automotive manufacturing.

Table 10.1 Advantage and disadvantages of natural fibres (Sreekumar 2008)

Advantages	Disadvantages
Low specific weight results in a higher specific strength and stiffness than glass	Lower strength especially impact strength
Renewable resources, production require little energy and low CO_2 emission	Variable quality, influence by weather
Production with low investment at low cost	Poor moisture resistant which causes swelling of the fibres
Friendly processing, no wear of tools and no skin irritation	Restricted maximum processing temperature
High electrical resistant	Lower durability
Good thermal and acoustic insulating properties	Poor fire resistant
Biodegradable	Poor fibre/matrix adhesion
Thermal recycling is possible	Price fluctuation by harvest results or agricultural politics

10.4 Medical and Dental Applications

10.4.1 Overview

Many clinical applications have been using Fibre Reinforced Composite (FRC), especially in dentistry applications as shown in Fig. 10.1. This is because it is able to manipulate the composite's properties to match and suit structural aspects. Other than that, the good mechanical properties of the composites can optimise to medical applications such as artificial bone or dentin. The unique properties are needed in dental applications; designing framework such as crowns, posterior, and anterior fixed prostheses a (Freilich et al. 1998). In addition, it is also designed for replacement of tooth (Belvedere 1998), used as periodontal splints, posts (Yeluri and Munshi 2012), space maintainers (Tayab et al. 2011), and orthodontic retainers (Scribante et al. 2011).

In dental applications, a few challenges faced include replacing missing and lost teeth. Furthermore, these must be performed according to the provisional of restorations, laboratory costs, and also multiple appointments. Despite implants and diversity of bridge techniques are well proven according to few studies and serve well, but these problems are needed to be overcome immediately because researchers found a few newer options for dealing with these problems. In many researches, it had been found that fibre reinforced composites (FRC) is proven to solve some of the problems due to their properties. In 1960, it was found that the fibre reinforced composites (FRC) were used as acrylic denture bases. This technique was improved due to the FRC's mechanical properties, however, there existed disadvantages when it came to the clinical acceptance, in which the fibre volume was reduced and contributed to inadequate wetting of fibres. However, in the late'80s, these problems were solved

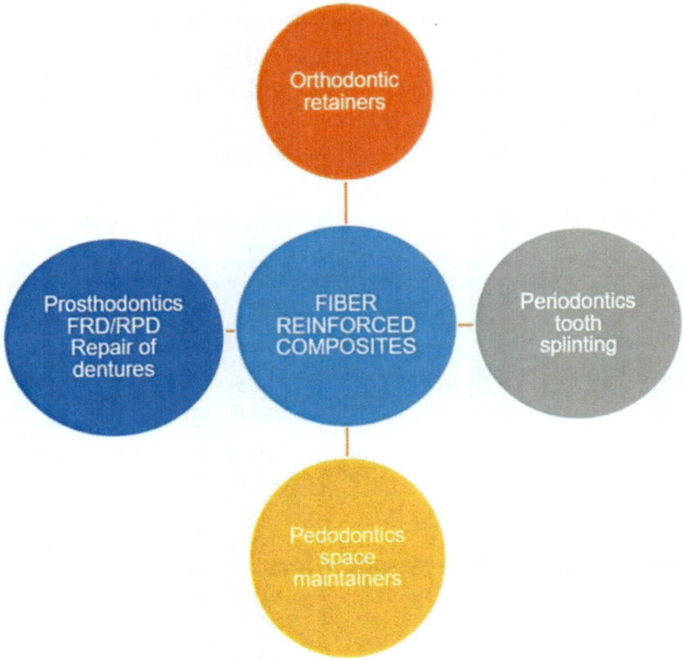

Fig. 10.1 Versatility of FRC in various dental specialities (Tayab et al. 2015)

by researchers by developing a fibre in a complete impregnation with resin (Kacir et al. 1977).

In addition, fibre reinforced biopolymer composites are very useful in biomedical applications. Therefore, it was believed that the use of biodegradable polymers was started in the late 1960s with the implementation of the first used bioabsorbable sutures. In the biomedical field, it was found that numerous applications have been used including body implants, materials of drug delivery and in vivo sensing, wound enclosures, and materials of tissue engineering (Suzuki and Ikada 2011).

This biodegradable polymers term refers to those polymers in medicine which are able to convert slowly to non-toxic products where the body can easily degrade without being eliminated from the body. Furthermore, there are a few alternative terms of polymers which are "bioabsorbable," "bioresorbable," and "bioerodible" with the same meaning as biodegradable polymer's definition. However, researchers found a bit confusion on the actual meaning of the each of terms, respectively (Martina and Hutmacher 2007).

- "Bioabsorbable" is referring to the biodegradable polymers that can be dissolved in body fluids. If the dissolved polymer is being metabolised and/or excreted by living organisms, thus it is categorised as bioresorbable.
- "Bioresorbable" is referring to the biodegradable polymers that are degraded by undergoing bulk erosion and resorbed within the body.

- "Bioerodible" is referring to water-insoluble biodegradable polymers which can be converted to water-soluble products by undergoing surface erosion due to the physical and chemical processes such as dissolution as well as resorbed within the body.

In addition, these polymers were studied broadly for the applications in biomedical. For example, they are used for fracture fixation, sutures, bone augmentation, wound closure by using mesh and staples, ligation clips, ligament reconstruction in orthopaedics, and dental repairs (Engelberg and Kohn 1991). These major applications are further explained in detail in the subsequent sections.

10.4.2 Dental and Medical Applications: Health Recovery

Structural materials of FRC have two constituents. For example, the surrounding matrix of FRC gives support to reinforcement and workability while another component is known as the reinforcing component. It provides strength and stiffness constituents, where most fibres being used in dental applications are carbon, polyethylene, glass, aramid, or polypropylene (Goldberg and Freilich 1999).

Fibres are well-known for their types and each of them has their own unique characteristics such as glass fibres, light-polymerised FRC substructure, and many more. In dental applications, glass fibre currently seems to be one of the choices because of its better aesthetic properties while possessing good adherence between these fibres and dimethacrylates (Kacir et al. 1977).

Other than that, fibres also exist naturally whether in animals or plants. These techniques and usage of these fibres must be understood to suit various applications (Sommerfeldt et al. 2001). Structure of wood contains a polymer fibre known as cellulose with a combination of fibre and lignin. Thus, wood is one of the examples of natural composites. However, when these elements individually are weak when separated. So, for producing strong components, it needs these combinations with the right structure of material's arrangements to develop composites with an impactful strength.

For instance, cotton is a material that consists of cellulose. This cotton is weaker than wood because wood structures are more organized (Cao et al. 2016). Other than that, bone is one of the components or elements in the human body which is also a natural biocomposite. Other than bone, tooth also has a structure which contains three different parts of natural biocomposites, such as cementum, enamel, and dentine. Main constituents of these structures are collagen and hydroxyl apatite (Hap) accompanied by minerals, specialised cells, and proteins. Thus, dentin has a soft structure compared to the enamel which has hard and brittle structures. The outer surface of the tooth is formed by enamel covering the inner dentine bulk. Furthermore, enamel exists and appears as a translucent material that once damaged is unable to reconstruct or regenerate again. However, it can be build up with other components called synthetic biocomposites such as ceramic, dental resin, or nanocomposite. This is due to some conditions such as pathological stimuli in the human body caused by the ageing process and physiological condition. However, a dentist must aware when using the biocomposites that different biocomposites have different appearances, properties, characteristics, and texture. Due to their specific percentages and also arrangements of these elements which are different in each of them, they can give a huge impact according to the characteristics and properties of each biocomposite. Figure 10.2 shows various applications of biocomposites in the human body.

In addition, there has a fibre called the light-polymerized FRC substructure, where this fibre can retain and inhibit the sticky oxygen on its outer surface which, allowing the bonding of chemical with composites. However, each type of fibres has their pros and cons when dealing with dental applications. Therefore, awareness and knowledge of these pros and cons with limitations are essential for selecting the best fibre according to the situations. Hence, each dentist must clearly understand and able to use these materials with well understanding of the basic structures of each type of materials (Kacir et al. 1977).

Furthermore, fibre reinforced composites being used as one of the treatments in dentistry due to the fibre that gives a reliable result with an immediate solution to overcome a certain problem related to the dentition. Thus, the treatment techniques have become well-known and demanded in dental applications for having advantages according to the types of fibre used. For example, some fibres are safely used in the narrow canal which acts as a post and gives reliable and excellent results. Other than that, the fibres also can reduce the use and wear of the dental tools and did not cause skin irritation. Most importantly, the fibre can be easily degraded and is non-toxic to human health. However, long term clinical studies are important and needed for evaluating the effects in terms of health after the prolonged use of fibre for treatment (Berthold et al. 2009).

In conclusion, clinicians have to make the best choice of fibre during the treatment of their patients. This is because, the fibres have different characteristic, pros, and cons that must be understood before their use in treatments. The unique characteristics of each fibre also give opportunities for their applications in all dentistry related fields, especially in clinical or laboratory dentistry.

10.5 Food Packaging

10.5.1 Overview

Food packaging is used to ensure the best quality of food is preserved during transportation and storage before getting to the customers. To achieve this aim, bionanocomposites are the most suitable materials to enhance the existing biopolymer-based packaging materials, and also as replacements of the non-biodegradable plastic packaging materials (Ramos et al. 2018).

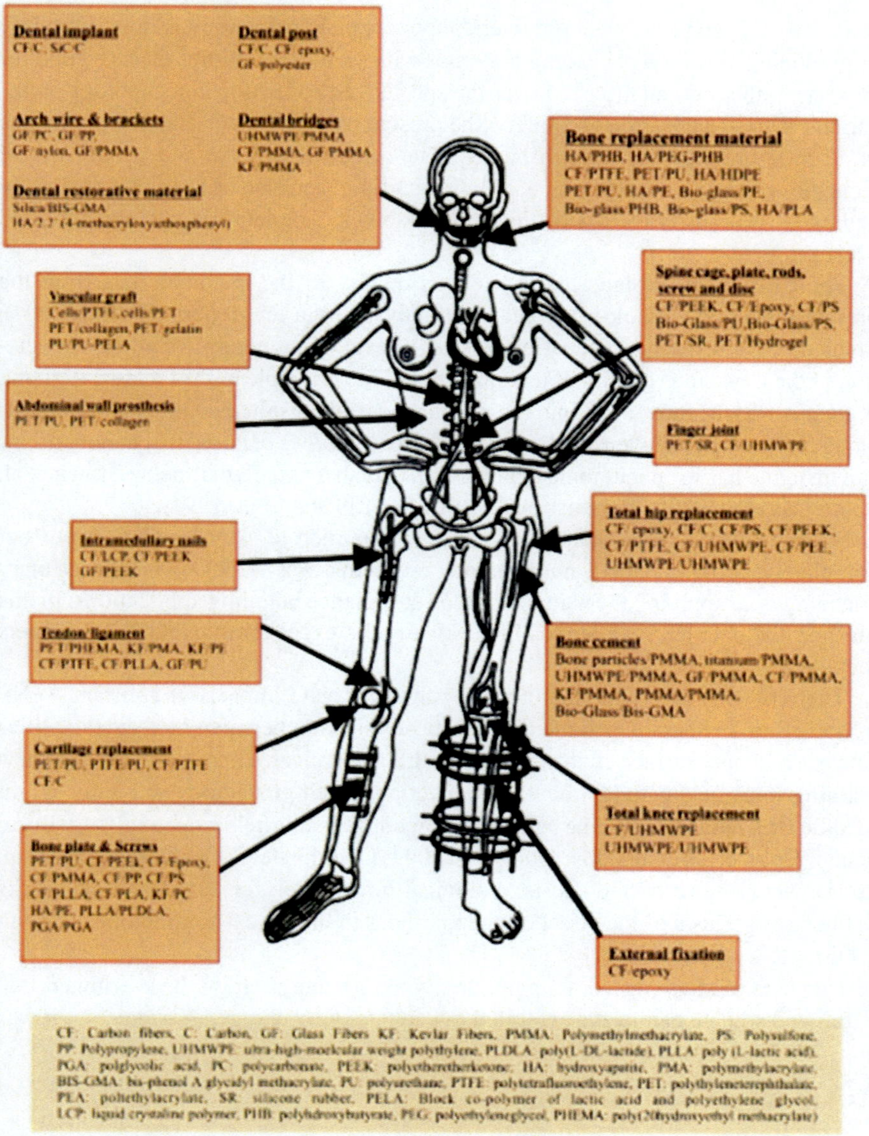

Fig. 10.2 Applications of biocomposites in the body of the human

10.5.2 Food Packaging

These days, most materials utilised in the bundling business are made from petroleum products (Shchipunov 2012). They are non-biodegradable, hence, the boundless utilisation of such materials motivated the awareness of their effect on the earth (Tang et al. 2012). Using bio-based polymers/polymers nanotechnology is in the certainty of producing new food packaging materials with various properties that are good for preserving the food quality. To cover the specialized parts of nanotechnology applications in food packaging and safety, the present business status and health impacts of all these innovations were discussed. The issues of concern and administrative strategies on the manufacturing, handling, bundling, and also utilisation were briefly addressed (Ramos et al. 2018). Figure 10.3 shows complete life cycle of the food packaging.

The capacity to evaluate the poisonous quality of the nanomaterials from the present molecule toxicological data was analysed by a few reports (Dreher 2004). Their discoveries demonstrated quite a high level of vulnerability in the expectation of the poisonous quality of nanomaterials. The principle hazard related to sizing segments was the movement into the diet that possibly resulted in unfavourable wellbeing impacts (Echegoyen and Nerin 2013). Some studies featured that nanoparticles can instigate harms, pneumonic aggravation, and also vascular ailment (Brown et al. 2000; Das et al. 2008; Nemmar et al. 2002; Oberdörster et al. 1994).

Since wellbeing and security properties of numerous nanomaterials were not completely comprehended, nourishment wellbeing was ought to be the primary concern while applying the nanomaterial in sustenance bundling applications. In this manner, the detailed toxicological investigation is expected to explain the dangers included.

There have been no attempts to study the aggregate impacts of constant AgNP introduction, deliberate examinations of the connection between molecule qualities (measure, shape, surface charge, and so forth), and danger. Thus, a great deal of new learning must be created on how nanomaterials based procedures and items might meddle with human wellbeing before any directions in this field can be built up (Borm et al. 2006; Derfus et al. 2004; Hoet et al. 2004; Oberdörster 2004; Oberdörster et al. 2005). Next, the restricted eco-toxicological information for nanomaterials blocks deliberate appraisal of the effect of nanoparticles in biological communities (Colvin 2003).

However, a study tested a single clay type and morphology. It determined that exfoliated silicate nanoclays exhibited low cytotoxicity and genotoxicity, but it was

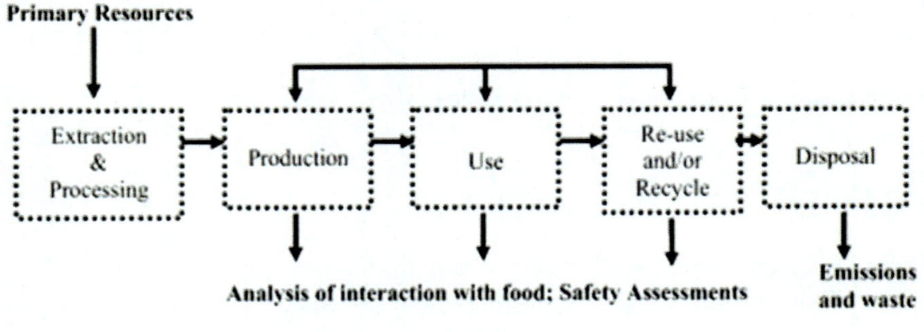

Fig. 10.3 Complete life cycle of the food packaging (Silvestre et al. 2011)

unclear whether it can be applied in a general sense. While there is a developing number of in vitro contemplates demonstrating that silver nanoparticles are cytotoxic to an assortment of mammalian cell types, in vivo studies exploring the precise impacts of AgNP introduction by oral routes of exposure are progressively uncertain. For instance, though AgNPs were discovered appropriated practically in every organ of rodents sustained an enduring nanoparticle diet, there were a couple of poisonous impacts seen aside from the highest concentration (Kim et al. 2008). An oral admission study in weaning pigs demonstrated AgNP amassing in the liver yet no intense lethal impacts (Fondevila et al. 2009). Then again, lymphocyte invasion and aggravation were observed in the livers of mice fed with nano and smaller scale silver particles and the impact was exacerbated when particles diameter was on the nanoscale (Cha et al. 2008). Figure 10.4 shows average diameter and number of measured microfibers and nanofibers.

Generally, there are three diverse methods for passage infiltrations of nanoparticles in organisms, which are inward breath, ingestion, and also skin penetration. The increasing scientific evidence reported that the free nanoparticles can cross cell boundaries and the exposure to a portion of these nanoparticles might lead to oxidative harm and inflammatory effects (Siegrist et al. 2008; Osman et al. 2003;

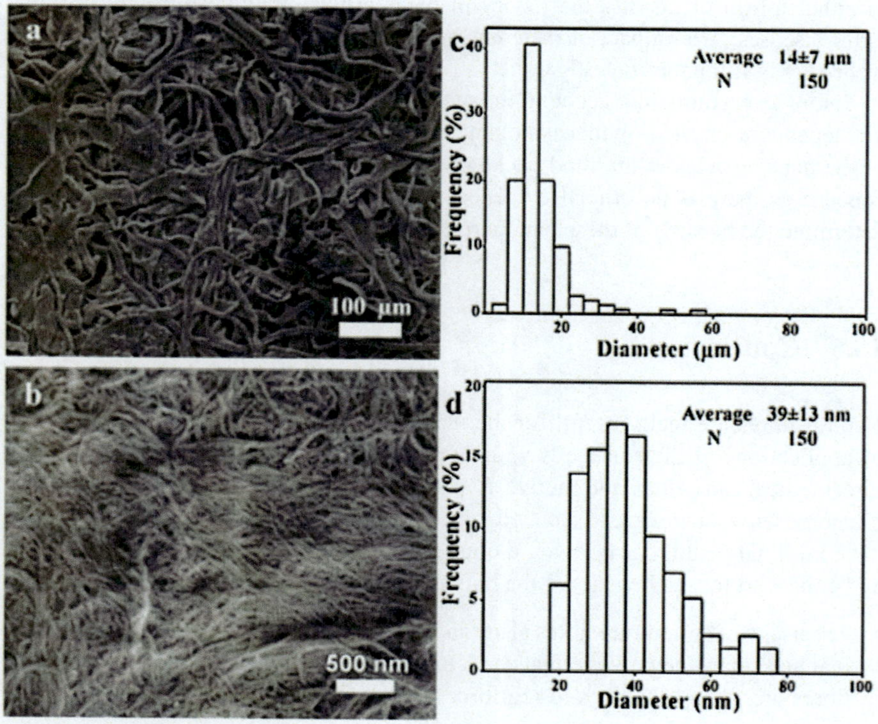

Fig. 10.4 The average diameter and number of measured microfibers (**a** and **c**) and nanofibers (**b** and **d**) (Ghaderi et al. 2014)

Li et al. 2003; Su et al. 2003; Kim and Rajapakse 2005; Kawasumi 2004; Akbari et al. 2007; Lagaron and Lopez 2010; Rhim and Ng 2007; Okada and Usuki 2006; Sanchez-Garcia and Lagaron 2010; Ciardelli et al. 2008). The workers in nanomaterials manufacturing and processing factories are highly exposed to the infiltrations of nanoparticles through inhalation and skin penetration. There is insufficient information on the effects of nanomaterials entry into the body. Despite the fact that nanomaterials function to propel sustenance allergen the board (Pilolli et al. 2013; Kumar et al. 2012), one can't disregard the way that specific nanomaterials might advance unfavourably and susceptible respiratory irritation (Yoshida et al. 2011; Ilves and Alenius 2016; Syed et al. 2013). A review of the literature revealed that exposure to nanomaterials promoted inflammatory response as common immune responses (Syed et al. 2013). Ryman-Rasmussen et al. (2009) also stated that mice with asthma would get airway fibrosis when exposed to multivalued carbon nanotubes with pre-existing inflammation.

Then, for all organs examined up until this point, either the compound segment of the nanoparticles or the nanoparticles themselves could be detected, demonstrating their dispersion to organs including the brain and testis/the reproductive system. There is particular concern about the possible relocation of nanoparticles into the cerebrum and foetus. Research in this territory must be conducted with the end goal to either affirm or dismiss the theory of nanoparticles relationship with different brain diseases. The impact of different particles utilised in food packaging on the wellbeing is still under investigation.

Taking everything into account, the most recent developments in food packaging are dependent on the polymer nanomaterials depicted in this paper. The utilisation of the nanoparticles in the food packaging can be said free from safety and health concern as there is no crucial evidence from the numerous studies conducted to determine the hazards of these nanoparticles on animals and human beings.

10.6 Conclusion

Natural fibres are really useful for the manufacturing of products in various fields of applications as their usage have now gone worldwide in the sector of agriculture, medical and dental, automotive, food packaging, engineering, and environment. Numerous advantages are associated with these fibres as they are biodegradable and have minimal health hazards. So, the usage of natural fibres needs to be promoted and enhanced for the benefits of the humankind in terms of safety and health.

- The usage of agriculture fibres also can reduce the usage of biopolymer composite that have many negatives effects on human and also the environment. Agriculture fibres are good initiatives to reinforce biopolymer that are free from safety and health concern.

- The use of natural fibre reinforced biopolymer composites in the automotive industry is able to avoid health and safety issues for the users and employees involved with the manufacturing.
- Natural fibres have the potential for reinforcement in dentistry, but the clinicians must understand very well their characteristic, pros, and cons prior to selecting the types of fibres to be used in the treatment.
- The natural fibres are very useful in food packaging and numerous studies have been conducted to determine the hazards of these nanoparticles on the animals and human beings.
- Therefore, it can be said that natural fibre reinforced biopolymer composites have various benefits in the aspect of global environmental pollution reduction.

Acknowledgements This chapter was written with M. D. N. Izzati, M. S. S. Syazwani, K. N. H. Aini, Y. Munirah, and M. Y. N. A. Danisya.

References

Akbari G, Sanavy SA, Yousefzadeh S (2007) Effect of auxin and salt stress (NaCl) on seed germination of wheat cultivars (Triticum aestivum L.). Pakistan J Biological Sci: PJBS 10(15):2557–2561

AL-Oqla F, Sapuan S (2014) Natural fiber reinforced polymer composites in industrial applications: feasibility of date palm fibers for sustainable automotive industry. J Cleaner Prod 66:347–354

Anandjiwala RD, Blouw S (2007) Composites from bast fibres-prospects and potential in the changing market environment. J Natural Fibers 4(2):91–109

Athalye A (2018) High tensile strength—jute is the golden fibre today. Retrieved from https://www.fibre2fashion.com/industry-article/7577/golden-fibre-still-glitters

Augustyn V, McDowell MT, Vojvodic A (2018) Toward an atomistic understanding of solid-state electrochemical interfaces for energy storage. Joule 2(11):2189–2193

Bakirci N, Niven RM, Tumerdem N (2004) Health effects of cotton dust and byssinosis. Turkish J Public Health 2:123–132

Belvedere PC (1998) Single-sitting, fiber-reinforced fixed bridges for the missing lateral or central incisors in adolescent patients. Dental Clinics of North America 42(4):665–682

Berthold MR, Cebron N, Dill F, Gabriel TR, Kötter T, Meinl T et al (2009) KNIME-the Konstanz information miner: version 2.0 and beyond. AcM SIGKDD Explorations Newsletter 11(1):26–31

Borm PJ, Robbins D, Haubold S, Kuhlbusch T, et al (2006) The potential risks of nanomaterials: a review carried out for ECETOC. Particle Fibre Toxic 3(1):1–35

Brown TN, Williams DR, Jackson JS, Neighbors HW, Torres M, Sellers SL, Brown KT (2000) "Being black and feeling blue": the mental health consequences of racial discrimination. Race Soc 2(2):117–131

Burton J (2018) The world leaders in coconut production. Retrieved from https://www.worldatlas.com/articles/the-world-leaders-in-coconut-production.html

Cao F, Zhao M, Yu Y, Chen B, Huang Y, Yang J, et al (2016) Synthesis of two-dimensional CoS1. 097/nitrogen-doped carbon nanocomposites using metal–organic framework nanosheets as precursors for supercapacitor application. J Amer Chem Soc 138(22):6924–6927

Cha SK, Ortega B, Kurosu H, Rosenblatt KP, Kuro-o M, Huang CL (2008) Removal of sialic acid involving Klotho causes cell-surface retention of TRPV5 channel via binding to galectin-1. Proc National Academy Sci 105(28):9805–9810

Ciardelli F, Coiai S, Passaglia E, Pucci A, Ruggeri G (2008). Nanocomposites based on polyolefins and functional thermoplastic materials. Polymer Int 57(6):805–836

Colvin VL (2003) The potential environmental impact of engineered nanomaterials. Nature Biotech 21(10):1166–1170

Cooke TF (1979) Chemical composition of cotton dust and its relation to byssinosis: a review of the literature. Textile Res J 49(7):398–404

Cotton fibers—The king of fibers—Textile School (2018). Retrieved from https://www.textilesc hool.com/129/cotton-fibers-the-king-of-fibers/

Das A, Pisana S, Chakraborty B, Piscanec S, Saha SK, Waghmare UV, Sood AK (2008) Monitoring dopants by Raman scattering in an electrochemically top-gated graphene transistor. Nature Nanotech 3(4):210–215

Derfus AM, Chan WC, Bhatia SN (2004) Intracellular delivery of quantum dots for live cell labeling and organelle tracking. Adv Mat 16(12):961–966

Dreher TW (2004) Turnip yellow mosaic virus: transfer RNA mimicry, chloroplasts and a C-rich genome. Mol Plant Pathol 5(5):367–375

Dungani R, Karina M, Sulaeman A, Hermawan D, Hadiyane A (2016) Agricultural waste fibers towards sustainability and advanced utilization: a review. Asian J Plant Sci 15(1):42–55. http://doi.org/10.3923/ajps.2016.42.55

Echegoyen Y, Nerín C (2013) Nanoparticle release from nano-silver antimicrobial food containers. Food Chemi Toxicol 62:16–22

Engelberg I, Kohn J (1991) Physico-mechanical properties of degradable polymers used in medical applications: a comparative study. Biomaterials 12(3):292–304

Ferdus Alam M (2018) Properties of coconut/coir fiber | Manufacturing process of coconut fiber | Application of coconut fiber. Retrieved from http://textilelearner.blogspot.com/2014/01/proper ties-of-coconutcoir-fiber.html

Fondevila M, Herrer R, Casallas MC, Abecia L, Ducha JJ (2009) Silver nanoparticles as a potential antimicrobial additive for weaned pigs. Animal Feed Sci Tech 150(3–4):259–269

Freilich MA, Karmaker AC, Burstone CJ, Goldberg AJ (1998) Development and clinical applications of a light-polymerized fiber-reinforced composite. J Prosthetic Dentistry 80(3):311–318

Ghaderi M, Mousavi M, Yousefi H, Labbafi M (2014) All-cellulose nanocomposite film made from bagasse cellulose nanofibers for food packaging application. Carbohyd Polym 104(1):59–65. https://doi.org/10.1016/j.carbpol.2014.01.013

Goldberg AJ, Freilich MA (1999) An innovative pre-impregnated glass fiber for reinforcing composites. Dental Clinics of North America 43(1):127–133

Hoet PH, Brüske-Hohlfeld I, Salata OV (2004) Nanoparticles–known and unknown health risks. J Nanobiotechnology 2(1):1–15

Jawaid MHPS, Khalil HA (2011) Cellulosic/synthetic fibre reinforced polymer hybrid composites: A review. Carbo Poly 86(1):1–18

Kacir L, Narkis M, Ishai O (1977) Oriented short glass fiber composites. III. Structure and mechanical properties of molded sheets. Polymer Eng Sci 17(4):234–241

Kalia S, Kaith BS, Kaur I (2011) Cellulose fibers: bio- and nano-polymer composites green chemistry and technology. Springer, Berlin, Heidelberg, UK

Kawasumi M (2004) The discovery of polymer-clay hybrids. J Polym Sci Part A Polym Chem 42(4):819–824. https://doi.org/10.1002/pola.10961

Kim SK, Rajapakse N (2005) Enzymatic production and biological activities of chitosan oligosaccharides (COS): a review. Carbohydrate Poly 62(4):357–368

Kim SK, Ravichandran YD, Khan SB, Kim YT (2008) Prospective of the cosmeceuticals derived from marine organisms. Biotechnol Bioprocess Eng 13(5):511–523

Kozłowski R (2012) Handbook of natural fibres. Woodhead Publishing, Oxford

Kumar V, Gu Y, Basu S, Berglund A, Eschrich SA, Schabath M, et al (2012) Radiomics: the process and the challenges. Magnetic Resonance Imag 30(9):1234–1248

Lagaron JM, Lopez-Rubio A (2010) Nanotechnology for bioplastics: opportunities, challenges and strategies. Trends Food Sci Techn 22(11):611–617

Li L, Stoeckert CJ, Roos DS (2003) OrthoMCL: identification of ortholog groups for eukaryotic genomes. Genome Res 13(9): 2178–2189

Ilves M, Alenius H (2016) Modulation of immune system by carbon nanotubes. Biomedical Appl Toxic Carbon Nanomat 397:e428

Mahaveer S, Jadhav Hemant R (2014) Melatonin: functions and ligands. Drug Discov Today 19(9):1410–1418

Marsh GP (2003) Man and nature. University of Washington Press

Martina M, Hutmacher DW (2007) Biodegradable polymers applied in tissue engineering research: a review. Polymer Int 56(2):145–157

Marques HMC (2010) A review on cyclodextrin encapsulation of essential oils and volatiles. Flavour Frag J 25(5):313–326

Monteiro S, Lopes F, Ferreira A, Nascimento D (2009) Natural-fiber polymer-matrix composites: cheaper, tougher, and environmentally friendly. JOM 61(1):17–22

Nemmar A, Hoet PM, Vanquickenborne B, Dinsdale D, Thomeer M, et al (2002) Passage of inhaled particles into the blood circulation in humans. Circulation 105(4):411–414

Oberdörster E (2004) Manufactured nanomaterials (fullerenes, C60) induce oxidative stress in the brain of juvenile largemouth bass. Environ Health Persp 112(10):1058–1062

Oberdörster G Ferin J, Lehnert BE (1994) Correlation between particle size, in vivo particle persistence, and lung injury. Environ Health Persp 102(suppl 5):173–179

Oberdörster G, Oberdörster E, Oberdörster J (2005) Nanotoxicology: an emerging discipline evolving from studies of ultrafine particles. Environ Health Persp 113(7):823-839

Okada A, Usuki A (2006) Twenty years of polymer-clay nanocomposites. Macromolecular Mat Eng 291(12):1449–1476

Osman I, Young A, Ledingham MA, Thomson AJ, Jordan F, Greer IA, Norman JE (2003) Leukocyte density and pro-inflammatory cytokine expression in human fetal membranes, decidua, cervix and myometrium before and during labour at term. MHR: Basic Science of Reproductive Medicine 9(1):41–45

Pilolli R, Monaci L, Visconti A (2013) Advances in biosensor development based on integrating nanotechnology and applied to food-allergen management. TRAC 47:12–26

Rahman S, Sharma A (2018) A complete MRCP (UK) parts 1 and 2 written examination revision guide: a systems-based competencies approach. CRC Press

Ramos R, Comas-Cufí M, Martí-Lluch R, Balló E, Ponjoa, A, Alves-Cabratosa L, et al (2018) Statins for primary prevention of cardiovascular events and mortality in old and very old adults with and without type 2 diabetes: retrospective cohort study. BMJ 362

Rhim JW, Ng PK (2007) Natural biopolymer-based nanocomposite films for packaging applications. Critical Rev Food Sci Nutrition 47(4):411–433

Rijswijk J (2018) Oil-based contrast cost effective in hysterosalpingography. PharmacoEconomics Outcomes News 813:21–26

Roy S, Lutfar LB (2012) Handbook of natural fibres. Woodhead Publishing, Oxford, pp 24–46

Ryman-Rasmussen JP, Cesta MF, Brody AR, Shipley-Phillips JK, Everitt JI, Tewksbury EW et al (2009) Inhaled carbon nanotubes reach the subpleural tissue in mice. Nature Nanotech 4(11):747–751

Sanchez-Garcia MD, Lagaron JM (2010) On the use of plant cellulose nanowhiskers to enhance the barrier properties of polylactic acid. Cellulose 17(5):987–1004

Scribante A, Sfondrini MF, Broggini S, D'Allocco M, Gandini P (2011) Efficacy of esthetic retainers: clinical comparison between multistranded wires and direct-bond glass fiber-reinforced composite splints. Int J Dentistry

Shchipunov Y (2012) Bionanocomposites: green sustainable materials for the near future. Pure Applied Chem 84(12):2579–2607

Siegrist M, Stampfli N, Kastenholz H, Keller C (2008) Perceived risks and perceived benefits of different nanotechnology foods and nanotechnology food packaging. Appetite 51(2):283–290. https://doi.org/10.1016/j.appet.2008.02.020

Silvestre C, Duraccio D, Cimmino S (2011) Food packaging based on polymer nanomaterials. Prog Polym Sci (oxford) 36(12):1766–1782. https://doi.org/10.1016/j.progpolymsci.2011.02.003

Snyder A (2018) Sodium chlorite: what is it & can it be medicinal? Retrieved from https://www.healthline.com/health/sodium-chlorite

Sommerfeldt DW, McLeod KJ, Rubin CT, Hadjiargyrou M (2001) Differential phosphorylation of paxillin in response to surface-bound serum proteins during early osteoblast adhesion. Biochem Biophy Res Comm 285(2):355–363

Sreekumar PA (2008) Matrices for natural-fibre reinforced composites. In: Pickering KL (ed) Properties and performance of natural-fibre composite. Woodhead Publication Limited, Brimingham, UK, p 541

Su KH, Wei QH, Zhang X, Mock JJ, Smith DR, Schultz S (2003) Interparticle coupling effects on plasmon resonances of nanogold particles. Nano Lett 3(8):1087–1090

Suzuki S, Ikada Y (2011) Biomaterials for surgical operation. Springer Science & Business Media

Syed S, Zubair A, Frieri M (2013) Immune response to nanomaterials: implications for medicine and literature review. Curr Allergy Asthma Rep 13(1):50–57. https://doi.org/10.1007/s11882-012-0302-3

Tang F, Li L, Chen D (2012) Mesoporous silica nanoparticles: synthesis, biocompatibility and drug delivery. Adv Mat 24(12):1504–1534

Tayab T, Vizhi K, Srinivasan I (2011) Space maintainer using fiber-reinforced composite and natural tooth–a non-invasive technique. Dental Traumatology 27(2):159–162

Tayab T, Shetty A, Kayalvizhi G (2015) The clinical applications of fiber reinforced composites in all specialties of dentistry an overview. Int J Compos Mater 5(1):18–24. https://doi.org/10.5923/j.cmaterials.20150501.03

United Nation (2018) The Sustainable Development Goals Report 2018. Retrieved from https://www.un.org/development/desa/publications/the-sustainable-development-goals-report-2018.html. 28 October 2021

van Rijswijk K, Brouwer WD, Beukers A (2018) Application of natural fibre composites in the development of rural societies. Retrieved from http://www.fao.org/docrep/007/ad416e/ad416e06.htm

Weigmann H (2018) Cotton | Description, cultivation, diseases, & facts. Retrieved from https://www.britannica.com/topic/cotton-fibre-and-plant

Wen J, Ma J (2012) Modulating morphology of thiol-based monolayers in honeycomb hydrogen-bonded nanoporous templates on the Au(111) surface: simulations with the modified force field. J Phys Chem C 116(15):8523–8534. https://doi.org/10.1021/jp211206n

Wittig R (1994) Invisible rendezvous: connection and collaboration in the new landscape of electronic writing. Wesleyan University Press

Yan L, Chouw N, Jayaraman K (2014) Flax fibre and its composites—a review. Compos B Eng 56:296–317

Yeluri, R., & Munshi, A. K. (2012). Fiber reinforced composite loop space maintainer: An alternative to the conventional band and loop. Contemporary clinical dentistry, 3(Suppl1), S26.

Yoshida T, Yoshioka Y, Fujimura M, Yamashita K, Higashisaka K, Morishita Y et al (2011) Promotion of allergic immune responses by intranasally-administrated nanosilica particles in mice. Nanoscale Res Lett 6(1):195. https://doi.org/10.1186/1556-276X-6-195

Chapter 11
Occupational Safety and Health Administration in Composite Industry

Abstract This book chapter reviews the issues of occupational safety and health administration in the composites industry in three countries, the United States, the United Kingdom, and Malaysia. In the United States, Occupational Safety and Health Administration (OSHA) was established in 1971 under the Occupational Safety and Health Act of 1970. It was the Department of United States of Labour agency that enforced the law and standards and responsible for ensuring safe and healthy working conditions and environments for the workers of both genders. In 1802, Sir Robert Peel introduced the first piece of Occupational Safety and Health legislation in the United Kingdom, The Health and Morals of Apprentices Act. The historical aims of OSH have been mostly safety matters, measurement problems, enhancing difficulty between well-being and health, and associated financial costs. In Malaysia, The Occupational Safety and Health Act 1994 (OSHA) was developed due to increased concern in enormous movement and transformation from year to year and was well-established in Feb 1994 in Malaysia. For further understanding, this paper discusses Occupational Safety and Health management in those countries. Based on the literature, different countries have different management systems in terms of administration. Besides, this paper also confers about the safety and health legislation that has been practiced in these three countries. The United States was concerned with protecting the workers from hazardous materials and chemicals, improving workplace performance, reducing workplace injuries and death, providing adequate information, and training to all employees especially those who engage in hazardous and dangerous works. In Malaysia, the purpose of the bureau is to encourage employers and workers to practise effective safety and health measures at the geographic point. This paper also proposes the safety and health management of composite industries in the United States, the United Kingdom, and Malaysia. Effective safety will solely be achieved once there is proper management of interaction between technological systems and other people.

Keywords OSH management · Legislation · Safety · Health · Composite industry

S. M. Sapuan et al., *Safety and Health in Composite Industry*,
Composites Science and Technology ,
https://doi.org/10.1007/978-981-16-6136-5_11

11.1 Introduction

In 1970, an administrative related to health and safety at work was established by the president of the United States which known as OSHA (Labor 2020). OSHA and its counterparts have improved safety at the workplace dramatically, reducing more than 65% of accidents at work through the cooperation between employers, employees, safety and health professionals, unions, and supporters (Labor 2020). The primary purpose of OSHA was to create surroundings healthy workplace for employees by conducting safety training and education on health and safety at work in accordance with the laws related to occupational safety and health (Labor 2020).

The Bureau of Labour Statistics confirmed that about 14,000 workers were killed due to workplace accidents or about 38 deaths per day in 1970, while the mortality was dropped to 4500 or about 12 deaths per day in 2010 (Labor 2020). Besides that, OSHA standards also clarify the employer's need to provide the safety, health, and welfare for workers at the workplace by conducting safety training, preventing and minimizing exposure to hazardous chemicals, providing appropriate protective equipment for workers while working, and conducting medical surveillance (Labor 2020).

The rapid development of the composite industry has shown that organic composites are widely used as the uniqueness of these materials allow them to continue to grow (Olson 1992). For example, in the United States, composites are in high demand and are being widely used in construction, infrastructure, industrial, transportation, sports, and recreation (Mazumdar et al. 2018). Composites are being used in many industries due to their desirable properties that include durability, design flexibility, corrosion resistance, strength, lightweight, low maintenance, consume less energy, strong, and many more (Mazumdar et al. 2018). However, to maintain the safety and security of these materials use, several factors should be taken into account for the control of dangerous risks in the composites manufacturing industry (Olson 1992).

Thus, this book chapter reviews several main points related to the topic of occupational safety and health administration in the composites industry, which are occupational safety and health in the United States, the United Kingdom, and Malaysia, health components in the composites industry, and safety components in the composites industry.

11.2 Occupational Safety and Health in the United States

11.2.1 Background

In the United States, composites are in high demand and are being widely used in construction, infrastructure, industrial, transportation, sports, and recreation ("Composites Manufacturing's 2018 State of the Industry Report" 2018). One of the composite industries in the United State is the American Composites Manufacturers Association (ACMA) ("Composites Manufacturing's 2018 State of the Industry Report" 2018). Composites' broad applications are due to their benefits which are durability, design flexibility, corrosion resistance, strength, lightweight, low maintenance, consume less energy, high strength, and many more ("Composites Manufacturing's 2018 State of the Industry Report" 2018).

According to market research firm Lucintel, by 2022 the global composites market is predicted to get $113.2 billion ("Composites Manufacturing's 2018 State of the Industry Report" 2018). Even though having interruptions in raw material supplies in terms of lack of shipping containers, extended shipping periods, and plant shutdowns in the Gulf Coast and Southeast because of Hurricanes Harvey and Irma, the US composites industry still had a robust performance in 2017 ("Composites Manufacturing's 2018 State of the Industry Report" 2018).

Glass fiber is the main strengthening material in the composites industry. According to the CEO Lucintel, the U.S. glass fibre market grew by four percent in 2017 to get 2.5 billion pounds in terms of volume and $2.1 billion in terms of value ("Composites Manufacturing's 2018 State of the Industry Report" 2018). By 2023, the market is predicted to reach 3.1 billion pounds with a compound annual development rate of 3.4% ("Composites Manufacturing's 2018 State of the Industry Report" 2018). Figure 11.1 shows the increasing demand and capacity of glass and carbon fibres composites.

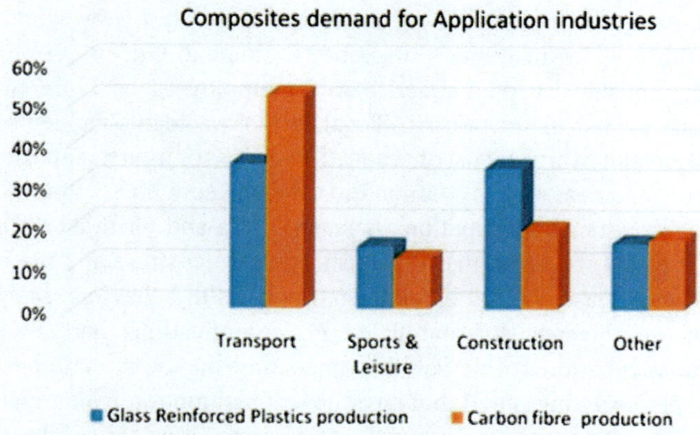

Fig. 11.1 Percent demand of glass and carbon fibres composites for industry sectors (Psomopoulos et al. 2019)

11.2.2 Scenario and Health Issues in the United States

Employees have the probability to be exposed to a wide range of physical and chemical hazards even with adequate safety training and devotion to procedures (Geldart 2014). Geldart (2014) stated that in the United States, fatal work injuries were over 5500 per year in the mid-2000s and roughly 4550 per year over the last few years, meanwhile, non-fatal injuries with lost time days from work were around 120 per 10,000 full-time employees in 2010 (Geldart 2014).

The U.S. Bureau of Labor Statistics reported that there were almost 2.8 million nonfatal job site injuries and illnesses reported by private industry employers in 2017, which occurred at a rate of 2.8 cases per 100 full-time equivalent (FTE) workers (U.S. Bureau of Labor Statistics 2018). According to estimates from the Survey of Occupational Injuries and Illnesses (SOII), Private industry employers reported almost 45,800 fewer nonfatal injury and illness cases in 2017 compared to a year earlier (U.S. Bureau of Labor Statistics 2018).

The total recordable cases (TRC) incidence rate among private industry employers declined from 2.9 cases per 100 full-time workers in 2016 to 2.8 cases in 2017 (U.S. Bureau of Labor Statistics 2018). For days away from work, job transfer or restriction (DART) cases dropped from 1.6 cases per 100 full-time workers in 2016 to 1.5 cases in 2017 (U.S. Bureau of Labor Statistics 2018). For cases with days of a job transfer or restriction only (DJTR), the recorded cases were the same from 2011 to 2017, while other recordable cases (ORC) were the same from 2016 to 2017 (U.S. Bureau of Labor Statistics 2018). Figure 11.2 shows the distribution of nonfatal occupational injuries and illnesses by private industry sector, in 2005 (McDiarmid 2014).

Based on Fig. 11.2, healthcare and social assistance are the main reported non-fatal occupational injuries and illnesses in the private sector with 8.1 injuries and 34.7 illnesses reported. Meanwhile, the manufacturing sector is the second-highest reported non-fatal occupational injuries and illnesses in the private sector with 394.6 injuries and 34.3 illnesses reported. For construction, there were 194.3 injuries and 3.8 illnesses reported.

One of the most challenging jobs is ensuring the safety of the workers when performing a job near or involving a machine (Danna and Griffin 1999). In many cases, it is an industry-accepted practice to let employees be close to moving machinery to accomplish their works (Ruff et al. 2011). Accidents/incidents are those occurrences that result in higher cost of production, illness or injury, damage to equipment or property, and near misses (Vernon Eidson and Reese 2006). One of the most concerning health issues is occupational stress (Danna and Griffin 1999). Cooper and Cartwright (1994) revealed the relationships between stress and the incidence of coronary heart disease, mental breakdown, poor health behaviours, job displeasure, accidents, absenteeism, lost output, family complications, and certain forms of cancer. The occupational illnesses that impacting the construction employees have not been precisely measured, but an educated assumption is that construction employees suffer both acute (short-term) and chronic (long-term) illnesses from

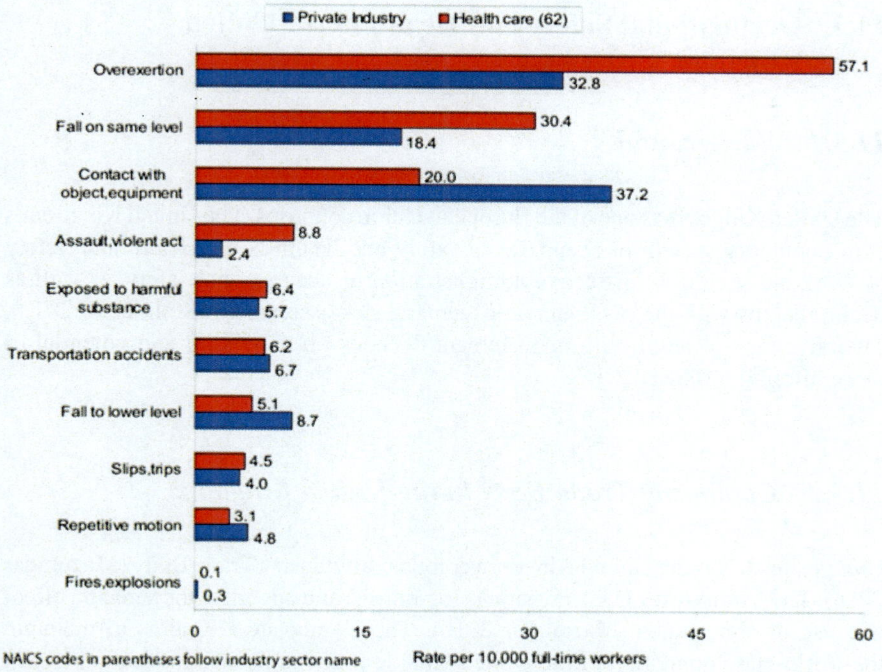

Fig. 11.2 Incidents rates of nonfatal occupational injuries and illnesses involving days away from work by selected nature of injury or illness, healthcare and social assistance sector (62)* and private industry, 2005 (McDiarmid 2014)

their exposure to chemicals, dust, fibres, noise, radiation, vibration, and extreme temperatures (Vernon Eidson and Reese 2006).

The most common health issues faced by workers in industrial works are exposure to excessive noise, falls, fractures, musculoskeletal disorders (upper limb disorders), low back pain, respiratory ailments, asthma, and psychosocial problems (Geldart 2014).

In a conclusion, Occupational Safety and Health Administration is a national health public health that emphasises on safety and health of employees including men and women in the workplace. Even though workplace injury/accident or illnesses cannot be prevented in the workplace, having the standards and regulations, could reduce the number of cases of accident or injury at the workplace and creating awareness to the employees and employer.

11.3 Occupational Safety and Health in the United Kingdom

11.3.1 Background

The United Kingdom is one of the European Union countries. The United Kingdom's core regulatory system in occupational safety and health is the Health and Safety at Work etc. Act 1974. Effective administration of occupational safety, as well as health, begins with the involvement of general risk assessment (Stolk et al. 2012). The efficiency of any health management depends on the ability and potential of those offering it (IOSH 2016).

11.3.2 Composite Trade Body in the United Kingdom

Composites UK is the United Kingdom composites industry Trade Body (Mazumdar 2018). The Composites UK's mission is to improve and advertise the safe and effective use of composites (Mazumdar 2018). The Composite UK aims to maintain the world-class reputation of the United Kingdom in composites innovation (Anon 2016). The Technology Working Gather incorporates the United Kingdom's composites aimed at science and technology society, subsidising organisations, and is functioning to convey arrangements to overlap trade division encounters inside the industrial groups (Anon 2016). The Composite UK also set sights on the sequence of the process involved in delivering the composite commodity to ensure global trade (Anon 2016).

Other than that, Composite UK also seeks for understanding and conveys the requirement of the present and upcoming workforce by the United Kingdom composite sequence of the process involved in the production of the composite commodity (Anon 2016). The Composite UK also aims to provide a sustainable composites industry into the future and to eliminate barriers to the new finding applications in United Kingdom composite products (Anon 2016). The strategy classified and planned issues that were essential for every single division and then finalized mutual requirements and acts which would enhance the benefit to various divisions (Anon 2016).

11.3.3 Occupational Health and Safety Management System

The efficiency of any health management depends on the ability and potential of those offering it, for example, in the case of occupational health services, there are numerous important elements to be concerned about (IOSH 2016). The prominence of applying appropriate Occupational Safety and Health measures and administration systems is established in the writings and as evidence of top-level management dedication, a documented policy is vital (Collins 2018). As expected, most developments in the European Union (76%) with elevated frequency were observed in the enormous foundation of a documented Occupational Safety and Health Policy (Collins 2018). There were significant variations between the European Union countries, with predominantly high levels in Ireland and the United Kingdom (Collins 2018).

In Health and Safety at Work etc. Act 1974, Part I Section 2(3) stated that it shall be the responsibilities of a person or organization that employs people to organize and improve a documented statement of this general policy of health and safety from time to time. This might be applicable to the employees and the organisation that procedures of health and safety should be informed to the employees for its implementation, and to bring the statement and any revision of it (WORSLEY 1978). In selected countries of northern Europe, the utmost number of OSH management methods establishments was observed, for instance, in Sweden and the United Kingdom, eight aspects of establishments were reported (Stolk et al. 2012). The average amount of viewpoints for risk management in manufacturing is around six aspects of establishments as reported in the scope of the risk management associated with the industry (Stolk et al. 2012). Table 11.1 shows the different approaches to occupational safety and health management systems in European Union countries.

Table 11.1 Different approaches to OSH management systems (Stolk et al. 2012)

Modalities	Cascade	Innovative	Applied	Ideological
Origin of decision	Senior management	Supervisory level management	OHSE dept. (quality, health, safety, environment)	Senior management
Expected goal	Integration of OSH into local policies	Integration of OSH into practices	Formalisation of OSH management	Integration of OSH into individual's behaviour
Leaders and partners	National (or regional) management and safety line managers	Supervisory level management and staff together with safety line managers	Supervisory level management and safety line managers	Senior and supervisory level management
Method of dissemination	Information and awareness-raising meetings	Working groups with staff	Supervisory level management meetings	Human resources and individual assessments
Resources provided	Limited	Negotiable	Limited	Extensive
Employee involvement	Low	High to start with	Limited	High at the end
Link with CHST*	Information	Participation and validation	Consultation	Information

*CHSCT: French Health, Safety and Working Conditions Committee

11.4 Occupational Safety and Health in Malaysia

11.4.1 Background

The Bright Sparklers fireworks explosion in 1991 of a manufacturing plant was thought of as the worst industrial incident within Malaysia's history. It's asserted 26 lives and injured over a hundred folks. A royal commission was founded to investigate this incident that alerted the government and therefore the public on the vitality of the industrial safety rules for the manufacturers. Because of this, an additional step was taken by the government by the introduction of the Safety and Health Act (OSHA) 1994. The authority body was established in 1994 to deal with safety and health problems in industrial within the producing district (Hee 2014).

11.4.2 Management Practices in Safety Culture

The Occupational Safety and Health Act 1994 (OSHA) concerned an enormous movement and transformation from year to year and was well-established in Feb 1994 in Malaysia. The purpose of the bureau is to encourage employers and workers to practise effective safety and health measures at the geographic point (Awang and Kamil 2014). Companies are possible offenders according to the Occupational Health and Safety Act (OSHA) 1994 (Malaysia) as employers (possibly entrepreneurs), and companies might carry out tasks that could injure or kill their workers and harm the public (Hassan 2015). In every organization, work-related accidents cause serious problems and place enormous costs on the industry and the country. In general, workplace accidents were attributed to safety engineering aspects that arise from the poor use and handling of machinery and tools (Ali et al. 2009). Table 11.2 shows the number of prosecutions under Section 15(1) and Section 17(10) of OSHA from 2008 to 2011 (Hassan 2015).

Table 11.2 Number of prosecutions under Section 15(1) and Section 17(10) of OSHA from 2008 to 2011 (Hassan 2015)

Year	Number of prosecutions under Section 15 (employer has breached his duty against its own employer)	Number of prosecutions under Section 17 (employer has breached his duty against no-employer)
2008	3	2
2009	15	27
2010	21	10
2011	36	27

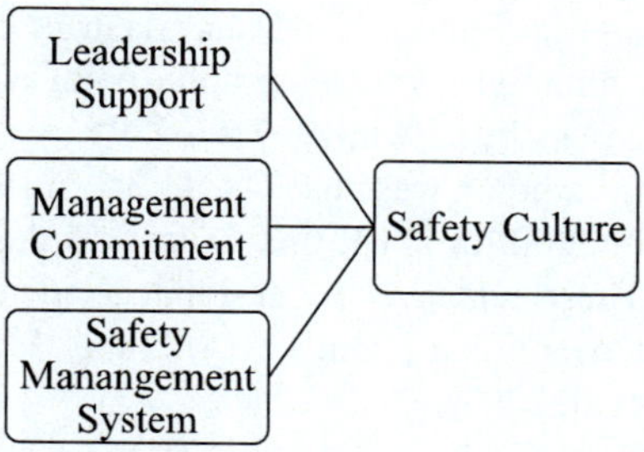

Fig. 11.3 Factors that can lead to safety culture in the manufacturing industries (Hee 2014)

The number of accidents in industrial and occupational diseases are growing at an alarming rate and OSHA plays a major role in helping employers and employees to become more aware of safety. The act lays down administrative actions such as awareness of campaigns, awards, and certification of standards, although they continue to be implemented as subjected to the enforcement of the legal (prosecution) (Hassan 2015). Therefore, in order to understand work-related accidents, correct understanding of the person who works independently or in groups that operate in a technological system is becoming increasingly important. Effective safety will solely be achieved once the interaction between technology systems and other people is properly managed (Ali et al. 2009).

Manufacturing businesses ought to have an on-site security management system for the efficient management of all protective problems. A safety management system shall accommodate security parts that are important in risk management. It should come up with the complete implementation and management of safety at various stages and features within the organization. It is the responsibility of different levels of staff within the safety management system to participate explicitly in safety programs. In general, the programs cover the entire organization and require staff to participate in activities such as identification of hazards and risk assessment (Hee 2014). Figure 11.3 shows the factors that can lead to safety culture in the manufacturing industries (Hee 2014).

11.4.3 Composite Industry Scenario in Malaysia

From year to year, statistics show an increasing number of accident cases in the industrial sector. All accidents can be eliminated with a correct and efficient health and safety policy or regulation layout by prime management (Awang and Kamil 2014). The average range of occupational accidents in Malaysia over nine-year from 1995 to 2003 was 249 cases annually, which solved an average daily rate of 250 occupational accidents. The Malaysia Social Insurance Organization (SOCSO) paid an estimated RM305 million in compensation for work in 2003. Based on the information revealed within the SOCSO annual report, there were 114,134 cases of workplace injuries rumoured in Malaysia in 1995-the whole range of occupational injuries. This rumoured case of geographic point accidents was decreased by 67% in 1996 (106,508 cases), by 87% in 1997 (86,589 cases), and by 1.4% in 1998 (85,338 cases). However, the amount was raised by 11.3% in 1999 to 92,074 cases and by 3.2% to 95,006 cases in 2000. In 2001, this figure was decreased another time by 9.6% to 85,926 cases. These evidences suggest that accidents at work in Malaysia are perpetually unsteady (Awang and Kamil 2014). Table 11.3 shows the injuries data for three years (Ali et al. 2009).

In 2011, there were 35,088 cases (58.58%) of commercial accidents reported and revealed by the Social Security Organization (SOCSO) of Malaysia (Zahrim Bin Osman et al. 2016). In the year 2014, there was a complete of 35,294 industrial accident rumoured cases. It's a tiny low reduction from 35,898 cases the year before. However, the cases on average did not vary a lot since 2007 (Chee and Rampal 2004). Accidents on industrial sites were typically caused by factors sourcing from a scarcity of organizational commitment, uncontrolled operations, lack of safety laws, personal protective equipment, coaching, poor technical superintendence, and unsafe instrumentation. Some initiatives were introduced by the Malaysian government to scale back the number of cases of activity accidents in trade (Awang and Kamil 2014).

11.4.4 Occupational Health Issues in Malaysia

Rapid industrialisation has resulted in a modification of the economic activity distribution among the Asian nation. High growth in employment within the producing services and construction have replaced agriculture and primary alternative industries with relatively sluggish growth. These changes have taken place in conjunction with the epidemiological changes of many Malaysian diseases. The prevalence of infectious diseases had decreased as non-communicable diseases had increased (Sadhra et al. 2012).

11.4.4.1 Muscular Disorder Diseases

A cross-sectional study was carried out because the highest prevalence was pain within the upper back, shoulders/neck, and lower limbs, and the highest exposures were prolonged (more than or equal to four hours per work shift) hand/wrist movement, standing, and hand lifting. Once provision regression, lower-back pain was significantly associated with standing, shoulder or neck pain with lifting and sitting, rising steps with upper-back pain, low back pain with movement of the hand or wrist, lifting pain of the hand or wrist. For employees with shorter operating times, neck

Table 11.3 Injuries data for three years (Ali et al. 2009)

				Non-weighted		Weighted	
Types of injury	Severity ranking	Total injuries (3 yrs)	Total weighted injury	Average injury per year	Percentage of injury	Average weighted injury per year	Percentage of injury
Scratch	1	140	140	46.67	14.30	46.67	2.39
Abrasion	2	107	214	35.67	10.93	71.33	3.65
Bruise	3	196	588	65.33	20.02	196.00	10.03
Blister	4	64	256	21.33	6.54	85.33	4.37
Laceration	5	149	745	49.67	15.22	248.33	12.70
Contusion	6	42	252	14.00	4.29	84.00	4.30
Strain	7	9	63	3.00	0.92	21.00	1.07
Sprain	8	82	666	27.33	8.38	218.67	11.19
Concussion	9	0	0	0	0	0	0
Dislocation	10	9	90	3.00	0.92	0	1.53
Fracture (U. limb	11	4	44	1.33	0.41	387.33	0.75
Poisoning (splash)	12	0	0	0	0	55.00	0
Poisoning (ingestion)	13	0	0	0	0	00.33	0
Eye injury	14	83	1162	27.67	8.48	226.67	19.82
Fracture (L. limb)	15	11	166	3.67	1.12	6.00	2.81
Poisoning (inhalation)	16	13	208	4.33	1.32	44.33	3.55

(continued)

Table 11.3 (continued)

				Non-weighted		Weighted	
Types of injury	Severity ranking	Total injuries (3 yrs)	Total weighted injury	Average injury per year	Percentage of injury	Average weighted injury per year	Percentage of injury
Superficial burn (<50%)	17	40	18	13.33	4.09	34.00	11.60
Superficial burn (>50%)	18	1	133	0.33	0.10	0	0.31
Crushing (U. limb)	19	7	120	2.33	0.72	110.00	2.27
Crushing (L. limb)	20	6	0	2.00	0.61	6.00	1.74
Amputate (L. limb)	21	0	330	0	0	0	0
Amputate (U. limb)	22	15	23	5.00	1.53	30.00	5.63
Deep burn (<50%)	23	1	0	0.33	0.10	14.67	0.31
Deep burn (>50%)	24	0	680	0	0	0	0
Total		979	5899	326.32	100	1954.66	100

Table 11.4 Outcomes of ergonomic in different duration of exposure (Chee and Rampal 2004)

Ergonomic exposure	Respondents (n = 906)	
	Number	Percentages (%)
Movement/position assumed for more than four hours per work shift		
Moving hands and wrists	632	68.8
Standing	499	55.1
Lifting with hands	464	51.2
Sitting	307	33.9
Pushing and pulling	240	26.5
Twisting the body	205	22.6
Bending	145	16.0
Climbing steps	32	3.5
Body pain experienced in the last year		
Any part of the body	729	80.5
Neck/shoulders	465	51.3
Neck	231	25.5
Shoulder	388	42.8
Upper limbs	317	35.0
Arms	246	27.2
Hands/wrist	184	20.3
Back (upper, lower)	464	51.2
Upper back	364	40.2
Low back	220	24.3
Lower limbs	471	52.0
Upper legs (hips, thighs, knees)	289	31.9
Lower leg (ankles, feet)	380	41.9
Temperature too cold	248	27.4
Experienced stress in the last year	70	7.7

or shoulder pain was significantly higher whereas lower back pain was considerably higher for staff with longer operating times. End of–line assembly staffs had considerably higher odds ratios of pain in any body parts, whereas middle-of-line employees had higher odds ratios for neck pain or shoulders and upper back, and wafer-fabrication staff had higher odds ratios for low back pain and lower limbs (Chee and Rampal 2004). Table 11.4 shows the outcomes of ergonomic in different duration of exposure.

11.4.4.2 Hearing Impairment

Noise-induced hearing disorder (NIHL) has risen to be the main cause of industrial diseases in the last few years, particularly in industrialized countries. However, in Malaysia, there are NIHL hindrance programs and legislation with poor implementation and social control. NIHL is a hearing disorder due to excessive noise exposure. Noise will cause hearing disorder either through acoustic trauma or through chronic exposure, the centre ear (tympanum) is broken from sudden or temporary exposure to extraordinary noise in acoustic trauma. Chronic noise exposure causes NIHL because it will injure whisker cells at intervals of the labyrinth structures (Sadhra et al. 2012).

In conclusion, a strong safety culture with nice management practices is important to reduce injury in a very meaty way. This can facilitate organisations to profit financially by reducing lost working hours and accident compensation. Additionally, reducing work injuries could increase staff motivation, productivity, product quality, and mutually reduce employees' compensation cost. Even so, the high rates of serious injuries in organisations that are similar to the performance of organisations will affect the image of the organisation.

11.5 Health Components in Composite Industries

11.5.1 Background

Composites consist of composites (PMCs), ceramic matrix composites (CMCs), and metal matrix composites (MMCs), which are also known as advanced composites materials. The characteristics of advanced composite materials are high in strength and stiffness, low in weight, resistant to corrosion, and in certain cases of special electrical properties. Initially, advanced composite materials were governed by the properties of the reinforcing fibres. Advanced composite materials like carbon fibre or epoxy composites have becoming very attractive and important structural parts in aircraft, military, and aerospace (Gandhi et al. 1999).

11.5.2 Fibre Reinforced Polymer Composites

Aircrafts highly use the fibre reinforced polymer for their primary and secondary structural components. When post-crash aircraft fires occur, gases, organic vapour particulate, fibre, and a complex mixture of combustion products will be released. Some potential health hazards are arising from the exposure of these combustion products that concern firefighters, investigators, and clean up personnel (Gandhi and Lyon 1998). Figure 11.4 illustrates the total of materials used by weight in the Boeing 787 body.

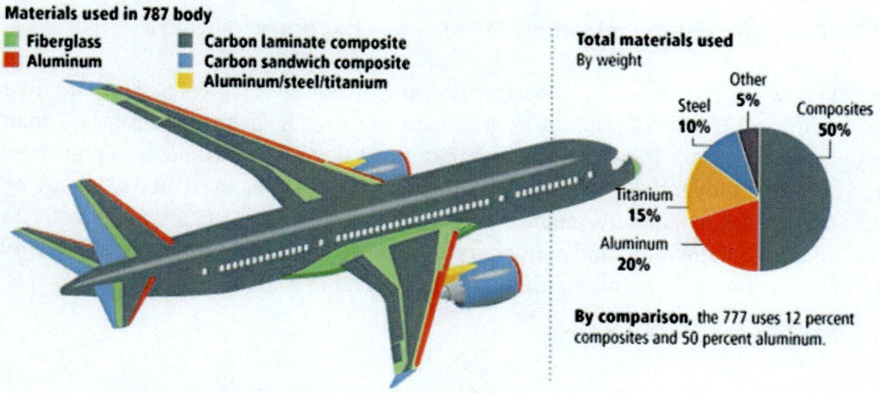

Fig. 11.4 Total materials used in Boeing 787 body (M. Aly 2017)

Furthermore, Gandhi and Lyon (1998) stated that there were several reported incidents of irritation and toxic effects involving exposure to fibrous matter on the workers due to post-crash aircraft fires. Exposure to fire emission among workers can affect human physiological which can lead to serious short term and long term adverse effects. The workers from the incidents suffered different degrees of serious health effects, e.g. skin and eye irritations and severe respiratory problems. Some factors affecting human health problems were influenced by the size, solubility, and chemical composition of the aerosols (Gandhi and Lyon 1998).

Dermal and inhalation are major routes of exposure to the fibres. Dermatitis can occur from skin irritation or sensitisation. This is due to surface abrasion response and puncture by sharp, needle-like fibres of diameters greater than 4–5 μm. However, the irritation effects are not permanent. Inhalation of fibres, depending on their physical and chemical compositions, contributes to the highest potential of adverse effects in humans. The deposition of small particles within the deep lung or the alveolar areas can cause chronic toxicity. Several studies showed that asbestos exposure can cause pulmonary fibrosis, lung cancer, and mesothelioma. The fibres' dimensions can cause toxic effects due to inhalation and exposure to asbestos (Gandhi and Lyon 1998). Table 11.5 shows critical fibre dimensions for asbestos-related diseases.

Table 11.5 Critical fibre dimensions for asbestos-related diseases (Gandhi and Lyon 1998)

Disease	Diameter (μm)	Length (μm)
Asbestosis	0.2–2	2–5
Lung cancer	0.3	Less than 10
Mesothelioma	Less than 0.1	More than 5

11.5.2.1 Medical Examinations for Asbestos Exposure

Workers who are exposed to asbestos dust at concentrations greater than one fibre per millilitre of TWA-8 should be provided with medical examinations by their employer (Congress 1994). The medical examination should be done by a registered medical practitioner that performs the full-size chest X-ray examination (350 mm by 430 mm), takes personnel's medical statement, occupational, and smoking histories and clinical examinations and pulmonary tests using spirometry, forced vital capacity (FVC), and forced expiratory volume in one second (FEV1) (Congress 1994).

11.5.3 Beryllium Reinforced Composite Materials

Beryllium is being used as an agent of reinforcing epoxy resin and metal matrix in composite materials. The properties of beryllium that include high tensile strength and elastic modulus relative to its low density have made it more preferable for structures that need high rigidity and low weight (Taylor and Hawk 1970). The development of beryllium-based metal matrix composites, AlBeMet, AlBe-Cast, and E materials is to meet the essentials and performance encouragement of advanced aerospace and commercial applications. Figure 11.5 shows the application of composite materials in aerospace.

Like many industrial materials, without safe handling practices of beryllium, the workers might pose a risk of health conditions. Airborne transmission of beryllium

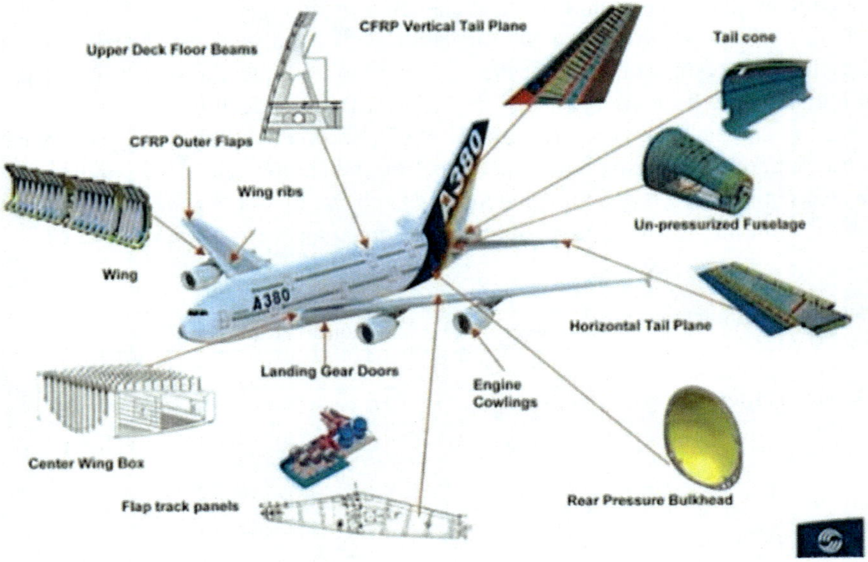

Fig. 11.5 Applications of composites in aerospace sector (Maria 2013)

might cause a serious lung disorder in susceptible individuals (World Health Organization 1997). Based on a study, the U.S. and international facilities workers in manufacturing nuclear weapons have been exposed to aerosols of beryllium which lead to the development of health conditions, e.g. beryllium sensitization (BeS) and chronic beryllium diseases (CBD) (Taylor and Hawk 1970). In certain individuals, BeS progresses to CBD, a progressive lung disease characterized by noncaseating granulomas (Van Dyke et al. 2011). The beryllium lymphocyte proliferation test (BeLPT) is a test to demonstrate the proliferation of beryllium-specific cells in the blood. This test is used broadly in the industry as a screening test for BeS and CBD (Newman et al. 2001).

11.5.3.1 Medical Examinations for Beryllium Exposure

Constant exposure of beryllium towards workers would increase the risk of respiratory hazards.

They should be diagnosed using chest X-rays diffuse, bilateral, granulomatosis, or in early stages are only enlarged lymph nodes. In this case, a radiologist report is necessary too. Moreover, the workers also should perform pulmonary function tests and impairment of respiratory function by a reduction in diffusion capacity of the lungs, which is detectable in the early stage of the disease. Furthermore, discontinued contact with beryllium should be done to workers who are exposed to acute berylliosis. The patient must be admitted to the hospital if mild symptoms precede a severe attack (Congress 1994).

11.5.4 Reinforced Plastics and Composite Products

Polymer and plastic composites are plastics that are reinforced with fibres, fillers, particulates, powders, and other matrix reinforcements to provide improved strength and/or stiffness. The reinforcing fibres and various chemicals used for the resinous matrix or in the finishing and assembling work of composite products are exposed to the workers in the reinforced plastics and more advanced plastic composite industries. The mechanical irritancy of mineral fibres and various solvents used mostly cause occupational dermatoses of irritant type in plastic composite work.

Based on a study, workers who are exposed to these composite products tend to have irritant contact dermatitis (ICD), allergic contact dermatitis (ACD), and immunological contact urticarial. ICD is caused by the glass fibre mechanical irritation, solvents, and sawing dust from cured plastic products. ACD was caused by diglycil ether bisphenol A epoxy resin, sensitisation to the epoxy compounds triglycidyl-*p*-aminophenol and tetraglycidyl-4–4′-methylene dianiline, cobalt naphthenate in unsaturated polyester resin, natural rubber latex and epoxy resin hardener, and methyltetrahydrophthalic anhydride (MTHPA). Some individuals pose bronchial asthma caused by epoxy resin or its hardener MTHPA and allergic rhinitis caused by

MTHPA. Whenever chemical knowledge of the safe test concentration and chemicals analysis is available, patch tests with plastic compounds to the exposed workers are commended (Tarvainen et al. 1995). Table 11.6 shows organ system target by resins, reinforcing materials, hardeners, curing agents, and other aromatic amines.

In a conclusion, exposure to composite materials might lead to acute and chronic health effects liable on how long the workers are exposed to the substances. Exposure to composite components might result in human target organs, for instance, skin,

Table 11.6 Organ system target by resins, reinforcing materials, hardeners, curing agents, and other aromatic amines (OSHA 1995)

Composite component	Organ system target (possible target)	Known (possible) health effect
Resins		
Epoxy resins	Skin, lungs, eyes	Contact and allergic dermatitis, conjunctivitis
Polyurethane resins	Lungs, skin, eyes	Respiratory sensitization, contact dermatitis, conjunctivitis
Phenol formaldehyde	Skin, lungs, eyes	As above (potential carcinogen)
Bismaleimides (BMI)	Skin, lungs, eyes	As above (potential carcinogen)
Polyamides	Skin, lungs, eyes	As above (potential carcinogen)
Reinforcing materials		
Aramid fibres	Skin (lungs)	Skin and respiratory irritation, contact dermatitis (chronic interstitial lung disease)
Carbon/graphite fibres	Skin (lungs)	As noted for aramid fibres
Glass fibres (continuous filament)	Skin (lungs)	As noted above
Hardeners and curing agents		
Diaminodiphenylsulfone	–	No known effects with workplace exposure
Methylenedianiline	Liver, skin	Hepatotoxicity, suspect human carcinogen
Other aromatic amines		
Meta-phenylenediamine (MPDA)	Liver, skin (kidney, bladder)	Hepatitis, contact dermatitis (kidney and bladder cancer)
Aliphatic and cycloaliphatic amines	Eyes, skin	Severe irritation, contact dermatitis
Polyaminoamide	Eyes, skin	Irritation (sensitization)
Anhydride	Eyes, lungs, skin	Severe eye and skin irritation, respiratory sensitization, contact dermatitis

lungs, and eyes. Medical surveillance is a must and must be conducted in a workplace as stated in the legal requirement.

11.6 Safety Components in Composite Industries

11.6.1 Background

The rapid development of the composite industry has shown that organic composites are widely used as the uniqueness of their materials allows them to continue to grow (Olson 1992). Some important composite-related factors need to be addressed such as safety, health, and environmental factors because to ensure that the composite industry continues to thrive, the control of hazardous risks needs to be more robust (Olson 1992). Besides that, the primary purpose of OSHA is to create surroundings of a healthy workplace for employees by conducting safety training, health, and safety education program in accordance with the law related to occupational safety and health (Labor 2020).

For example, in the automotive industry, steel manufacturers commonly depend on test methods developed by ASTM International's oldest committee. Automotive manufacturers use the A01 standard to test the strength, readiness, and integrity of the steel products (A370).

11.6.2 Safety Training Program

A safe and healthy working area should be provided by the employers because none of the workers should be injured, sick, or die (OSHA 1995). Lippin et al. (2000) conducted a cross-sectional telephone interview survey with 362 respondents who already participated in a safety training program approximately six to twelve months prior to the interview. They found that over 89% (89.3–98.1%) of respondents from each type of facility indicated that they were more aware of health and safety conditions because of the training. Respondents were asked to qualify their increase in awareness into the following categories: a great deal, a lot, somewhat, and a little. 55% (55.4%) to 73% (72.9%) of respondents said that they were a great deal more aware in Table 11.7. Thus, the safety training program must be conducted at the workplace as it is one of the legal requirements to help the workers improving their knowledge and create awareness (Lippin et al. 2000).

Table 11.7 Change in health and safety awareness, use of information, and work practices (Lippin et al. 2000)

Question asked	Percentage who answered yes			
	Blue-collar public sector	Chemical processing plants	Hospitals	Nuclear facilities
Did the training make you more aware of your safety and health?	98.1	97.5	89.3	93.3
If yes, how much more aware? (Choices: A great deal, a lot, somewhat, or a little) Reported here a great deal and a lot	72.9	68.0	55.4	57.8
Have you or your co-workers referred to the workbook or other materials received at the training?	53.7	44.9	54.3	49.7
Are people in your work area following established safety procedures more since the training?	73.7	71.7	67.2	62.3

11.6.3 Hazard Communication Standard

To ensure that hazardous chemicals in the workplace are safe to use, OSHA's hazard communication is established (Stein 2018). Based on the article written by Elliot Stein, she explained that manufacturer should provide safety data sheets for each hazardous chemical as listed in schedule 11 Use and Standard of Exposure Chemical Hazardous to Health (USECHH) regulation so that employees can understand the characteristics and side effects of the chemicals they work with (Stein 2018). Additionally, chemical manufacturers and distributors or importers should provide a safety data sheet to buyers for any hazardous chemicals so that their information can be communicated and understood by users (OSHA 1995).

Based on the standard, every employer should conduct training programs related to hazardous chemicals to their employees who might be exposed to these chemicals so that they know the side effects of the chemical on physical health and the safety precautions to take in case of accidents (Robins and Klitzman 1988). Robins and Klitzman (1988) conducted a study examining the results of educational programs conducted on workers in five different plans in the United States to meet the standard federal hazard communication requirements.

11.6.4 Medical Emergency Preparedness

Medical emergency preparedness is one of the safety components at the workplace. Regarding this topic, Jefferelli et al. (2014) conducted a study reviewing the availability of medical emergencies in BASF Hong Kong in 2012 and the same training was conducted in 2011 with the involvement of 650 workers in Hong Kong. This study was aimed to compare the medical emergency response in 2012 and 2011 in a company in Hong Kong (Jefferelli et al. 2014). The findings showed that there was a significant increase in medical practice in 2012 where the workers were showing a better performance in medical emergency preparedness than in 2011 (Jefferelli et al. 2014). They found that with the First Aid Pouch introduction, 20% improvement was observed as compared to 2011, and getting help from other first aiders was also enhanced by 22%, remembering that CPR involved 30 chest compressions (45%) and two rescue breaths (61%). There were no significant differences in action check responsiveness (−5%), open airway and breathing check (−7%), call for help (9%), get help to call an ambulance (0%), and start CPR(0%) as shown in Table 11.8.

In conclusion, it is important to train employees working with composite materials in terms of the health effects of exposure to fibres and chemical compounds. The health risks of working with composites are significant. Mitigating risks can be achieved by ensuring they receive the proper training in health and safety procedures.

Table 11.8 Recall of appropriate actions (Jefferelli et al. 2014)

Action	Spontaneous 2012 (%)	Spontaneous 2011* (%)	Difference (%)
Arrive with first aid box/pouch	93	73	+20
Check responsiveness	86	91	−5
Open airway and breathing check	93	100	−7
Call for help	100	91	+9
Get help from other first aiders	86	64	+22
Get help to call an ambulance	100	100	0
Get help to get AED	79	0	+79
Start CPR	100	100	0
Thirty compressions	100	55	+45
Two rescue breaths	79	18	+61

*To be consistent, two participants who had participated in medical drill in 2011 but were not certified were excluded from these figures

There are numerous resources available through government agencies and private industries to obtain training services in workplace health and safety practices.

11.7 Conclusion

Composites have been used extensively in industries such as aircraft, military, automotive, construction, marine, and aerospace. Different countries have different administration of occupational safety and health. This paper reviews the issues of occupational safety and health administration in the composite industry based on three countries, the United States, the United Kingdom, and Malaysia.

- Occupational safety and health in the United States have their own safety management practice and safety culture involving employers and employees and it is also concerned about their health-related issues among the workers.
- In the United Kingdom, they also have their own legislative system, management system, and risk assessment to incapacitate the hazards and risk themselves. This country's core regulatory system in occupational safety and health is the Health and Safety at Work etc. Act 1974.
- There is also no difference with Malaysia that also has its own management practice in safety cultures by introducing the Safety and Health Act (OSHA) 1994. A strong safety culture with good management practice is important in the reduction of injuries within the work.
- Exposure to composite materials might lead to acute and chronic health effects depending on how long the workers are exposed to the substances. Medical surveillance needs to be conducted at the workplace to diagnose the current state of health of workers that comply with the legal and other requirements.
- To prevent any accidents and occupational injuries among workers at the workplace, precautions and prevention steps should be implemented and introduced to the workers. The employer also should give enough supervision and training program about correct operating procedures to them, e.g. employee training program, hazard communication standard, and medical emergency preparedness. The employer has to make sure the employees follow those standard operating procedures at the workplace.

Acknowledgements This chapter was written with M. H. N. Baini, A. N. Sakinah, Z. N. Z. Ashykin, P. Z. N. Farahiah, Z. N. Farina.

References

Ali H, Abdullah AC, Subramaniam C (2009) Management practice in safety culture and its influence on workplace injury: an industrial study in Malaysia. Disaster Prev Manag Int J 18(5):470–477. https://doi.org/10.1108/09653560911003660

Aly NM (2017) A review on utilization of textile composites in transportation towards sustainability. IOP Conf Ser Mater Sci Eng 254:42002

Anon (2016) 2016 UK composites strategy, 24. Composites Leadership Forum

Awang H, Kamil IM (2014) Execution of the Occupational Safety and Health Act (1994) in the construction industry from contractors' point of view. SHS Web Conf 11:01003. https://doi.org/10.1051/shsconf/20141101003

Chee HL, Rampal KG (2004) Work-related musculoskeletal problems among women workers in the semiconductor industry in Peninsular Malaysia. Int J Occup Environ Health 10(1):63–71. https://doi.org/10.1179/oeh.2004.10.1.63

Collins A, Shaw K, Porawski C, Cann O, Geiger T, Tedeneke A, Tan J-FT, Verin S (2018) The global risks report 2018 13th edition. In: World economic forum, vol 3, Issue 1. https://doi.org/10.1056/nejm196802152780701

Composites manufacturing's 2018 state of the industry report (n.d.)

Congress IR (1994) Guidelines on (1)

Cooper CL, Cartwright S (1994) Healthy mind; healthy organization—a proactive approach to occupational stress. Human Relat 47(4):455–471. 10.1177/001872679404700405

Danna K, Griffin RW (1999) Health and well-being in the workplace: a review and synthesis of the literature. J Manag 25(3):357–384. https://doi.org/10.1177/014920639902500305

European Risk Observatory Report (n.d.) European agency for safety and health at work. European survey of enterprises on new and emerging risks. European Risk Observatory Report. http://doi.org/10.2802/30026

Gandhi S, Lyon RE (1998) Health hazards of combustion products from aircraft composite materials. Security

Gandhi S, Lyon R, Speitel L (1999) Potential health hazards from burning aircraft composites. J Fire Sci 17(1):20–41. https://doi.org/10.1177/073490419901700102

Geldart S (2014) Health and safety in today's manufacturing industry. In: Comprehensive materials processing, vol 8. Elsevier. http://doi.org/10.1016/B978-0-08-096532-1.00816-5

Hassan KH (2015) Corporate liability under Malaysian occupational safety and health legislation. Int J Bus Soc 16(2):281–294

Hee OC (2014) Factors contribute to safety culture in the manufacturing industry in Malaysia. Int J Acad Res Bus Soc Sci 4(4):63–69. https://doi.org/10.6007/IJARBSS/v4-i4/753

IOSH (2016) Occupational health management in the workplace, pp 1–43

Jefferelli et al (2014) J Occup Saf Health 11

Labor of U. S. D. (2020) All about OSHA, vol 3, Issue 2017. http://repositorio.unan.edu.ni/2986/1/5624.pdf

Lippin TM, Eckman A, Calkin KR, McQuiston TH (2000) Empowerment-based health and safety training: evidence of workplace change from four industrial sectors. Am J Ind Med. https://doi.org/10.1002/1097-0274(200012)38:6%3c697::AID-AJIM9%3e3.0.CO;2-T

Maria M (2013) Advanced composite materials of the future in aerospace industry. INCAS Bull 5(3):139–150. http://doi.org/10.13111/2066-8201.2013.5.3.14

Mazumdar S (2018) The 2018 State of the Composites Industry Report. Retrieved from http://compositesmanufacturingmagazine.com/2018/01/2018-composites-manufacturing-state-of-the-industryreport/

Mazumdar S, Benevento M, Pichler D, Witten E, Hinrichsen J (2018) The 2018 State of the Composites Industry Report. Retrieved from https://compositesmanufacturingmagazine.com/2018/01/2018-composites-manufacturing-state-of-the-industryreport/

McDiarmid MA (2014) Hazards of the health care sector: looking beyond infectious disease. Ann Glob Health 80(4):315. https://doi.org/10.1016/j.aogh.2014.08.001

Newman LS, Mroz MM, Maier LA, Daniloff EM, Balkissoon R (2001) Efficacy of serial medical surveillance for chronic beryllium disease in a beryllium machining plant. J Occup Environ Med 43(3):231–237. https://doi.org/10.1097/00043764-200103000-00011

Olson LJM (1992) Safety, health, and environmental hazards associated with composite: a complete analysis, vol 298

Psomopoulos CS, Kalkanis K, Kaminaris S, Ioannidis GC, Pachos P (2019) A review of the potential for the recovery of wind turbine blade waste materials. Recycling 4(1). http://doi.org/10.3390/recycling4010007

Robins TG, Klitzman S (1988) Hazard communication in a large U.S. manufacturing firm: the ecology of health education in the workplace. Health Educ Behav. http://doi.org/10.1177/109019818801500406

Ruff T, Coleman P, Martini L, Ruff T, Coleman P, Martini L (2011) Machine-related injuries in the US mining industry and priorities for safety research. 7300(May 2010):10–20. http://doi.org/10.1080/17457300.2010.487154

Sadhra S, Beach JR, Aw T (2012) Occupational health research priorities in Malaysia: a delphi study. Occup Environ Med 58(7):426–431

Stein E (2018) Radioactive materials and the OSHA hazard communication standard. J Chem Health Saf. https://doi.org/10.1016/j.jchas.2017.11.003

Tarvainen K, Kanerva L, Jolanki R, Estlander T (1995) Occupational dermatoses from the manufacture of plastic composite products. Am J Contact Dermatitis 6(2):95–104

Taylor W, Hawk JA (1970) Beryllium-reinforced composite materials. JOM 22(6):45–53

United States. Occupational Safety and Health Administration. Office of Science and Technology Assessment (1995) OSHA technical manual. University of Michigan Library

U.S. Bureau of Labor Statistics (2018) 2017 survey of occupational injuries & illnesses industry-level estimates case circumstances and worker characteristics. U.S. Bureau of Labor Statistics, p 19

Van Dyke MV, Martyny JW, Mroz MM, Silveira LJ, Strand M, Fingerlin TE et al (2011) Risk of chronic beryllium disease by HLA-DPB1 E69 genotype and beryllium exposure in nuclear workers. Am J Respir Crit Care Med 183(12):1680–1688. https://doi.org/10.1164/rccm.201002-0254OC

van Stolk C, Staetsky L, Hassan E, Kim CW (2012) Manag Occup Saf Health. https://doi.org/10.2802/90924

Vernon Eidson J, Reese C (2006) Handbook of OSHA construction safety and health, 2nd edn. https://doi.org/10.1201/9781420006230

World Health Organization (1997) Environmental health criteria nitrogen oxides. In: Encyclopedia of toxicology

Worsley JL (1978) Health and safety at work act. J Soc Dyers Colour 94(11):471–486. https://doi.org/10.1111/j.1478-4408.1978.tb03383.x

Zahrim Bin Osman H, Hasan Samad A, Dato Y, Ibrahim Hussein I, Bahari Lynas Sdn Bhd I, Jeffereli Shamsul Bahrin M et al (2016) J Occup Saf Health 13(2). http://doi.org/10.1093/oxfordhb/9780199928286.013.0036

Chapter 12
The Role of Biocomposites in Health Issues During COVID-19 Pandemic

Abstract 2019 Novel Coronavirus (2019-nCoV) or widely called as COVID-19 has spread vigorously throughout 2019 until present. This Wuhan-originated virus had been the cause of global pandemic and thus, impaired global economy, industry as well as health issues. Various approaches, non- and pharmaceutical remedies, had been practiced by the publics. The best way to avoid virus spread is social distancing as asserted by World Health Organization (WHO). However, in order to assist and facilitate the healthcare workers, researchers had focused on development of composite applications, including biocomposite and nanocomposite via technologies. From our study, recent research on composite contributions in health issues during the COVID-19 pandemic had been compiled and discussed. It is found that the application of bio- and nanocomposites with the help of efficient tech-based approach had produced useful product for the pharmaceutical uses, such as antimicrobial materials for medical devices manufacturing, bio-based facemasks with high filtration properties and 3D-printing biocomposite-derived face shields and testing tools. The different innovations derived from biocomposites applied for the war against COVID-19, which benefit the health systems, government and the public is highlighted. Furthermore, the technological swift during the pandemic is addressed and its effect towards the environment and society is reviewed.

Keywords Health and safety · Composites · COVID-19 · Pandermic

12.1 Introduction

Coronavirus disease 2019 or COVID-19 is defined as illness caused by a novel coronavirus which called severe acute respiratory syndrome coronavirus 2 (SARS-CoV-2 or 2019-nCoV). It was first identified in Wuhan, China and spread vigorously throughout the globe. As updated on 24th February 2021 from Woldometers.com website, for almost 112,680,152 recorded coronavirus cases globally with around 2.4 million deaths reported from the first day of pandemic (Worldometer 2021). Coronavirus cases piled up drastically and thus, contributed to a shortage of facemasks and protective personal equipments (PPE). Economically, the pandemic had

S. M. Sapuan et al., *Safety and Health in Composite Industry*,
Composites Science and Technology ,
https://doi.org/10.1007/978-981-16-6136-5_12

led to increase the demand and supply of the petrochemical-derived PPEs, which commonly used by the healthcare workers (HCW). The most worrying scenario is the health of Earth is affected by the massive production of non-degradable products and waste. Improper plastic waste management play important role in endangering the environment while poor COVID-19 related waste management plan will help the transmission of coronavirus (Harussani et al. 2020).

As concerned by World Health Organization (WHO) (2020), various steps to encounter the rapid transmission of the virus had been taken into care. According to WHO (2020), the best nonpharmaceutical remedies against the disease which rapidly spread via airborne and surface transmission are social distancing methods (Chu et al. 2020). Whereas, for self-protection measures, HCWs treating their COVID-19-infected patients while protected with complete PPEs. Facemasks and respirators-wearing are paramount as its act as our main physical barrier. Respiratory droplets secreted from mucosalivary system of infected individuals may spread out and enter through nose and mouth. Facemasks are particularly important in encountering this epidemic spread as the virus can be shed while the infected individuals are in asymptomatic or presymptomatic (Gandhi et al. 2020; He et al. 2020).

The existing PPE, masks and protection equipments still not suitable to be utilized in various conditions. For example, HCWs are facing with uncomfortability when using PPEs especially in hot weather, it will make them sweat a lot. In addition, global shortage of supply and raw materials resulted in self-made masks with less protection than actually needed. Therefore, researchers and industries had done many research in order to produce and engineer alternative materials with filtering properties, designing and developing multifunctional masks and protection equipments as well as introducing new technologies to enhance the production of masks and PPEs. In composite applications, many attempts and research to improvise the existed masks and PPEs as well as discover new materials and products had been done. As a results, the studies will pivot on strong understanding on COVID-19 related products and technologies, a new important field needed to be explored.

12.2 Composites and Nanocomposites Contribution During COVID-19 Pandemic

A number of polymer and non-polymer composites have been developed, with unique surface chemistry, mechanical and thermal properties, for the investigations of air (virus) filtration and purification as well as hindering virus infections. Multifunctional COVID-19 facemasks with improvised poly (vinyl alcohol)/cellulose nanocrystals (PVA/CNC) composite nanofibrous filter had been engineered by Zhang et al. (2019). The electrospun composite nanofibrous filter is significant in providing low air resistance with high-efficiency particulate matter filtration. The work introduced us with the role of CNCs in facemasks, which act as mechanical reinforcing agents and also used to improve the surface charge density of the electrospinning solution, thus, reducing fibre diameter. The thinner fibres greatly decreased the pressure drop and improved particulate matter removal efficiency (Chua et al. 2020). To give better emphasis, this work had been supported by several previous and recent works from Zhang et al. (2016), Li et al. (2018) and Yang et al. (2020).

Zhang et al. (S. Zhang et al. 2016) had reported an anti-deformed poly(ethylene oxide)@polyacrylonitrile/ polysulfone (PEO@PAN/PSU) composite membranes which comprised of efficient air and particulate matter filtration, $PM_{2.5}$, with stable porous membranes. With small pore size, good mechanical properties with high porosity, the polymer membrane exhibits enhanced $PM_{2.5}$ filtration efficiency (99.992%) with low pressure drop (95 Pa) Based on the investigations from Li et al. (2018), the membranes made of graphene oxide/polyacrylonitrile (GOPAN) composite nanofibrous presents high $PM_{2.5}$ removal efficiency with low pressure drop at 8 Pa (99.97%). The role of GO and macrostructure design of olive-like bead had led to air filter properties improvements, see Fig. 12.1. The basic illustration of PAN- and GOPAN-based nanofibrous membranes shown in Fig. 12.1, where the surface of the GOPAN-based nanofibre membrane is rougher and had olive-like bead structure due to the GO addition. Thus, the addition of GO has led to low pressure drop as their high surface area and the slip effect of air flow on the surface. Whereas, Yang et al. (2020) had developed barium titanate@polyurethane/ polysulfonamide ($BaTiO_3$@PU/PSA) composite nanofibrous membrane which exhibits better mechanical properties and flexibility. The introduced composite also has high thermal stability, favourable flame-retardancy and high stability when exposed to acid and alkali.

Nanocomposites derived from nanostructured materials have been used widely as an alternative for antibiotics, additives and antimicrobial agents. Bhushan (2017) stated that only materials with size range of 1–100 nm can be considered as nanostructured materials and its size may exceed 100 nm when hybridised with other materials, including polymers and biomolecules, to form composite of nanostructured materials. Whereas, nanostructure materials can be categorised into three main

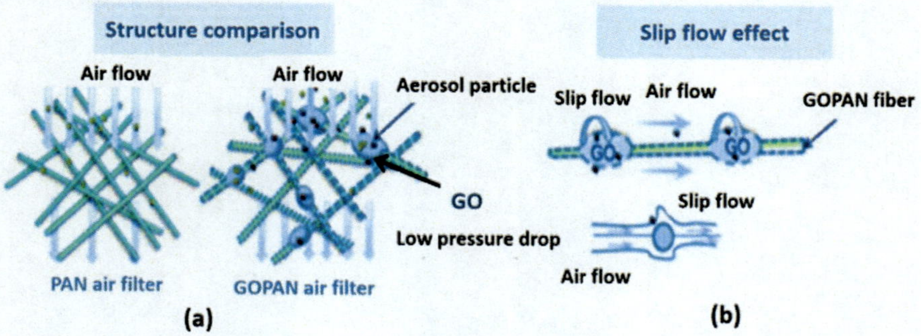

Fig. 12.1 **a** Structure and performance comparison diagram of PAN and GOPAN, representing the high efficiency and low pressure GOPAN nanofibrous membrane. **b** Visualization of aerosol particle-fibre interactions, where the particle sticks on the fibre and particle slips along the fibre (Zhang et al. 2016)

classes and sub-classes as shown in Fig. 12.2. Recently, graphene-based nanocomposites, due to its distinct surface properties, have attracted the researchers' interest for their suitability and applicability to act as antimicrobial agent (Jilani et al. 2018), and according to Oscoy et al. (2017), their microbicidal activity has been contributed by the "sheet effect", oxidative stress and cell membrane dysfunction inside the cell. In a review paper by Baranwal et al. (2018), there are several nanocomposites which actively applied in biomedical devices with highly efficacious antimicrobial activity. Nanocomposites of carbon nanotubes (CNT)-doped TiO_2 thin film had been developed as antimicrobial surface coatings (Akhavan et al. 2010).

Antibacterial application is one of the nanocomposite contributions, thus, aided in encountering COVID-19 epidemic. Pei et al. (2013) introduced antibacterial activated carbon (AC) nanocomposites with silver nanoparticles hybridised with ACs (Ag/ACs) nanocomposite which exhibit high antibacterial activity especially against airborne bacteria and toluene adsorption. Generally, the nanocomposite was utilized in indoor air quality (IAQ) applications due to its antibacterial AC, which can remove volatile organic compounds and kill bacteria. In addition, Ag particles aggregation reduced antibacterial activity since homogenous dispersion contributed

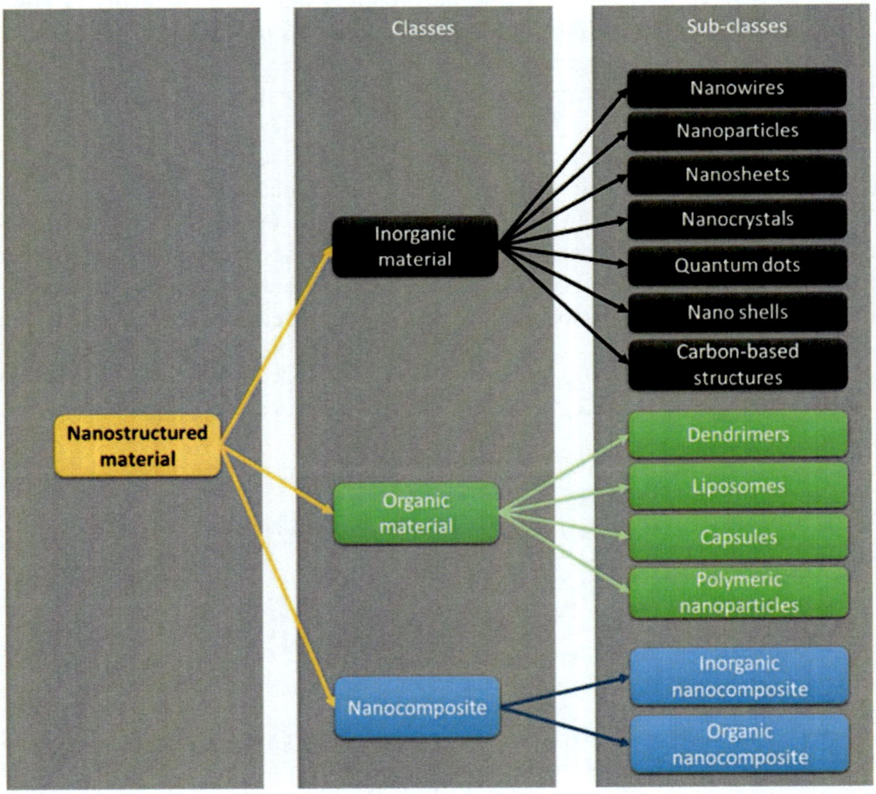

Fig. 12.2 Classification of nanostructured materials used as antimicrobials (Bhushan 2017)

to Ag-microbial species interaction and block micropores. Tamayo et al. (2016) had compiled several works on copper-polymer nanocomposites due to its enhanced antibacterial properties and facile process. Numerous copper-polymer nanocomposites had been utilized in water-treatment (disinfections), medical devices such as wound dressing and implants, and in food packaging applications. Here, the authors had explained and illustrated the biocidal effects happened in copper-polymer nanocomposites which was divided into three events: release of copper ions, release of copper nanoparticles from composites and biofilm inhibition, as shown in Fig. 12.3. Nowadays, there are many nanocomposites designed for medical uses as antibacterial wound-dressing, such as poly (ethylene glycol), cellulose, chitosan and PVP-alignate containing silver based nanocomposites (Liu and Kim 2012; Singh and Singh 2012; Wu et al. 2014). Whereas, copper nanoparticles widely reinforced, into composites, with these polymeric matrices: epoxy resin (Barua et al. 2014), low-density polyethylene (LDPE) (Hundáková et al. 2014), nylon (Komeily-Nia et al. 2013), polyaniline (PANI) (Thampi et al. 2015), poly (*D,L*-lactide-*co*-glycolide) (Amina et al. 2014), poly (ethylene glycol diacrylate) hydrogel (Cometa et al. 2013), polymethylmethacrylate (PMMA) (Weickmann et al. 2005), polypropylene (PP) (Delgado et al.

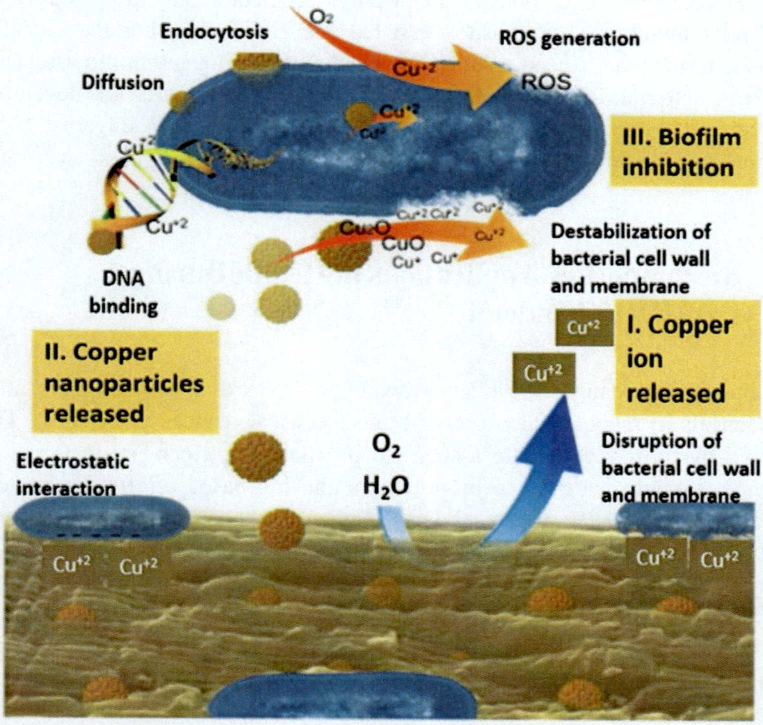

Fig. 12.3 Illustration diagram of antibacterial effects existed in copper-polymer nanocomposites (Tamayo et al. 2016)

2011; Palza et al. 2010), polythiophene (PTh) (Ma et al. 2014) and polyvinyl chloride (PVC) (Becerra et al. 2013).

Shen et al. (2020) developed metal–organic composites which comprised of multi-dentate organic ligands and metal ions. It is reported that metal–organic composites have unique properties such as high porosity and structural flexibility. Hence, these composites are suitable to be used as antibacterial agents. Metal nanoparticles reinforced metal–organic frameworks (MOF) composites demonstrated high effect as the released of metal ions from the composites contributed by the disruption of cytomembrane permeability led to intracellular contents leakage (Abd El Salam et al. 2018). In this applications, various researchers had embedded silver (Ag) nanoparticles with the metal–organic frameworks composites including Ag-CuTPP MOFs (Ximing et al. 2017), Ag-NPs@Ni-MOF (Abd El Salam et al. 2018) and Ag@MOF-5 (Thakare and Ramteke 2017) composites. Turcheniuk et al. (2015) had worked on pegylated reduced graphene oxide nanoparticles (rGO-PEG) and core/shell nanocomposites of gold nanorods/rGO-PEG (rGO-PEG-Au NRs) for the selective killing of uropathogenic *E. coli* UTI89. It is stated that rGO-PEG and Au NR particles have good light absorption properties in near-infrared region in order to kill the pathogens of *E. coli* photothermally.

Therefore, these applications of metal-polymer, metal nanoparticles- MOFs and nanofibrous membrane composites assisted the HCWs treating the COVID-19 patients in terms of enhanced medical device manufacturing and improvised antimicrobial raw materials for facemasks and PPEs production. Research and development efforts do offer improved protection for users especially HCWs against airborne pathogens and chemicals.

12.3 Biocomposites Applications in Health During COVID-19 Pandemic

Biocomposite has caught the many researchers' eyes as it acts as main alternative materials in mechanical, structural and electrical devices production. This is attributed by its biodegradable assets and parallel with green products policy as biocomposites are made up of biopolymers and bio-based reinforcing materials. Compilations of studies and research had been done on natural fibre-reinforced biopolymers, biocomposites for various applications (AL-Oqla and Omari 2017; Das et al. 2020b; Ilyas and Sapuan 2020; Sakthiguru and Sithique 2020; Sanyang et al. 2016), including in health.

Increasing COVID-19 active cases everyday must be proportional to the amount of available rooms in hospitals and isolation centres. In Wuhan city, they had to build almost 20 temporary isolation-and-care centres with total of 16 thousand beds to occupy a massive amount of positive-COVID-19 patients at the beginning of the epidemic (Fang et al. 2020). Thus, innovation of easily-build healthcare centre is an excellent idea to solve the problem of increasing number of patients. Thus, a group of researcher from CETMA (Brindisi, Italy) had proposed an idea of SOS—Smart Operating Shelter—project (CETMA 2020), which introducing rapid and facile set up of mobile health-and-care centres for the treatment of infected patients. Eco-friendly biocomposites panels made from bio-based polyurethane and recycled polyethylene terephthalate (PET) foam as structural panels for the smart shelters (SOS 2020). In addition, Natural fibre and bio-based polymers had been utilized as composite laminates for external sandwich panel skins.

Antibacterial biocomposite masks from biocellulose materials were widely improvised by researchers (Oladapo et al. 2021). In order to assist HCWs as the front-liners during COVID-19 epidemic, 3D-printing technology had been practiced extensively to produce masks, face shields and PPE as well. As reviewed by Oladapo et al. (2021), 3D-printing approach had helped in manufacturing 3D antibacterial bio-cellulose swabs (testing tools) as shown in Fig. 12.4 (Gupta and Misra 2020; Oladapo et al. 2020a), masks and respirator valves (Oladapo et al. 2020b; Zehra et al. 2020).

Das et al. (2020a), in his work, had developed bio-based facemask equipped with lanosol added gluten biocomposites and electrospun filter membranes. Biopolymers including poly (lactic acid), poly (vinyl alcohol), chitin, chitosan, cellulose or blends of these biopolymers had been used to produce the electrospun nanofibre membranes which were then utilized as filter for the facemasks (Kadam et al. 2018). To encounter the coronavirus, several improvements are needed for the ordinary facemasks. The facemask cloth or electrospun carbon membranes must exhibit smaller pore size range, 120–160 nm, to filter pathogens especially COVID-19 virus (Nurazzi et al.

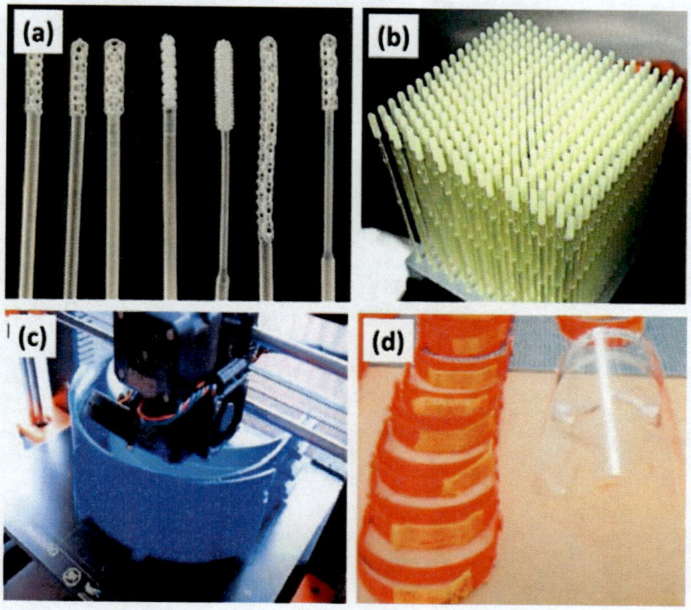

Fig. 12.4 **a–b** 3D-printed biocellulose swabs and **c–d** face shield for the COVID-19 test and demands (Oladapo et al. 2021)

2021; Pellett et al. 2014). Next, the electrospun carbon mat also must comprises with higher mechanical strength, and good moisture and fire resistance, as shown in Fig. 12.5. Moreover, there are many studies, previously, had been done on the development of carbon nanofibre mat equipped on facemasks such as addition carbon to gluten composites (Das et al. 2019), addition of polyaminoamide epichlorohydrin (PAE) to the gluten (Das et al. 2020a) and electrospun carbon nanofibre mat laminated with glycerol plasticised gluten (with lanosol) biocomposite films (Norizan et al. 2020).

According to scientists from Cardiff University (2020), they had discovered that bio-based materials derived from waste crab shells, chitosan can be utilised into natural medical products. This is attributed by the anti-microbial and anti-viral properties by the chitosan extracted from the crab shells (Fig. 12.6). It helps in reducing the risks of contamination from the coronaviruses by killing any virus closely contact with it (Joseph et al. 2020). Previously, chitin (chitosan derivatives) had been vastly applied in composite materials, surgical sutures, drug delivery and tissue engineering (Nakajima et al. 1986; Wu et al. 2018). Based on Lee et al. (2008), chitin exhibits unique properties as it can stimulate immune cells in anticancer activity and can cure fungal disease candidiasis (Rementería et al. 1997) as well as it has anti-aging

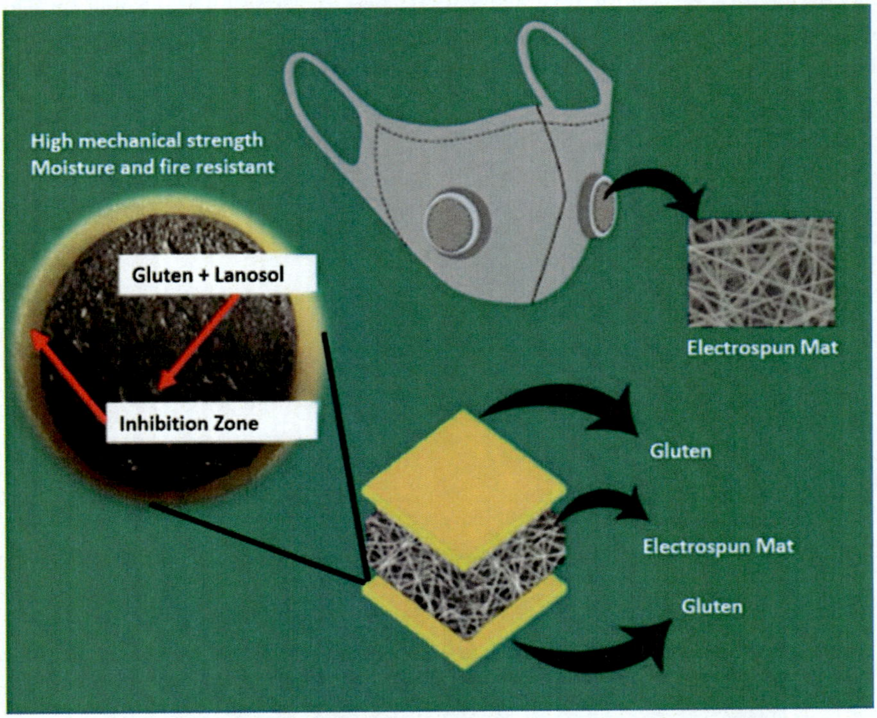

Fig. 12.5 Schematic of potential facemasks developed with enhanced mechanical and thermal strength (Das et al. 2019)

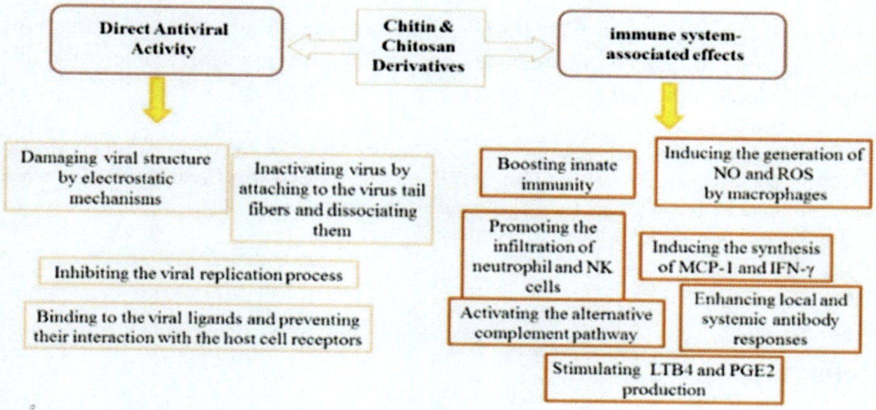

Fig. 12.6 Chitosan, chemical-derived from crab shells could help in anti-COVID-19 activity (Safarzadeh et al. 2021)

properties for skin protection (Ito et al. 2014). Therefore, it is recorded by Global Industry Analysis Inc., USA, that the global market for chitosan derivatives is about 282 thousand metric tons in the year 2020, and will keep increasing until 2027 (Global Industry Analysts 2020).

In addition, various biopolymers had been applied as matrices to support copper nanoparticles and generate biocomposite materials with effective antimicrobial properties, including agar (Shankar et al. 2014), carboxymethylcellulose (CMC) (Yadollahi et al. 2015; Zhong et al. 2013), cellulose (Cady et al. 2011; Llorens et al. 2012), bamboo-rayon (Teli and Sheikh 2013), chitosan (Gouda and Hebeish 2010; Mallick et al. 2012; Zain et al. 2014), cotton (Anita et al. 2011; El-Nahhal et al. 2012), cotton-cellulose (Bhowmik et al. 2017; Grace et al. 2009) and cotton silica (Li et al. 2020). Conclusively, bio- and nano-composites did play strong roles in encountering COVID-19 transmission and contamination which already caused to almost 2.5 million deaths globally. Therefore, extensive research and investigations are needed in order to improve the existing anti-COVID-19 equipments and medical devices.

12.4 Conclusions and Future Outlooks

To conclude, it is certain that mankind attitude will fight its best and heal from the epidemic. However, to assist and ease the HCWs, various development of composites and biocomposites based products had been introduced either in medical device manufacturing or antiviral medicine. The utilization of technologies such as 3D-printing have showed their strengths and abilities in producing massive amount of facemasks and PPEs production in order to fulfil the market demands. Besides that, the government also must play bigger role to manage and deploy assets towards research and development field as it is a right-step in countering and solving the problem of rapid spread of COVID-19. Therefore, deep studies and investigations are compulsory to understand the behaviour of this virus and next, to overcome this pandemic-derived problem.

Acknowledgements The authors would like to appreciate Ministry of Higher Education (MoHE) for the financial support through the Post COVID-19 Special Research Grant Scheme 2020 and Universiti Putra Malaysia for the facilities provided throughout this research.

This chapter was written with M. M. Harussani.

References

Abd El Salam HM, Nassar HN, Khidr ASA, Zaki T (2018) Antimicrobial activities of green synthesized Ag nanoparticles@ Ni-MOF nanosheets. J Inorg Organomet Polym Mater 28(6):2791–2798

Akhavan O, Azimirad R, Safa S, Larijani MM (2010) Visible light photo-induced antibacterial activity of CNT–doped TiO_2 thin films with various CNT contents. J Mater Chem 20(35):7386–7392

AL-Oqla FM, Omari MA (2017) Sustainable biocomposites: challenges, potential and barriers for development. In: Green biocomposites. Springer, pp 13–29

Amina M, Hassan MS, Al Musayeib NM, Amna T, Khil M-S (2014) Improved antibacterial activity of HAP garlanded PLGA ultrafine fibers incorporated with CuO: synthesis and characterization. J Sol-Gel Sci Technol 71(1):43–49

Anita S, Ramachandran T, Rajendran R, Koushik CV, Mahalakshmi M (2011) A study of the antimicrobial property of encapsulated copper oxide nanoparticles on cotton fabric. Text Res J 81(10):1081–1088

Baranwal A, Srivastava A, Kumar P, Bajpai VK, Maurya PK, Chandra P (2018) Prospects of nanostructure materials and their composites as antimicrobial agents. Front Microbiol 9:422

Barua S, Chattopadhyay P, Phukan MM, Konwar BK, Karak N (2014) Hyperbranched epoxy/MWCNT-CuO-nystatin nanocomposite as a high performance, biocompatible, antimicrobial material. Mater Res Express 1(4):45402

Becerra A, Rodríguez-Llamazares S, Carrasco C, Díaz-Visurraga J, Riffo C, Mondaca MA (2013) Preparation of poly (vinyl chloride)/copper nanocomposite films with reduced bacterial adhesion. High Perform Polym 25(1):51–60

Bhowmik S, Islam JMM, Debnath T, Miah MY, Bhattacharjee S, Khan MA (2017) Reinforcement of gelatin-based nanofilled polymer biocomposite by crystalline cellulose from cotton for advanced wound dressing applications. Polymers 9(6):222

Bhushan B (2017) Springer handbook of nanotechnology. Springer

Cady NC, Behnke JL, Strickland AD (2011) Copper-based nanostructured coatings on natural cellulose: nanocomposites exhibiting rapid and efficient inhibition of a multi-drug resistant wound pathogen, *A. baumannii*, and mammalian cell biocompatibility in vitro. Adv Functional Mater 21(13):2506–2514

Cardiff (2020) Crab shell bid to tackle COVID-19

CETMA (2020) CETMA: Contro Il COVID-19 Un Progetto Per L'Allestimento Di Mini Ospedali Mobili

Chu DK, Akl EA, Duda S, Solo K, Yaacoub S, Schünemann HJ, El-harakeh A, Bognanni A, Lotfi T, Loeb M, Hajizadeh A (2020) Physical distancing, face masks, and eye protection to prevent person-to-person transmission of SARS-CoV-2 and COVID-19: a systematic review and meta-analysis. The Lancet 395(10242):1973–1987

Chua MH, Cheng W, Goh SS, Kong J, Li B, Lim JY, Mao L, Wang S, Xue K, Yang L, Ye E (2020). Face masks in the new COVID-19 normal: materials, testing, and perspectives. Research, 2020

Cometa S, Iatta R, Ricci MA, Ferretti C, De Giglio E (2013) Analytical characterization and antimicrobial properties of novel copper nanoparticle–loaded electrosynthesized hydrogel coatings. J Bioact Compat Polym 28(5):508–522

Das O, Hedenqvist MS, Johansson E, Olsson RT, Loho TA, Capezza AJ, Raman RS, Holder S (2019) An all-gluten biocomposite: Comparisons with carbon black and pine char composites. Compos A Appl Sci Manuf 120:42–48

Das O, Kim NK, Hedenqvist MS, Bhattacharyya D, Johansson E, Xu Q, Holder S (2020a) Naturally-occurring bromophenol to develop fire retardant gluten biopolymers. J Clean Prod 243:118552

Das O, Neisiany RE, Capezza AJ, Hedenqvist MS, Försth M, Xu Q, Jiang L, Ji D, Ramakrishna S (2020b) The need for fully bio-based facemasks to counter coronavirus outbreaks: a perspective. Sci Total Environ 736:139611

Delgado K, Quijada R, Palma R, Palza H (2011) Polypropylene with embedded copper metal or copper oxide nanoparticles as a novel plastic antimicrobial agent. Lett Appl Microbiol 53(1):50–54

El-Nahhal IM, Zourab SM, Kodeh FS, Selmane M, Genois I, Babonneau F (2012) Nanostructured copper oxide-cotton fibers: synthesis, characterization, and applications. International Nano Letters 2(1):1–5

Fang D, Pan S, Li Z, Yuan T, Jiang B, Gan D, Sheng B, Han J, Wang T, Liu Z (2020) Large-scale public venues as medical emergency sites in disasters: lessons from COVID-19 and the use of Fangcang shelter hospitals in Wuhan, China. BMJ Global Health 5(6):e002815

Gandhi M, Yokoe DS, Havlir DV (2020) Asymptomatic transmission, the Achilles' heel of current strategies to control Covid-19. Mass Medical Soc

Global Industry Analysts, I. (GIA) (2020) Chitin and chitosan derivatives: a global strategic business report. Global Industry Analysts, Inc., (San Jose, CA, USA)

Gouda M, Hebeish A (2010) Preparation and evaluation of CuO/chitosan nanocomposite for antibacterial finishing cotton fabric. J Ind Text 39(3):203–214

Grace M, Chand N, Bajpai SK (2009) Copper alginate-cotton cellulose (CACC) fibers with excellent antibacterial properties. J Eng Fibers Fabr 4(3):155892500900400300

Gupta R, Misra A (2020) Contentious issues and evolving concepts in the clinical presentation and management of patients with COVID-19 infection with reference to use of therapeutic and other drugs used in Co-morbid diseases (Hypertension, diabetes etc.). Diabetes Metab Syndr 14(3):251–254

Harussani MM, Sapuan SM, Khalina A, Ilyas RA, Hazrol MD (2020) Review on green technology pyrolysis for plastic wastes. In: 7th postgraduate seminar on natural fibre reinforced polymer composites 2020, pp 50–53

He D, Zhao S, Lin Q, Zhuang Z, Cao P, Wang MH, Yang L (2020) The relative transmissibility of asymptomatic COVID-19 infections among close contacts. Int J Infect Dis 94:145–147

Hundáková M, Valášková M, Samlíková M, Pazdziora E (2014) Vermiculite with Ag and Cu used as an antibacterial nanofiller in polyethylene

Ilyas RA, Sapuan SM (2020) Biopolymers and biocomposites: chemistry and technology. Curr Anal Chem 16(5):500–503. https://doi.org/10.2174/157341101605200603095311

Ito I, Osaki T, Ifuku S, Saimoto H, Takamori Y, Kurozumi S, Imagawa T, Azuma K, Tsuka T, Okamoto Y, Minami S (2014) Evaluation of the effects of chitin nanofibrils on skin function using skin models. Carbohyd Polym 101:464–470

Jilani A, Othman MHD, Ansari MO, Oves M, Alshahrie A, Khan IU, Sajith VK (2018) A simple route to layer-by-layer assembled few layered graphene oxide nanosheets: optical, dielectric and antibacterial aspects. J Mol Liq 253:284–296

Joseph B, Mavelil Sam R, Balakrishnan P, J Maria H, Gopi S, Volova T, Fernandes S, Thomas S (2020) Extraction of nanochitin from marine resources and fabrication of polymer nanocomposites: recent advances. Polymers 12(8):1664

Kadam VV, Wang L, Padhye R (2018) Electrospun nanofibre materials to filter air pollutants—a review. J Ind Text 47(8):2253–2280

Komeily-Nia Z, Montazer M, Latifi M (2013) Synthesis of nano copper/nylon composite using ascorbic acid and CTAB. Colloids Surf, A 439:167–175

Lee CG, Da Silva CA, Lee J-Y, Hartl D, Elias JA (2008) Chitin regulation of immune responses: an old molecule with new roles. Curr Opin Immunol 20(6):684–689

Li H, Granados A, Fernández E, Pleixats R, Vallribera A (2020) Anti-inflammatory cotton fabrics and silica nanoparticles with potential topical medical applications. ACS Appl Mater Interfaces 12(23):25658–25675

Li J, Zhang D, Yang T, Yang S, Yang X, Zhu H (2018) Nanofibrous membrane of graphene oxide-in-polyacrylonitrile composite with low filtration resistance for the effective capture of PM2.5. J Membr Sci 551:85–92

Liu Y, Kim H-I (2012) Characterization and antibacterial properties of genipin-crosslinked chitosan/poly (ethylene glycol)/ZnO/Ag nanocomposites. Carbohyd Polym 89(1):111–116

Llorens A, Lloret E, Picouet P, Fernandez A (2012) Study of the antifungal potential of novel cellulose/copper composites as absorbent materials for fruit juices. Int J Food Microbiol 158(2):113–119

Ma G, Liang X, Li L, Qiao R, Jiang D, Ding Y, Chen H (2014) Cu-doped zinc oxide and its polythiophene composites: preparation and antibacterial properties. Chemosphere 100:146–151

Mallick S, Sharma S, Banerjee M, Ghosh SS, Chattopadhyay A, Paul A (2012) Iodine-stabilized Cu nanoparticle chitosan composite for antibacterial applications. ACS Appl Mater Interfaces 4(3):1313–1323

Nakajima M, Atsumi K, Kifune K, Miura K, Kanamaru H (1986) Chitin is an effective material for sutures. Jpn J Surg 16(6):418–424

Norizan MN, Moklis MH, Demon SZ, Halim NA, Samsuri A, Mohamad IS, Knight VF, Abdullah N (2020) Carbon nanotubes: Functionalisation and their application in chemical sensors. RSC Adv 10(71):43704–43732. https://doi.org/10.1039/d0ra09438b

Nurazzi NM, Harussani MM, Siti Zulaikha ND, Norhana AH, Imran Syakir M, Norli A (2021) Composites based on conductive polymer with carbon nanotubes in DMMP gas sensors—an overview. Polimery 66(2):85–97. https://doi.org/10.14314/polimery.2021.2.1

Ocsoy I, Temiz M, Celik C, Altinsoy B, Yilmaz V, Duman F (2017) A green approach for formation of silver nanoparticles on magnetic graphene oxide and highly effective antimicrobial activity and reusability. J Mol Liq 227:147–152

Oladapo BI, Daniyan IA, Ikumapayi OM, Malachi OB, Malachi IO (2020a) Microanalysis of hybrid characterization of PLA/cHA polymer scaffolds for bone regeneration. Polymer Testing 83:106341

Oladapo BI, Ismail SO, Zahedi M, Khan A, Usman H (2020b) 3D printing and morphological characterisation of polymeric composite scaffolds. Eng Struct 216:110752

Oladapo BI, Ismail SO, Afolalu TD, Olawade DB, Zahedi M (2021) Review on 3D printing: fight against COVID-19. Mater Chem Phys 258:123943

Palza H, Gutiérrez S, Delgado K, Salazar O, Fuenzalida V, Avila JI, Figueroa G, Quijada R (2010) Toward tailor-made biocide materials based on poly (propylene)/copper nanoparticles. Macromol Rapid Commun 31(6):563–567

Pei L, Zhou J, Zhang L (2013) Preparation and properties of Ag-coated activated carbon nanocomposites for indoor air quality control. Build Environ 63:108–113

Pellett PE, Mitra S, Holland TC (2014) Basics of virology. Handb Clin Neurol 123:45–66

Rementería A, Abaitua F, García-Tobalina R, Hernando F, Pontón J, Sevilla MJ (1997) Resistance to candidiasis and macrophage activity in chitin-treated mice. FEMS Immunol Med Microbiol 19(3):223–230

Safarzadeh M, Sadeghi S, Azizi M, Rastegari-Pouyani M, Pouriran R, HajiMollaHoseini HM (2021) Chitin and chitosan as tools to combat COVID-19: a triple approach. Int J Biol Macromol 183:235–244. https://doi.org/10.1016/j.ijbiomac.2021.04.157

Sakthiguru N, Sithique MA (2020) Fabrication of bioinspired chitosan/gelatin/allantoin biocomposite film for wound dressing application. Int J Biol Macromol 152:873–883

Sanyang ML, Sapuan SM, Jawaid M, Ishak MR, Sahari J (2016) Recent developments in sugar palm (*Arenga pinnata*) based biocomposites and their potential industrial applications: a review. Renew Sustain Energy Rev 54:533–549. https://doi.org/10.1016/j.rser.2015.10.037

Shankar S, Teng X, Rhim J-W (2014) Properties and characterization of agar/CuNP bionanocomposite films prepared with different copper salts and reducing agents. Carbohyd Polym 114:484–492

Shen M, Forghani F, Kong X, Liu D, Ye X, Chen S, Ding T (2020) Antibacterial applications of metal–organic frameworks and their composites. Compreh Rev Food Sci Food Safety 19(4):1397–1419

Singh R, Singh D (2012) Radiation synthesis of PVP/alginate hydrogel containing nanosilver as wound dressing. J Mater Sci - Mater Med 23(11):2649–2658

SOS (2020) S.O.S. Project: Smart Operating Shelter

Tamayo L, Azócar M, Kogan M, Riveros A, Páez M (2016) Copper-polymer nanocomposites: an excellent and cost-effective biocide for use on antibacterial surfaces. Mater Sci Eng, C 69:1391–1409

Teli MD, Sheikh J (2013) Modified bamboo rayon–copper nanoparticle composites as antibacterial textiles. Int J Biol Macromol 61:302–307

Thakare SR, Ramteke SM (2017) Fast and regenerative photocatalyst material for the disinfection of E. coli from water: silver nano particle anchor on MOF-5. Catal Commun 102:21–25

Thampi VVA, Rajan ST, Anupriya K, Subramanian B (2015) Functionalization of fabrics with PANI/CuO nanoparticles by precipitation route for anti-bacterial applications. J Nanopart Res 17(1):1–12

Turcheniuk K, Hage CH, Spadavecchia J, Serrano AY, Larroulet I, Pesquera A, Zurutuza A, Pisfil MG, Héliot L, Boukaert J, Boukherroub R (2015). Plasmonic photothermal destruction of uropathogenic *E. coli* with reduced graphene oxide and core/shell nanocomposites of gold nanorods/reduced graphene oxide. J Mater Chem B 3(3):375–386

Weickmann H, Tiller JC, Thomann R, Mülhaupt R (2005) Metallized organoclays as new intermediates for aqueous nanohybrid dispersions, nanohybrid catalysts and antimicrobial polymer hybrid nanocomposites. Macromol Mater Eng 290(9):875–883

WHO (2020) COVID-19 public health emergency of international concern (PHEIC) global research and innovation forum

Worldometer (2021) COVID-19 Coronavirus Pandemic

Wu H, Williams GR, Wu J, Wu J, Niu S, Li H, Wang H, Zhu L (2018) Regenerated chitin fibers reinforced with bacterial cellulose nanocrystals as suture biomaterials. Carbohyd Polym 180:304–313

Wu H, Williams GR, Wu J, Wu J, Niu S, Li H, Wang H, Zhu L (2014) In situ synthesis of silver-nanoparticles/bacterial cellulose composites for slow-released antimicrobial wound dressing. Carbohyd Polym 102:762–771

Ximing G, Bin G, Yuanlin W, Shuanghong G (2017) Preparation of spherical metal–organic frameworks encapsulating Ag nanoparticles and study on its antibacterial activity. Mater Sci Eng, C 80:698–707

Yadollahi M, Gholamali I, Namazi H, Aghazadeh M (2015) Synthesis and characterization of antibacterial carboxymethylcellulose/CuO bio-nanocomposite hydrogels. Int J Biol Macromol 73:109–114

Yang X, Pu Y, Zhang Y, Liu X, Li J, Yuan D, Ning X (2020) Multifunctional composite membrane based on BaTiO3@ PU/PSA nanofibers for high-efficiency PM2.5 removal. J Hazard Mater 391:122254

Zain NM, Stapley AGF, Shama G (2014) Green synthesis of silver and copper nanoparticles using ascorbic acid and chitosan for antimicrobial applications. Carbohyd Polym 112:195–202

Zehra Z, Luthra M, Siddiqui SM, Shamsi A, Gaur N, Islam A (2020) Corona virus versus existence of human on the earth: a computational and biophysical approach. Int J Biol Macromol

Zhang Q, Li Q, Young TM, Harper DP, Wang S (2019) A novel method for fabricating an electrospun poly (vinyl alcohol)/cellulose nanocrystals composite nanofibrous filter with low air resistance for high-efficiency filtration of particulate matter. ACS Sustain Chem Eng 7(9):8706–8714
Zhang S, Liu H, Yin X, Yu J, Ding B (2016) Anti-deformed polyacrylonitrile/polysulfone composite membrane with binary structures for effective air filtration. ACS Appl Mater Interfaces 8(12):8086–8095
Zhong T, Oporto GS, Jaczynski J, Tesfai AT, Armstrong J (2013) Antimicrobial properties of the hybrid copper nanoparticles-carboxymethyl cellulose. Wood Fiber Sci 45(2):215–222

Chapter 13
Safety Issues in Transportation Design

Abstract With the recent development of technology as well as rapid modernisation, transportation has been a vital part of them. Awareness about safety has been prioritised among people. Therefore, safety design in different transportations plays a huge role in ensuring the safety of everyone. In this review, several safety issues are being discussed which are safety in aircraft, train, automotive, maritime, and bus designs. For aircraft design, this paper reviews the fire safety and seat design. It focuses on the airworthiness design as a protective measure from the ignition of flammable fluids. In order to secure the occupants from getting thrown away, the seat design focuses on the crashworthiness design. For train design, this paper focuses on the safety of train design towards the human. It focuses on how every aspect counts in determining the safety of all people. The next safety issue in a bus is vibration, where it focuses on ways it affects human health and its source. For design in maritime transportation, this paper reviews aspects of stability and fire main system in ships by focusing on the factors that affect stability, operation, and installation of fire main system in ships.

Keywords Safety · Design · Aircraft · Train · Bus · Ship · Automotive

13.1 Introduction

Safety can be defined as the condition where someone is being protected from danger or unlikely or unpleasant conditions that able to cause injury or harm. According to World Health Organization (WHO), more than 1.25 million population involved in road traffic accidents in 2015. There are many types of transportation such as aircraft, trains, automotive, maritime, and also motorcycle. Nowadays, public transportation plays a huge role in enabling people to commute from one place to another. Besides its' cost-effectiveness, it is also advised by the government as a more environmentally friendly option. Priority to the public transport, improving the quality of service, and enhancing the transportation that brings comfort have attracted passengers to travel by public transport to ease traffic congestion and reduce environmental pollution. Due to that, a few safety issues are arising from transportation. For example, for

S. M. Sapuan et al., *Safety and Health in Composite Industry*,
Composites Science and Technology ,
https://doi.org/10.1007/978-981-16-6136-5_13

trains, the more amount of people using them, the more risk they pose. Therefore, there a few design improvements were done to improve the safety of people using trains. At the same time, aircraft accidents always occur due to many factors that are unpredictable and unpreventable. Hence, the main priority issue has always been the safety of the aircraft. Generally, during driving any transportation or specifically bus, riders or drivers will be constantly exposed to hazards, depending on the situation, e.g. vibration transmission to the human body. Nevertheless, safety issues are an essential factor affecting various industries, especially in maritime. Advanced technology leads to more functional, stable, and safer maritime transportations. There are 1.3 million children involved in motor vehicles crash (MVCs) each year. Therefore, a development of safer environment with better engineering design to reduce the number of accidents, injuries and casualties.

13.2 Safety Issues in Aircraft Design

Safety concerns are the most important in aircraft establishments. The safety of an aircraft can be affected by many factors. It can also be related to the aircraft's design, operation, and maintenance. Fire is one of the main concern factors related to an aircraft accident. Therefore, fire safety design is critical to ensure aircraft safety (Dong et al. 2014). Meanwhile, the aircraft's seat design is crucial too to satisfy ergonomics factors besides the occupants' comfort.

13.2.1 Aircraft Fire Safety Design

Commonly, fire eruption happens during many aircraft accidents (Dong et al. 2014). One of the high-risk threats an airplane is a fire. Fire is dangerous to aircraft on grounds and in flights. Hence, fire safety design for the aircraft is important. Aircraft zones need to be identified and the fire characteristics of each fire zone have to be reviewed for the aircraft fire safety airworthiness design. The features of fire of the aircraft zones are summarized in Table 13.1.

Table 13.1 Features of fire of the aircraft zones (Dong et al. 2014)

Compartment/zone	Left/right tanks/center
Flammable fluids/sources	Fuel in the fuel tank
Absorbents	Insulation blanket
Combustibles	None
Drainage paths/outlet location	Drip. tank lower skins
Ventilation	From wing leading edge
Possible ignition sources	Fuel pumps, fuel quantity

Airplane fire safety services help to decrease the possibility for ignition of flammable fluids and reduce the risks associated with ignition (Dong et al. 2014). In order to achieve the minimisation for ignition and resultant hazards, the practices and design characteristics must include such elements:

1. Lessen the potential leak path
2. Ventilation
3. Appropriate fluid drainage
4. Reduce the amount of fluid leakage
5. Compartmentalisation
6. Minimizing explosion sources through the use of eruption proof electrical equipment and limit the surfaces temperatures of apparatus and also air duct
7. Isolation and separation of electrical components from fluid-carrying components
8. The correct method of clamping and support of wires
9. To put the fire out of wire insulation
10. Parts of qualification: fire-resistant or fireproof
11. Grounding and attachment of electrical components
12. Containment in the occurrence of an over-speed of elements
13. Non-absorbent insulation element qualification: impenetrable fire
14. Monitoring of condition, a convenient arrangement, and detection
15. Decide for areas of possible impaired landing effect in the event of a wheels-up landing.

The aircraft zones have been classified into five zones as shown in Table 13.2. Based on the table, the fire safety specification levels of aircraft can be explained (Dong et al. 2014).

The initiatives taken for protective measures include fire protection and separating leakage from ignition sources (Yue et al. 2014). In general, the following provisions have been recognised as reducing the risk in the event of ignition, such as ventilation in areas to lower down the amount of flammable vapours in the zone and to reduce the likelihood of an explosion. Other than that, it is intended to provide a means of shutting down the ventilation airflow following the fire deck signal and supply the highest practical quantity and efficiency of drainage to cut down the available leakage capacity and detect multiple flight leakage exposure (Yue et al. 2014).

Table 13.2 The level fire safety specification of aircraft

Area	Fire provision
Fire	b-o
Flammable fluid leakage area	b-o
Inflammable area	b, c, f, k, l, o
Ignition area	f, g, h, i, k, l, n
Low hazard area	a, b, c, d, l, m
Non-hazard area	b-n

Design for isolation includes a cover to conceal the components carrying fluid and inflammable plumbing, double-walled fuel tanks, and a secondary layer of transparent on integrated tanks that block the leakage from sealant leaks or rupture of a structure, and not only that, it also can detect these leaks (Yue et al. 2014). For instance, the use of baffle ribs in the foremost wing edge enables the drainage of leakage without the leakage moves to different areas that could enhance the possibilities of ignition, fire size, and hazard when ignition happens (Yue et al. 2014).

13.2.2 Airworthiness Design

Airworthiness can be defined as the measurement of the suitability of an aircraft for safe flight. Airworthiness certificate for aircraft means that the entire aircraft meets design and production standards and is in safe flight conditions. The certificate of airworthiness shall be carried on the aircraft during all flight operations. The airworthiness certificate shall remain valid until the maintenance and inspections required for the aircraft are kept up to date.

Figure 13.1 shows a flow chart on the design of airworthiness for fire safety. The process-top down allocation specification which is on the left side, the aircraft level fire safety design requirements can be denoted from Section 6 of Guidelines

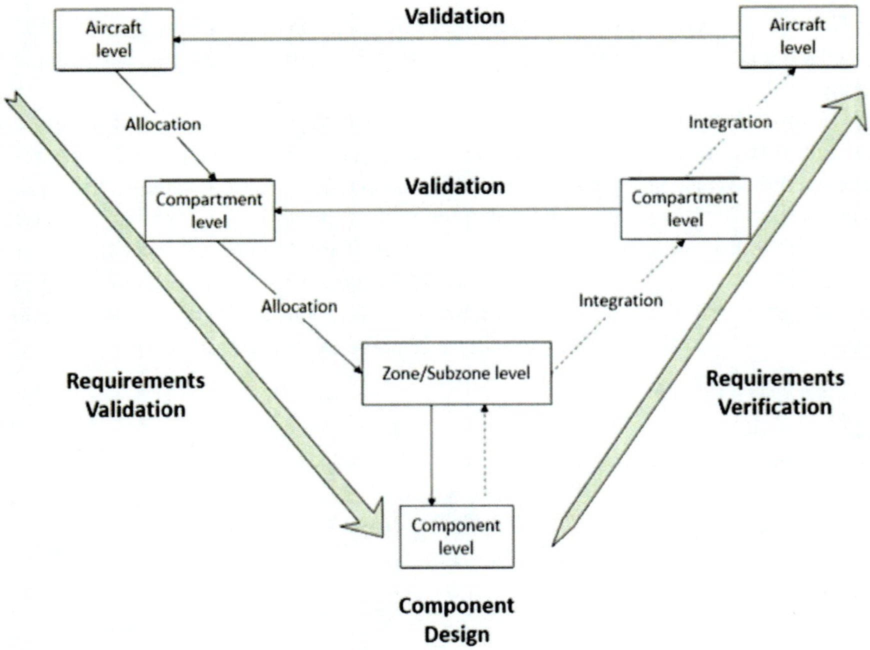

Fig. 13.1 Aircraft fire safety airworthiness design flow chart (Dong et al. 2014)

for Development of Civil Aircraft and System. The working cell is an area, hence that fire safety design requirements of the level of the aircraft are assigned to the Compartment level and Zone or Subzone level. Finally, to carry out the fire model of the component, the fire safety design necessity of components is acquired (Dong et al. 2014).

On the right side, verifying the requirements for process-bottom up integration, the fire component designs are verified to comply with the assigned Zone or Subzone on the left side. Likewise, the fire designs of the Zone or Subzone level and Compartment level are verified to fulfill their respective designs (Dong et al. 2014).

13.2.3 Design of Aircraft Seat

The safety of the passengers is essential for air travel. Besides, the people boarding in-flight need to maintain a correct seating posture. Seats on the aircraft need to be comfortable for all passengers, including the pilots and also the flight crews. The flight crews are required to not be just seated comfortably but also minimize the fatigue from being in the same position for some time. Not only that, but the pilots also need to have an ergonomic seat design in order for them to operate the aircraft efficiently too.

Some aspects need to be considered when designing the seats of an airplane. The pilot seat should be designed with strong adjustment features and easy control access. Next, the flight crew seats are often located near the exits and have always been designed to stow against the bulkhead when there are no passengers.

The features of the seat should be given more attention to the ergonomics factors. These include the headrest, backrest, and lower parts of the body. The headrest should be designed to support both the neck and skull. For the back part of the body, the seat backrest should support the back of the passenger when the upper body of the passenger is in a position of seating with the back straight (Katedra 1998). To support the lower parts of the body such as the waist or lower back area, the aircraft seat should have the Lumbar-Dorsal-Support (LDS) (Katedra 1998). The heights of the people also should be considered when designing the aircraft seat. The passengers should be able to alter the height of the LDS to enable the LDS to support the waist or lower back part (Katedra 1998).

Passengers in side-facing seats are at high risk of injury during aircraft landing in emergency braking (Wu et al. 2017). Based on the article by Michael (2001), a recent study concluded that economy-class air passengers do not have sufficient space to assume the correct "brace" position during an emergency landing. The hurdle for quick emergency evacuation of the cabin can also be from the seats in the aircraft. Attention was also given to the health implications of aircraft seats such as the occurrence and prevention of Deep Vein Thrombosis (DVT). In particular, it stated that the minimum dimensions of the Clean Air Act (CAA) must be increased by a least 3 in. regarding the seat pitch or space between rows of seats and up to 10 in. to adequately fit in taller people (Michael 2001).

Next, the usual occurrences are such as seat tears from its attachment and the re-strain system fails during aircraft crashes. Crashworthy seat design is frequently tested when it comes to the safety of an aircraft. Crashworthiness seat design is complex with the aid of the demands for comfort and appearance (Hooper and Ellis 1996). However, it is obvious that a crashworthy seat will boost the safety of the passengers when in use. Hence, to create an efficient crashworthy seat style, a thoughtful integration of many mechanical elements is needed (Hooper and Ellis 1996).

Crashworthiness can be stated as the capability of a structure to protect its occupants throughout an effect. It depends on the features of the impact and the vehicle involved because different criteria are used to verify the crashworthiness of the structure. Once addressing the lumbar load requirement, the cushion, seat sheet, and seat legs too shall be aware (Hooper and Ellis 1996).

A test on the seat cushion was done and discovered that the classic foam material response has lower stress in very slow loading rate (Hooper and Ellis 1996). Based on the graph (Fig. 13.2), at first, the material develops quite little resistance to the loading since it displaces at almost zero force. Then, the loading increases rapidly with the consolidation of the material (Hooper and Ellis 1996). For the upper curve, it was recorded during a higher loading rate. The changes in the curve are typical rate-sensitive foams and cause an orifice effect as the air is forced out of the foam or by the cell walls bulking (Hooper and Ellis 1996).

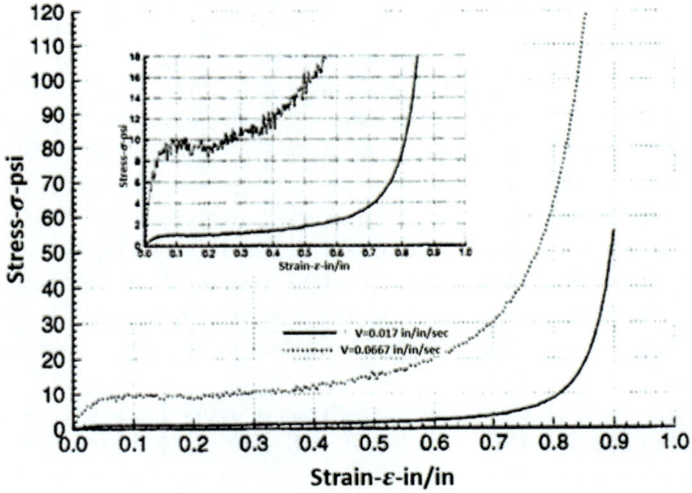

Fig. 13.2 Stress–strain curves for rate-sensitive foam (Hooper and Ellis 1996)

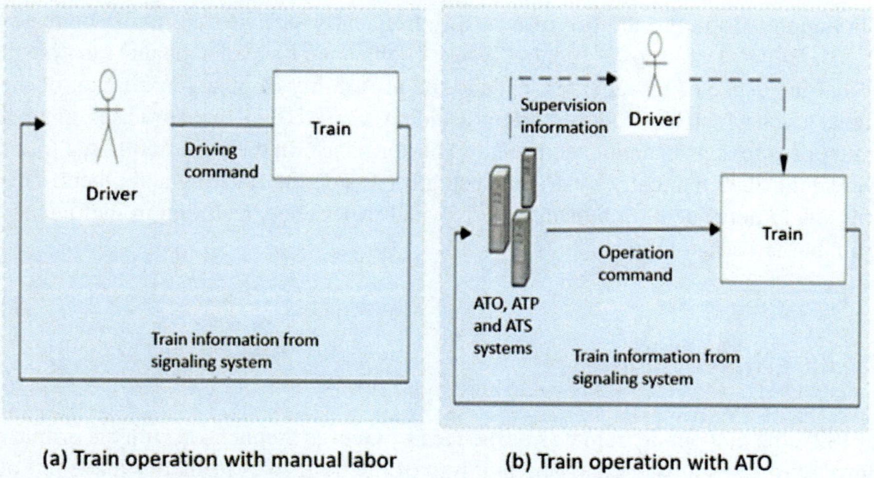

Fig. 13.3 Train operation with manual labour and with ATO (Yin et al. 2017)

13.3 Safety Issues in Train Design

Safety is defined as no danger or no conditions that can create a risk (Ling et al. 2017). Trains are currently being widely used due to its' effectiveness. The most important criterion in safety is its' product design. Nowadays, with the vast development of high-speed rail transport, high-speed trains tend to be more functional, integrated, and complex. The number of accidents, personal safety, providing safe and stability can be achieved by enhancing the safety of design.

13.3.1 Train (Automatic Train Operation)

One of the design safety aspects is the automatic train operation (ATO), to replace traditional manual driving so that drivers do not need to pay full attention to the driving to ensure safety of all people on board. (Yin et al. 2017). Figure 13.3 shows comparison of train operation between manual labour and ATO.

13.3.2 Safety in Automatic Train Operation (ATO)

During the earlier days of railways, safe movement is ensured by the usage of visual signals. To be specific, the driver has full responsibility for the control of the train, especially in the cabin. Track state information will be received by the driver from wayside equipment. This includes track circuits, on-track balises, and etc. (Midya and

Thottappillil 2008). In addition, drivers also frequently receive train control demands from a railway traffic control centre. However, the main issue of a manual ran system is that the usage of visual signs leads to the probability of misinterpretation of the signal given which can lead to a serious accident (Clark 2012). In many cases, manual train operations just cannot achieve good performance. In the past, traditional trains were controlled manually by drivers, using track-side interlocking and others. This was solely dependent on human expertise and experience, resulting in inefficiency from human errors (Yin et al. 2017).

13.3.3 Rail Vehicles

In 1825, the first public train was introduced by George Stephenson with the primary purpose of transporting coal. The build-up of the trains was primarily made out of steel and the trains were powered by steam from the coal. Nowadays, the primary goal of the rail industry is to provide passenger services and transporting many types of cargo. The advancement of the rail industry is due to many factors such as technology advancement, market demands, and government policies (Matsika et al. 2013).

13.3.4 Safety in Rail Vehicles

Security and safety are usually combined in the operation of rail vehicles because of their benefits. However, when considering the design of rail vehicles, it is better to separate them. Safety is defined as a way of reducing the consequences of unintentional system failure, meanwhile, security is a way to reduce the outcomes of an intentional drive of a system. From a safety perspective, it discusses the aspect of crashworthiness and fire. There are 2 standards used; EN 12663 and EN 15227. EN 12663 stated that the static (proof) is the load that a vehicle must be able to withstand without any deformation or breaks (Matsika et al. 2013). Table 13.3 shows the category of railway vehicle based on their structural design.

As an example, a vehicle from category P-II must withstand the amount of force shown in Fig. 13.4.

Table 13.3 Structural design categories of railway vehicles (EN 12663,2010)

Category	Description
P-I	Locomotives and coaches
P-II	Fixed units
P-III	Underground and rapid transit vehicles
P-IV	Heavy duty tramway vehicle and light duty metro
P-V	Tramway vehicle

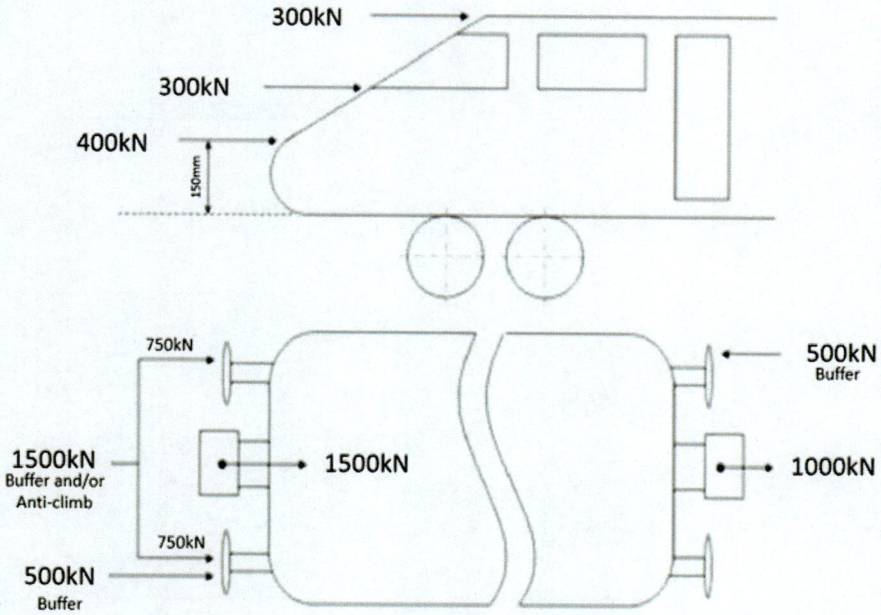

Fig. 13.4 Longitudinal loads for a category P-II vehicle (Matsika et al. 2013)

Other requirements such as lifting loads, maximum operating load, proof loads for equipment attachments, and many more are also considered. On the other hand, EN 15227 specified the number of collision scenarios that a vehicle must safely respond to. To achieve the number of collisions where a vehicle must respond to safely, a system called Crash Energy Management (CEM) is developed. In that system, a designated area is designed to absorb energy in case of accidents (Matsika et al. 2013). Figure 13.5 shows the Driver's cab design approach to CEM.

For fire safety, there are a few processes introduced to reduce the effect of fire. First is prevention that can be done by materials testing and selection. Next is detection which can be performed using smoke detection systems or any other alert mechanism. The last process is suppression that can be done by minimising the dispersion of flames, where water mist technologies are introduced that lead to a smoke reduction (Matsika et al. 2013).

Furthermore, other considerations must look into the design of a rail vehicle to be more flexible. Firstly, the materials chosen must have a huge impact in decreasing injuries and fatalities. Besides that, glazing films can be used to ensure that the glass is retained in one piece so that the shattering of the glass can be prevented and at the same time, assembling joints allows movement of structures without causing any detachment. This allows heavy objects from getting detached in any case (Matsika et al. 2013).

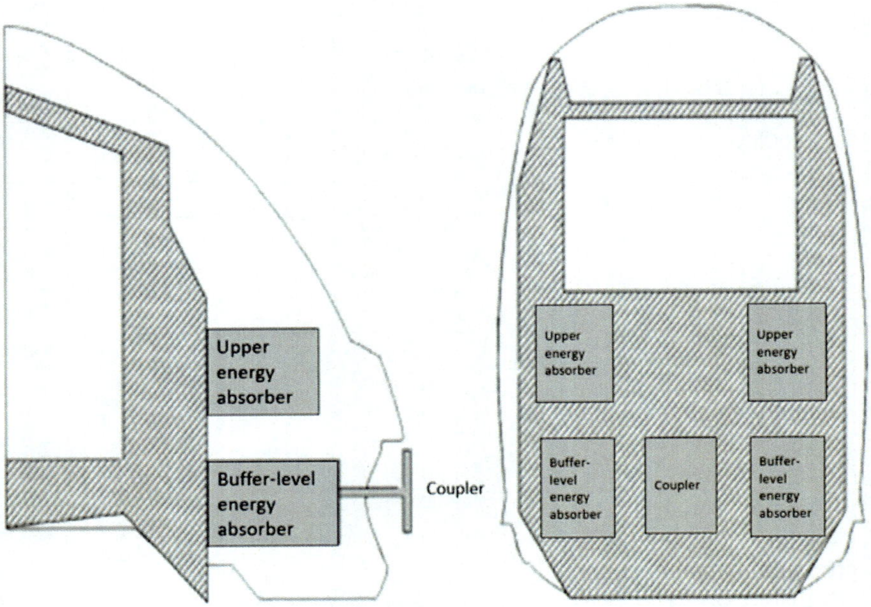

Fig. 13.5 Driver's cab design approach to CEM (Matsika et al. 2013)

13.3.5 Passive Road-Rail Crossing

One of the serious safety issues is road-rail collisions between the people riding the trains and heavy road vehicles. It is proven from the analysis given by the Federal Railroad Administration of the United States of America, in the year of 2001–2015 in Fig. 13.6. From 2001 to 2009, the number of collision incident had decreased by one third, however, the number of injuries and fatalities remain the same, which tell us that the occurrence of accidents had become more serious (Ling et al. 2017).

13.3.6 Safety in Passive Road Rail Crossing

On top of that, there are several measures taken to avoid road-rail crossing accidents, e.g. traffic laws, communication technology, and warming devices which consist of flashlights and boom gates (Tey et al. 2013).

However, passive civil infrastructure enables road vehicle drivers to slow down which reduces the severity of a collision. The most important point highlighted is that adding the height of road pavement in an area when a rail track is there or called a road hump. Researchers studied and found that the early design was effective, however, it possessed a certain weakness. Later, the design was improved, as shown in Fig. 13.7 (Ling et al. 2017).

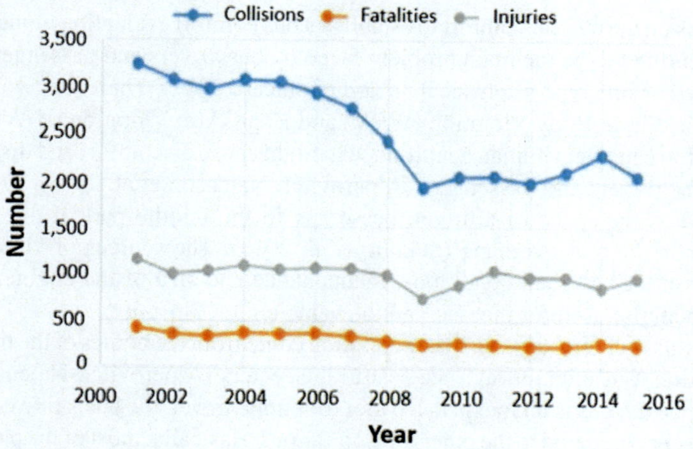

Fig. 13.6 Road-rail crossing collisions, fatalities, and injuries in the USA (Ling et al. 2017)

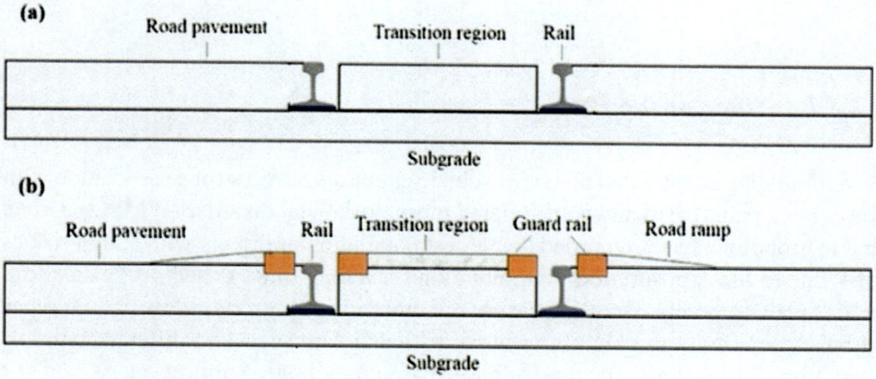

Fig. 13.7 Track structures at the road-rail crossing: **a** normal design; and **b** passive safety design (Ling et al. 2017)

In a nutshell, there are a few methods or designs made to ensure the safety of train design. Safety is something we must work hard to earn. It also requires awareness of the people to make sure that safety can be achieved efficiently.

13.4 Safety Issues in Bus Design

Eboli and Mazzulla (2011) proposed four independent impact factors that contribute to vehicle comfort, which are thermal comfort, noise, vibration comfort, and longitudinal acceleration, and they combined these factors with the influence of individual differences on the overall comfort to establish a bus comfort evaluation model. Vibration is identified to be the main problem faced by bus drivers and passengers which is classified as one type of physical hazard (Comcare 2017). There are two types of vibrations; Whole-Body Vibration (WBV) and Hand-Arm Vibration (HAV).

Vibration can affect human health since it might cause discomfort and disturbance towards the drivers and passengers (Alperovitch-Najenson et al. 2010). Comfort is elaborated as the space of a driver, passenger in the middle part of the bus, and passenger in the rear overhang (Sekulic et al. 2013). The sources of vibration are from uneven or bad road conditions, maintenance and size of the engine, and the types of material used for the seat and the vehicle.

But in this problem, the sources of vibration come from the bodies of the bus. Next, the vibration transmits through the seat to the seaters (Alperovitch-Najenson et al. 2010). Since the bus is an inexpensive tool for public travel, the bus's ride comfort is usually lower compared to the other transportations. Basically, most of the passengers sitting on the last row seats will receive a higher shock impact. Therefore, to achieve a balance system, the human acceleration exposure must be considered in designing a bus.

13.4.1 Floor of the Bus

Several studies investigated that most vehicle operators have a strong association with lower back pain (LBP) and work-related musculoskeletal disorder (WMSD). Urban and metropolitan buses recorded the largest population of drivers with a high risk of LBP due to the exposure to shock vibration for a long time. Based on Thamsuwan et al. (2013), a comparison to measure the vibration was produced using two types of buses; low-floor and high-floor buses with standard seats and different types of road. Figure 13.8 shows the type of buses based on vibration ability.

The instrument used by the researchers to measure personal vibration is an accelerometer. After performing the test with different types of routes, the results

Fig. 13.8 Type of buses based on vibration ability (Jonsson et al. 2014)

showed that WBV exposures on the high-floor coach bus were higher on the road with the speed humps compared to the road without humps. Based on the discussion, this was likely because of the bus driver being situated much higher over the road that concluded that stability depends on altitude. At the same time, hitting holes or any bumps on the road might influence shock vibration production (Thamsuwan et al. 2013). The main difference between these two buses was high-floor bus had greater suspension travel and a greater ability to absorb terrain-related perturbations compared with the low-floor bus (Johnsson 2014).

13.4.2 Type of Seat

The elasticity and softness of a seat influence seat comfort. This plays a vital factor in designing a seat that can cause vibration transmission to passengers' bodies and their intensity (Zhang et al. 2015). By focusing on that factors, designers are able to prevent passengers or drivers from getting 'static discomfort' and minimise 'vibration discomfort'. Different types of material of bus seats give different WBV exposure to seaters. Somehow, the way a person seat might influence the amount of vibration the person receives (Wilder et al. 1994). One study was conducted to identify the association between the types of seat used and vibration transmitted by using air-suspension and pedestal seats.

Both seats shared the same characteristics, where both have backrests, height armrests, and seat pan depths that were adjustable in angle. According to the researchers, the suspension travel was designed for absorbing road-related perturbations while pedestal seats, which contained a gas spring (Fig. 13.9) in the seat base was for height adjustment. Those two types of seats were compared with different types of buses, which were low- and high-floor buses. It was found that the air-suspension seat attenuated the vibration on the low-floor bus but amplified the vibration on the high-floor bus. It can be concluded that the amplitude of vibration was less at low-floor bus compared to the high-floor bus if using the air-suspension seat. Among the good properties of the air-suspension seat are able to minimise vibration, ensure smoother operations for driver or passengers, durability, and stability. The manufacturers also need to consider various sizes and weights of seaters for smooth and vibration-free bus ride.

13.4.3 Bus Tyres

Tyres play an important part in vehicle and driver safety and they need to be closely monitored and maintained. The lifespan of a tyre depends on the road surface conditions since speed bumps and humps are used widely to enforce drivers to reduce vehicle speeds (Krylov 2015); (North Western Maintenance Liaison Committee 2015). Somehow, the mass and load of passengers contribute to the increase of

Fig. 13.9 The pedestal seat (left) and air-suspension seat (right) used in the study (Jonsson et al. 2014)

production of friction on the tyres. The bus and bus suspension system act like a mass-spring-system (Thamsuwan et al. 2013). The grip and flexibility of tyres also take part in terms of safety. The effect of overinflated tyres will cause them to be stiff and rigid and at the same time reduces the tyre's contact surface. This could result in a rough ride that leads to uneven wear and tear. For under-inflated tyres, over flex and overheat could occur and might cause rapid depletion if continuously used (North Western Maintenance Liaison Committee 2015). Improper maintenance of tyres will lead to bus shaking and the vibration on the steering wheel is exhausting and discomfort to the driver.

What can be concluded for safety issues in a bus is designers should consider health issues into the design. To control vibration, engineering and administrative control approaches can be employed. For example, by choosing and improving the types of seats and to reduce the driver's exposure to continuous vibration, the company should at least provide work rotation for each worker, so they will not be exposed to the same work routine for a long time.

13.5 Safety Issues in Maritime Transportation Design

Nowadays, safety issues are an essential factor that affects all elements in various industries, especially the maritime industry. This is due to the increase in demands on maritime transportation, particularly cargo and cruise ships. The increasing number of daily cargo has resulted in many accidents that eventually leads to economic losses and tragic deaths. Advanced technology, increasing stability, and safety systems in ships are due to the blooming in naval transportation. In addition, marine safety can be divided into some elements which are technological and operational safety of ships, the safety of navigation, the safety of mankind during emergencies, and pollution prevention (Kopacz et al. 2001).

To enhance safety in maritime transportation, design plays an important aspect. Stability and fire protection system are among the aspects of the design safety issue as they can increase the safety and security of a ship in an open sea.

13.5.1 Stability Design in Maritime Transportation

Designing the structure of a ship or vessel is the most crucial part as it plays a key role in ensuring the ship returns to an upright position and maintain buoyancy. As time goes by, the revolution in the building and structure of vessels has boosted up marine safety. These betterments did not only attained in design, but also in robustness and size of ships that had become increasingly larger. Table 13.4 shows the revolution of ship size from the year 1912–2017. The trend shows that the increased surface area of the base of the ship exists as the year increases.

The stability is affected by gravitational force and buoyant force operating in opposite directions. By increasing the base of the ship, the stability and gravitational force of the ship increases. Over the years, the ship size is increasing to increase the surface area of the ship. According to Kulovaara (2015), "the new types of watertight doors, adding ducktails, removing weight up high or splitting tanks was applied to enhance stability." This shows that there are many ways to increase stability to eventually improve ship safety.

Some challenges will affect ships' stability which are upsetting forces; external heeling moments such as beam winds and grounding and internal heeling moments such as water trapped on deck. Waves from the seawater always got trapped on the deck and the motion will result in weight shift in both directions, producing a cyclic heeling occurrence.

13.5.2 Fire Protection Design in Maritime Transportation

Fire is one of the major causes of a total loss. Firefighting procedures should be established as it plays an essential role in minimising the risks as fire on ships is extremely dangerous to human lives. According to Zhang (2000), traditional fire protection on board can be divided into structural fire protection, fire detection, and fire extinction. Structural fire protection is passive protection to slow down the spread of fire on board to give time for evacuation. Apart from that, fire detection and fire extinction are active protection as their goal is to detect a fire and extinguish it.

Table 13.4 Size of Titanic compared to current modern ships throughout time (Galić et al. 2014)

Ship size comparison	Name	Type	Year
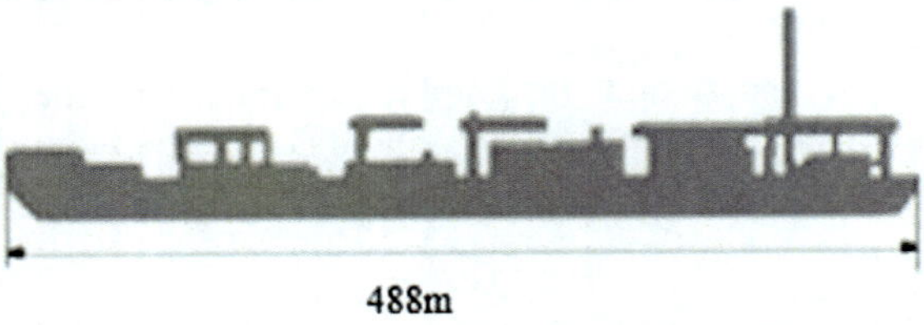	Prelude	FLNG	2017
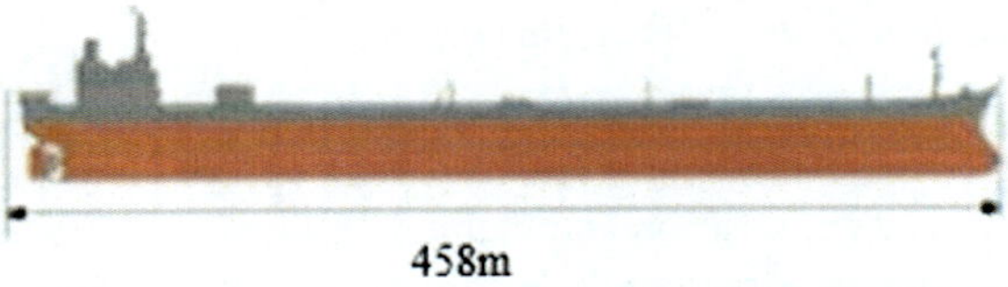	Knock Newis	Oil tanker	1979
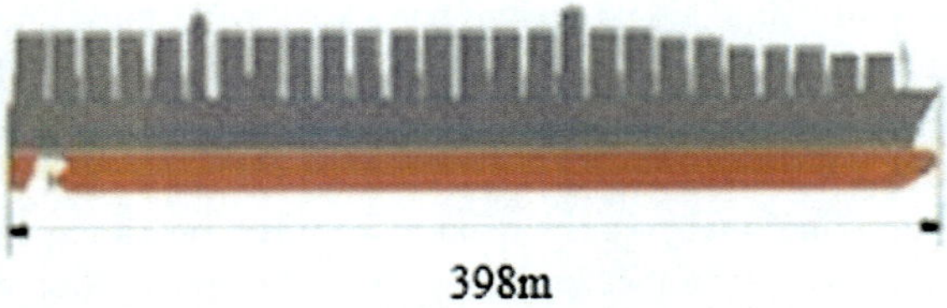	Maersk Mc-Kinney Meller	Container ship	2013
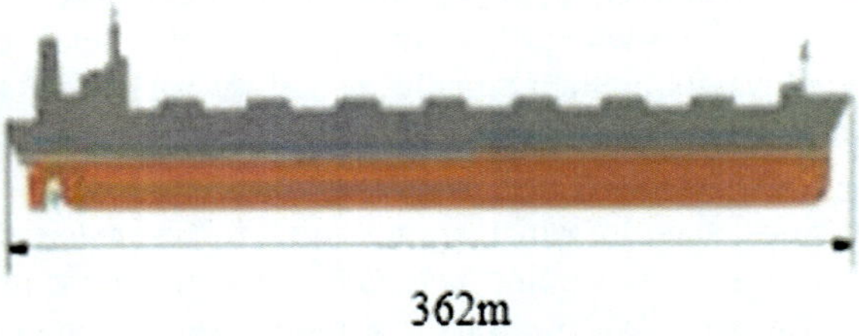	Valemax	Bulk carrier	2011
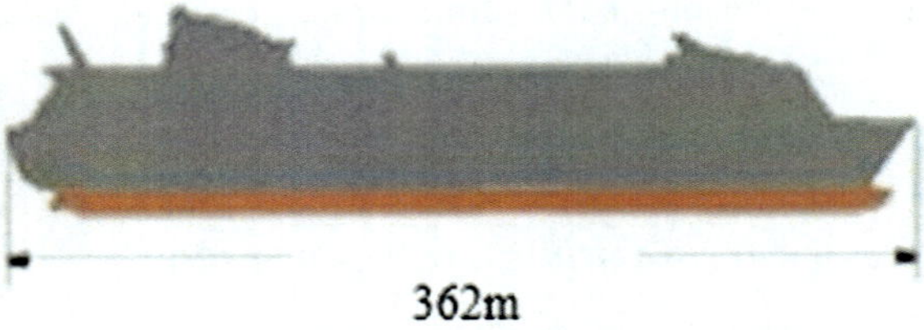	Alure of the seas	Cruise ship	2009
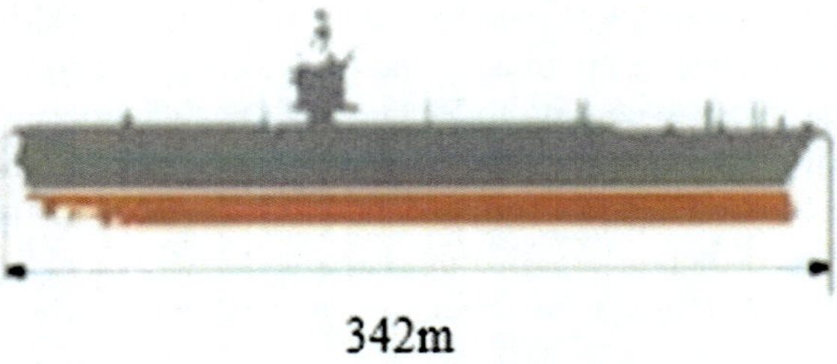	USS Enterprise	Aircraft carrier	1961

(continued)

Table 13.4 (continued)

Ship size comparison	Name	Type	Year
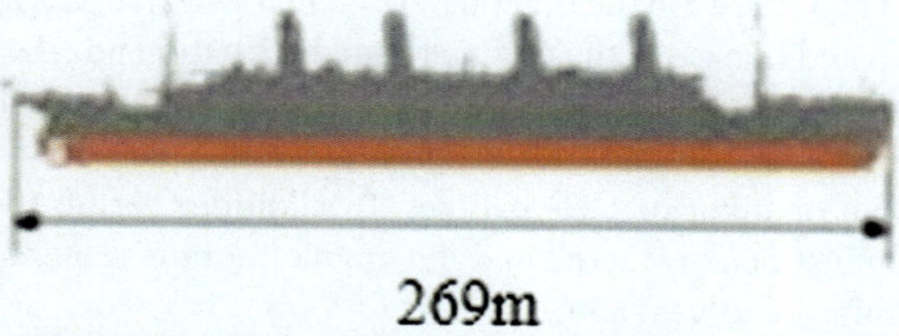	Titanic	Ocean liner	1912

Zhang (2000) also stated that some of the main lessons to be learned from the Morro Castle fire are the design of the lining should be improved to be fire-resisting, doors to compartments should be self-closing, automatic fire alarms should be installed throughout, more sprinklers should be installed, etc.

13.5.2.1 Ship Fire Main System for Cargo Ship

The utilisation of a water channel to put out a fire or fire main is essential when there is an occurrence of a fire outbreak and the fire main spread out covering the span of the vessel. Water is the core medium for firefighting and fire main is the basic installation for firefighting.

The system has two autonomous powered pumps that supply engine room hydrants and the deck main through the screw-down isolating valve. The latter is required to forestall any loss of water through impaired pipework in the engine room. The emergency fire pump positioned in an underpass by the engine room watertight door needs to be utilised to sustain the deck supply. Next, the deck main has a drain at the bottom position so that the pipeline can be emptied (particularly of freshwater) in cold weather. If this action is not executed, the pipeline can be destructed by the water freezing but what more crucial is, it will be obstructed by ice and became non-functional. It is a statutory specification that a fire main and deck wash system should be allocated. This water pipeline has hose outlets on different decks and is dispensed by power-driven pumps in the machinery spaces. The allocation might be made for drenching down the anchor chain from a connection to the fire main.

13.5.2.2 Carbon Dioxide (CO_2) Fire Extinguishing Installations for Cargo Ship Machinery Spaces

Another essential design aspect is carbon dioxide (CO_2) installation in ships such as cargo ships. CO_2 is an important element in extinguishing fire from the outbreak, especially in a ship on an open sea which is always exposed to many sources of fire.

CO_2 systems are voluminously used to safeguard ship's cargo compartments, boiler rooms, and machinery spaces.

Storage bottles of CO_2 will supply high-pressure CO_2. A servo-piston-operated gang release will open the storage bottles. On the other hand, the master valve acts as a safety factor against unanticipated release to the engine room distribution nozzles.

According to Dean (2016), the CO_2 structure is utilised if a fire is appalling enough to compel deportation of the engine room or to forestall entry. The alarm system will go on signalling that there is an opening of CO_2 cabinet. In addition, before the release of CO_2, the workforce must be considered and the engine room is required to be closed down with all openings and vent shut.

The challenges are as the world evolves, new technology has introduced composite materials in the manufacturing processes. Glass-reinforced polymer (GRP) composites are extensively used in the establishment of nautical craft such as life-boats and racing yachts. The main problem is that polymer matrix can be kindled within a short period when exposed to fire. Inadequacy of fire coating on the surface of a ship might affect firefighting design.

In a conclusion, the evolution of maritime transportation in improving and enhancing safety issues has increased as the world is moving towards modernization. Safety issues are not only for the transportation itself but also for the people who are working with risks as well. Therefore, it is very essential to build a strong base from scratch to produce a more efficient and safe mode of transportation, especially ships. From previous studies, the depth of stability and fire system aspects of ships were reviewed and revealed that it is crucial to curb any improper designs or installations to increase safety in maritime transportation.

13.6 Safety Issues in Automotive Design

MVCs are one of the prominent causes of death and injury among children of <1 year of age. Based on observations, infants involved in moderate to high-speed MVCs often sustain significant head injuries despite being properly controlled. Based on our observations, a high incidence of traumatic brain injuries (TBI) in properly controlled infants involved in higher speed motor vehicle crashes (MVCs). Hence, we assume that car safety seats are inadequately protecting infants from TBI. The car is the second most common mode of transportation in Malaysia after the motorcycle. In 2001 only, 11,302,545 vehicles registered with the Department of Transportation, near to half (4,557,992) were cars, representing 40% of the total vehicles registered in Malaysia (Kulanthayan et al. 2004).

Globally, the vehicle fleet on the road has shown stable growth from 130 million in the year 1940 to more than 450 million in 2013. It is convinced that, over the next 50 years, this number is about to grow 3–5 times greater. Besides fuel cost and CO_2 emission, the biggest challenge for regulators and engineers are finding ways to maintain the existing passive safety standards as the number of vehicles increases on the road.

13.6.1 Safety Issues of Infants Car Seat

In 2014, it was estimated that 167,000 children were injured in motor vehicle crashes (MVCs) in the United States. Among the 135 fatalities of children ages 1–3 years, 18% were unrestrained and 7% were restrained only by a seat belt. The second prominent cause of injury death for US children of 1–2 years of age are MVCs. It was roughly calculated that car safety seats reduced the risk of death and prevent substantial nonfatal traumatic injuries. Child fatality from MVCs had decreased over the past decade as child passenger restraint use has increased. However, a large amount of percentage of children was found to ride in misused car seats, placing them at high risk of injury or death in the event of a crash. Moreover, children of 12–23 months of age involved in MVCs have a 5 times greater risk of serious injury in forward-facing car safety seats when compared with rear-facing car safety seats (Jones et al. 2017).

Based on statistical analysis, motor vehicle crashes (MVCs) among infants usually involve damage or injury to the head. American Academy of Pediatrics suggests using rear-facing car safety seats among children up to 2 years of age. Rear-facings are much preferable to the forward facing car seats. Based on the observations, a high incidence of traumatic brain injuries (TBI) was reported in properly restrained infants involved in higher speed motor vehicle crashes (MVCs). Hence, we hypothesised that car safety seats are inadequate in protecting infants from TBI (Kulanthayan et al. 2004).

13.6.2 Safety Issues of Seat Belt

Despite the well-known benefits of wearing seat belts, most vehicle occupants still do not make use of them. WHO stated that wearing a seat-belt reduces the risk of fatality among drivers and front-seat passengers by 45–50%, and the risk of minor and serious injuries by 20–45%, respectively. Moreover, the seating position in the car has a strong correlation with seat belt use. Seat belt use was higher among drivers followed by front passengers and an alarming non-use level by rear passengers (Kulanthayan et al. 2004).

Controlled seat-belt systems continuously vary the belt force such that the motion of the occupant follows an optimal trajectory. Standard automotive seat belt systems nowadays have three points seat belts, a load limiter, and a pre-tensioner. The pre-tensioner is a pyrotechnical device that removes slack from the seat belt when a crash is detected. The load limiter upper-bounds mechanically force the seat belt. Some respondents felt that the seat belt is only needed to be used for those travelling at high speed to protect them in the car. Level of education and awareness might be one of the factors as well. Law enforcement towards compliance of wearing a seat belt can enhance safety and reduce the issues (Kulanthayan et al. 2004).

This study provides a snapshot of adult rear-seat passengers who do not frequently use their seat belt when riding in the rear seat, a population that, to date, has received little attention. Several reasons such passengers give for not buckling up were identified; these can be easily grouped into four categories: misperception of safety benefits, ambivalence, design and usability, and the law. While respondents note that issues with their outfit, comfort, and convenience, as well as legal requirements contribute to their behaviours, the most common responses reflected participants' ambivalence, and misperception about the consequences of not buckling up (Jermakian and Weast 2018).

13.6.3 Safety Issues in Bumper System

It is found that to reduce cost and save fuel, lightweight vehicle developments seem to be necessary. Lightweight design in vehicles comes with many advantages. The bumper design must be flexible enough to reduce the occupant injury and being firm enough to dissipate the kinetic energy at high speed intact (Belingardi et al. 2015).

The bumper system in a car is a safety issue that we can consider with. This is because many studies showed that frontal car crashes highly led to fatal. In that case, the bumper system should be improvised for better performance in preventing car crashes that cause fatal events. The design of the frontal crumple zone plays an important role as it manages crash energy, absorbing it within the frontal parts of the vehicle structure, and carefully avoiding its direct transfer to the vehicle occupants, the intrusion into, or the damage of the passenger compartment. In this issue, the bumper subsystem is one of the critical design spaces to boost the performance of the crumble zone (Davoodi et al. 2012).

Next, the bumper system is a set of components around the front and rear parts of the vehicle designed for damping the kinetic energy without any damage to the vehicle when there is low-speed impact and for energy dissipation in high-speed impact conditions besides serving aesthetic and aerodynamic purposes. The bumper system contains bumper beam, bumper cover (fascia), bumper foam (energy absorber) as the main components along with the rails, and cooling support system, also the foam cushions as shown in Fig. 13.10. Those components are highly associated with the safety of occupants and pedestrians (Belingardi et al. 2015).

The bumper beam is the main structural component of all as it is deformable enough to absorb shock and to reduce the risk of injury to the road users as well as the pedestrians. However, the bumper beam must have sufficient strength and stiffness in order to protect and absorb the impact energy. So, the material and design are crucial. Based on the studies, three materials can be considered for designing a bumper for the case of a die-formed integrated crash box–beam solution. Figure 13.11 shows the difference between GMT, GMTex, and GMT-UD on dynamic impact tests (Beyene et al. 2014).

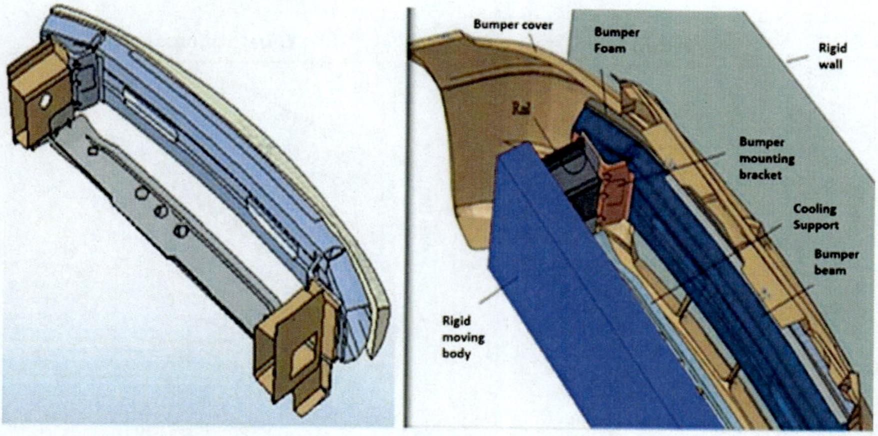

Fig. 13.10 Bumper component (Belingardi et al. 2015)

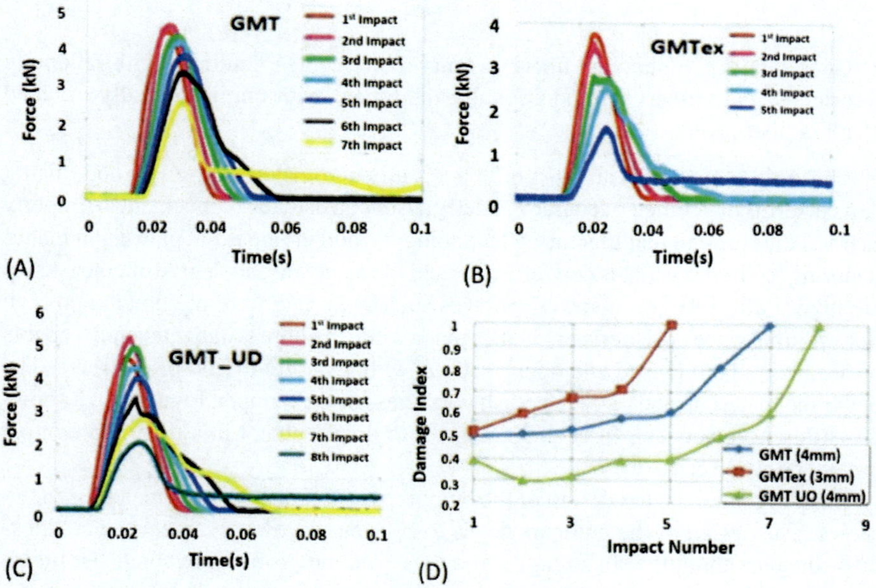

Fig. 13.11 **a–c** Load versus time curves for repeated dynamic impact tests and **d** damage index versus impact number (Belingardi et al. 2015)

- A classic glass-mat-reinforced thermoplastics (GMT), e.g. an end-less fiberglass mate reinforced PP with randomly oriented glass fibers
- GMTex, e.g. a chopped fiberglass mat reinforced PP laminate with randomly oriented glass fibers and an additional fabric reinforced inside and

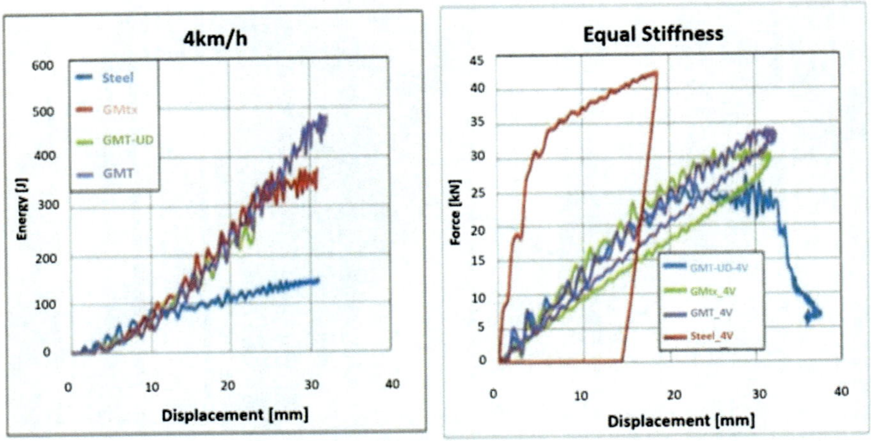

Fig. 13.12 Energy versus displacement and force versus displacement curve for the modified bumper system (Beyene et al. 2014)

- GMT-UD, e.g. a chopped fiberglass mat reinforced PP laminate with randomly oriented glass fibers and additionally reinforced with unidirectionally oriented glass fiber layers.

Those three materials are cost-effective, fully automated, and give good quality parts in terms of geometry accuracy and degree of consistency of mechanical property as it is a die forming manufacturing technology. Another one is the pultrusion manufacturing technology that is cost-effective and the accuracy and degree of consistency are high. Pultrusion has many advantages such as perfect fibre alignment and high fibre volume since polymerisation takes place while the fibre is under tension, capable of producing both closed and open sections with a variety of end profiles, etc. The better the materials, the more the crashworthiness of the bumper. Figure 13.12 shows the differences between force and energy with displacement in a modified bumper system (Beyene et al. 2014).

Besides materials, the design of the bumper is as important as the materials themselves. The design of the bumpers needs to consider several parameters. The thickness, the curvature, the rib strength, and cross-section profiles are some significant parameters that are useful to improve the absorption of energy. The curvature functions to improve the stability of the beam. Other than that, optimising the cross-section of a bumper beam magnifies its strength, dimensional stability, and damping capability. Moreover, it has crucial effects on the rate of energy damping and bending resistance compared to other parameters. Rigidity, distortion of resistance, and structural stiffness can be increased by strengthening the ribs. The optimal thickness of a bumper beam can construct a balance between the weight and strength of the structure as well as to provide more effective energy absorption (Beyene et al. 2014).

Hence, to overcome this issue, a bumper system needs to have the right materials selection and design to ensure the bumper can help to reduce the risk by absorbing the impact energy during crashes and protecting the road users, occupant as well as pedestrians. Overall, every part and component in a car plays a significant role regarding the safety issues of the occupants and the pedestrians. Those issues need to be taken into account to prevent them from happening. In order to protect infants, rear-facing car seats, the rear seat belt, as well as a car bumper need to be used to help to reduce the impact of a collision.

13.7 Conclusion

In conclusion, this paper reviews the safety in transports design. In an aircraft, fire safety and seat design are discussed since safety has become essential when it comes to aircraft. This paper focuses on the aircraft zones which help to reduce the possibility for ignition of flammable fluids. The seat design is, too, important to prevent the occupants from being thrown in case of an accident. On the other hand, the automatic train operation, rail vehicles, and passive road-rail crossing in trains are deliberated. Safety in trains has been improving from time to time. For buses, this chapter reviews the sources of vibrations based on the types of bus floor, seat, and tyres. For maritime design, the safety issues being reviewed are the stability and fire system. Last but not least, for automotive design, the safety issue discussed are car seats, bumper system, and seat belts. Every design has its own unique system to ensure the safety of its passengers.

Acknowledgements This chapter was written with Farah Farhanah Aziz, Nursarah Muhammad Anwar, Marissa Hafez, Thaqifah Rusydina Tarmizi, and Fatin Nadzirah Mohd Asri.

References

Alperovitch-Najenson D, Santo Y, Masharawi Y, Katz-Leurer M, Ushvaev D, Kalichman L (2010) Low back pain among professional bus drivers: ergonomic and occupational-psychosocial risk factors. Isr Med Assoc J 12(1):26–31

Belingardi G, Beyene A, Koricho E, Martorana B (2015) Alternative lightweight materials and component manufacturing technologies for vehicle frontal bumper beam. Compos Struct 120:483–495. https://doi.org/10.1016/j.compstruct.2014.10.007

Beyene A, Koricho E, Belingardi G, Martorana B (2014) Design and manufacturing issues in the development of lightweight solution for a vehicle frontal bumper. Procedia Eng 88:77–84. https://doi.org/10.1016/j.proeng.2014.11.129

Clark S (2012) A history of railway signaling (from the bobby to the balise). In: IET professional development course on railway signaling and control systems (RSCS), pp 6–25

Comcare (2017) Vibration. Retrieved from https://www.comcare.gov.au/preventing/hazards/physical_hazards/vibration

Davoodi M, Sapuan S, Aidy A, Osman NA, Oshkour A, Abas WW (2012) Development process of new bumper beam for passenger car: a review. Mater Des 40:304–313. https://doi.org/10.1016/j.matdes.2012.03.060

Dong J, Han B, Li X (2014) The study of transport category aircraft fire safety airworthiness design. Procedia Eng 80:44–48. https://doi.org/10.1016/j.proeng.2014.09.058

Eboli L, Mazzulla G (2011) A methodology for evaluating transit service quality based on subjective and objective measures from the passenger's point of view. Transp Policy 18(1):172–181

Galić S, Lušić Z, Skoko I (2014) The role and importance of safety in maritime transportation [Online]. Bib.irb.hr. Available at: https://bib.irb.hr/datoteka/700720.imsc2014.pdf. Accessed 13 Nov 2018

Hooper SJ, Ellis DR (1996) Aviation safety and crashworthy seat design. Int J Crashworthiness 2(1):39–54. https://doi.org/10.1533/cras.1997.0034

Jermakian JS, Weast RA (2018) Passenger use of and attitudes toward rear seat belts. J Safety Res 64:113–119. https://doi.org/10.1016/j.jsr.2017.12.006

Jones AT, Hoffman BD, Gallardo AR, Gilbert TA, Carlson KF (2017) Rear facing car safety seat use for children 18 months of age: prevalence and determinants. J Pediatr 189. http://doi.org/10.1016/j.jpeds.2017.06.020

Jonsson P, Rynell PW, Hasberg M, Johnson PW (2014) Comparison of whole-body vibration exposures in buses: effects and interactions of bus and seat design. Ergonomics 58(7):1–10

Kopacz Z, Morgas W, Urbanski J (2001) The maritime safety system; its components and elements. J Navig 199–211

Krylov VV (2015) Generation of ground vibrations by vehicles crossing exible speed bumps. Proc Inst Acoust 37(2):178–185

Kulanthayan S, Law T, Raha A, Radin UR (2004) Seat belt use among car users in Malaysia. IATSS Res 28(1):19–25. https://doi.org/10.1016/s0386-1112(14)60088-1

Kulovaara H (2015) Safety & stability through innovation in cruise ship design, STAB2015 (2015):3

Ling L, Dhanasekar M, Thambiratnam D (2017) A passive road-rail crossing design to minimise wheel-rail contact failure risk under frontal collision of trains onto stuck trucks. Eng Fail Anal 80:403–415. https://doi.org/10.1016/j.engfailanal.2017.07.003

Matsika E, Ricci S, Mortimer P, Georgiev N, O'Neill C (2013) Rail vehicles, environment, safety and security. Res Transp Econ 41(1):43–58. https://doi.org/10.1016/j.retrec.2012.11.011

Michael R (2001, November 06) Anthropometry and ergonomics in airline seating. Retrieved from https://ergoweb.com/anthropometry-and-ergonomics-in-airline-seating/

Midya S, Thottappillil R (2008) An overview of electromagnetic compatibility challenges in European rail traffic management system. Transp Res Part C 16(5):515–534

North Western Maintenance Liaison Committee (2015) Best practices guide for bus and coach tyre maintenance

Sekulic D, Dedovic V, Rusov S, Salinic S, Obradovic A (2013) Analysis of vibration effects on the comfort of intercity bus users by oscillatory model with ten degrees of freedom. Appl Math Model 37(18–19):8629–8644

Tey LS, Wallis G, Cloete S, Ferreira L (2013) Modelling driver behaviour towards innovative warning de. Vicesa trailway level crossings. Accid Anal Prev 51:104–111 (2013)

Thamsuwan O, Blood RP, Ching RP, Boyle L, Johnson PW (2013) Whole body vibration exposures in bus drivers: a comparison between high-floor coach and a low-floor city bus. Int J Ind Ergon 43(1):9–17

Wilder D, Magnusson ML, Fenwick J, Pope M (1994) The effect of posture and seat suspension design on discomfort and back muscle fatigue during simulated truck driving. Appl Ergon 25(2):66–76

Wu J, Lin S, Cao L, Adamou FN (2017) A numerical study on the injury prevention for the occupants seated in aircraft side-facing seats. Int J Crashworthiness 23(3):253–265. https://doi.org/10.1080/13588265.2017.1315896

Yin J, Tang T, Yang L, Xun J, Huang Y, Gao Z (2017) Research and development of automatic train operation for railway transportation systems: a survey. Transp Res Part C Emerg Technol 85:548–572. https://doi.org/10.1016/j.trc.2017.09.009

Yue N, Li Y, Bai M, Hou S (2014) Flammable fluid fire protection airworthiness design and verification method of civil transport aircraft. Procedia Eng 80:110–118. https://doi.org/10.1016/j.proeng.2014.09.067

Zhang S (2000) Fire protection onboard: enhance fire safety by design. World Maritime University Dissertations, 56. Retrieved from http://commons.wmu.se/all_dissertations/56

Zhang X, Qiu Y, Griffin MG (2015) Transmission of vertical vibration through a seat: effect of thickness of foam cushions at the seat pan and the backrest. Int J Ind Ergon 48:36–45